Mastering Linux

Mastering Linux

Paul S. Wang

CRC Press
Taylor & Francis Group
Boca Raton   London   New York

CRC Press is an imprint of the
Taylor & Francis Group an **informa** business

A CHAPMAN & HALL BOOK

Chapman & Hall/CRC
Taylor & Francis Group
6000 Broken Sound Parkway NW, Suite 300
Boca Raton, FL 33487-2742

© 2011 by Taylor and Francis Group, LLC
Chapman & Hall/CRC is an imprint of Taylor & Francis Group, an Informa business

No claim to original U.S. Government works

Printed in the United States of America on acid-free paper
10 9 8 7 6 5 4 3 2 1

International Standard Book Number: 978-1-4398-0686-9 (Paperback)

Library of Congress Cataloging-in-Publication Data

Wang, Paul S.
   Mastering Linux / Paul S. Wang.
     p. cm.
   Includes bibliographical references and index.
   ISBN 978-1-4398-0686-9 (pbk. : alk. paper)
   1. Linux. 2. Operating systems (Computers) I. Title.

QA76.76.O63W365143 2011
005.4'32--dc22
                                                  2010026042

Visit the Taylor & Francis Web site at
http://www.taylorandfrancis.com

and the CRC Press Web site at
http://www.crcpress.com

Preface

Linux is one of the great success stories of open-source, community-developed software. It is increasingly used as Web and application servers, software development platforms, personal workstations, and research machines. In computer science and engineering departments, you'll find Linux systems in classrooms, programming labs, and computer centers, not just because Linux is free, but also because it offers a rich computing environment for teaching and learning.

From its beginning in 1991, and with help from the GNU Project, Linux has evolved quickly and has brought new powers and conveniences to users. Competency on Linux will be important for any serious computer professional.

This book provides a comprehensive and up-to-date guide to Linux concepts, usage, and programming. The text helps you master Linux with a well-selected set of topics. Hands-on practice is encouraged; it is the only way to gain familiarity with an operating system. A primer gets you started quickly. The chapters lead you from user interfaces, commands and filters, Shell scripting, the file system, networking, to kernel system calls. There are many examples and complete programs ready to download and run. A summmary and exercises of varying degrees of difficulty can be found at the end of each chapter. A companion website provides appendices, information updates, an example code package, and other resources for instructors as well as students. See page 405 for details.

User Friendly and Comprehensive

There is both breadth and depth in this book's presentation. Chapter 1 contains a Linux primer to get the new user started as quickly as possible without awkwardness or confusion. Being able to play and experiment with the system adds to the user's interest and motivation to learn more. Once introduced and comfortable, the user is guided through a well-selected set of topics covering the type of detailed material appropriate for a one-semester course at the advanced undergraduate or beginning graduate level.

The first part of the textbook covers interactive use of Linux via the *Graphical User Interface* (GUI) and the *Command-Line Interface* (CLI), including comprehensive treatment of the Gnome desktop and the Bash Shell. Using different commands and filters, building pipelines, and matching patterns with regular expressions are major focuses.

Next comes Bash scripting, file system structure, organization, and usage, which bring us to about the middle of the book.

Chapters 7 and 8 present networking, the Internet and the Web, and data encryption, as well as Web hosting. The Linux Apache MySQL PHP (LAMP) Web hosting combination is presented in depth. Such practical knowledge can be valuable for many Linux programmers.

In Chapters 9–11, attention is then turned to C-level programming. Because the Linux kernel and most of its applications are implemented in it, C is considered the native language of Linux. In-depth knowledge of Linux requires understanding the *Standard C Libraries* and the *system calls* which form the interface to the Linux kernel. Topics covered include the C compiler, preprocessor, debugger, I/O, file manipulation, process control, inter-process communication, and networking. Many complete example programs, written in the standard ISO C99, are provided.

Chapter 12 deals with GUI programming in Ruby/GTK2, where we present topics such as widgets, layout, event-driven programming, and object orientation. Examples show how to build GUI applications for Linux.

Appendices and the bibliography for the book are kept on the book's website (`http://ml.sofpower.com`).

Flexible Usage

This book is for people who wish to learn Linux and to become good at using it and writing programs in it. The book does not assume prior knowledge of Linux or UNIX, but has the depth to satisfy even those with Linux experience.

Compared to other Linux books, this text is not a thick volume. However, it presents many topics comprehensively and in depth. Many examples are given to illustrate concepts and usage. It is well-suited for a one-semester course. An instructor can cover all the chapters in sequence or choose among them depending on the class being taught.

For an *Introduction to Linux* course, the chapters on C-level programming and perhaps on Web hosting can be skipped.

For a system programming-oriented course, the Linux primer, interactive use of Bash, and the GNU desktop material can be omitted or assigned for reading at the beginning of the class. This will provide more time for the hard-core topics on programming.

For an *introduction to operating system principles* course, this book is a good supplement. Discussion of Linux subjects—the Shell, file system structure, concurrent process management, I/O structure, signals/interrupts, and inter-process communication— provides concrete examples and adds to the students' understanding of the abstract operating system principles being studied.

For a server-side Web programming course, the coverage of Bash, file system, Internet and the Web, and Web hosting can make this book a great supplemental text.

For courses on network programming, graphics, C programming, distributed computing, etc., the book can be a valuable supplement as well.

For those who use Linux in school or at work, this book enables you to apply the system's capabilities more effectively, resulting in much increased productivity. Ready-to-use examples provide many immediate practical applications.

Going beyond, you can learn how to write programs at the Shell and the C levels. This ability enables you to build new capabilities and custom tools for applications or R&D.

Appendices Online

The table of contents lists all the appendices, but to reduce the volume of the book, we are keeping the appendices online at the book's website. The appendices in PDF are easier to use and search. The appendix "Secure Communication with SSH and SFTP" is actually a Web page so we can supply download and usage hyperlinks. This arrangement also allows us to add and improve the appendices in the future.

Example Code Package

Throughout the book, concepts and usages are thoroughly explained with examples. Instead of using contrived examples, however, every effort has been made to give examples with practical value and to present them as complete programs ready to run on your Linux system.

These programs are collected in an *example code package* ready to download from the companion website (`http://ml.sofpower.com`). See page 405 for instructions on downloading and unpacking the example code package. The description for each example program is cross-referenced to its file location with a notation such as (**Ex: ex05/argCheck.sh**).

Easy Reference

You'll find a smooth readable style uncharacteristic of a book of this type. Nevertheless, it is understood that such books are used as much for reference as for concentrated study, especially once the reader gets going on the system. Therefore, information is organized and presented in a way that also facilitates quick and easy reference. There are ample resource listings and appendices (at the website) and a thorough and comprehensive index. The in-text examples are also cross-referenced with files in the example code package. This book will be a valuable aid for anyone who uses tools, accesses the Internet, or writes programs, under Linux, even occasionally.

Acknowledgments

I would like to thank the editors Alan Apt, Sunil Nair, and David Tumarkin, as well as others at CRC Press for their help and guidance throughout the writing and production of this book. Their work is indeed much appreciated.

During the planning and writing of this book, several reviews have been conducted. Much appreciated are the input and suggestions from the reviewers:

- Skona Brittain, Santa Barbara City College

- Rob Kolstad, Consultant at Delos Enterprises

- Bill Leahy, Georgia Tech

- Weidong Liao, Shepherd University

Several chapters from the book draft have been used in computer science classes here at Kent State, and I would like to thank the students for reading the materials and providing feedback.

My daughter, Deborah S. Wang, helped proofread the early drafts. I know how busy young people can be, especially someone who just graduated from college and started working full-time. Deb, you make your parents very proud.

Finally, I want to express my sincere gratitude to my wife, Jennifer, whose support and encouragement have been so important to me through the years.

Paul S. Wang

王 士 弘

Kent, Ohio

Contents

Introduction

The term Linux refers to a free and open-source operating system that works, in most respects, exactly like UNIX. Linux became popular as a widely preferred *server platform*. However, with the introduction of the GNOME and KDE desktop user interface environments, together with many important applications, Linux is also gaining ground as a home/office system. Because it is free and open source,[1] Linux is a very attractive teaching tool for computer science and engineering departments. Also, because it is fast and reliable, corporations often choose Linux to run their Web and application servers. Companies in the United States such as *Novell*, *Redhat*, and *Penguin Computing*, as well as many others worldwide, provide products, training, and support for Linux, while the operating system itself remains free.

Lets take a brief look at the history of Linux, its versions and features, and the topics involved in learning how to use Linux.

A Brief History of Linux

The beginning of Linux can be traced back to 1991 when Linus Torvalds, a young student at the University of Helsinki, Finland, began to create an operating system more powerful than MINIX (MIni-uNIX).[2] Three years later, version 1.0 of the Linux *kernel*, the central part of the new UNIX-like system, was released.

The GNU open-source software movement would also later make many contributions to Linux, as remarked upon by Richard Stallman:

> When you are talking about Linux as a OS, you should refer to it as GNU/Linux. Linux is just the kernel. All the tools that make Linux an OS have been contributed by GNU movement and hence the name GNU/Linux.

Linux has come a long way since its early days. Today, it is a prime example of the success of open-source, community-developed software. Linux is used on Web/network servers, desktop computers, laptops, and netbooks. Even the *Google Chrome* operating system is based on the Linux kernel. Linux is

---

[1]Linux is distributed under the *GNU General Public License*.

[2]MINIX is the first open-source clone of UNIX for the IBM PC written by Professor Andrew S. Tanenbaum in 1987

also going mobile with such open-source efforts as *Moblin* (`moblin.org`) and *Maemo* (`maemo.org`). In 2010, Indian engineers designed a Linux-based tablet PC, similar in functionality to the Apple iPad, which was reported to cost as little as $35 USD—and the Indian government hopes to lower the price even further, to perhaps $10.

Linux Versions

Unlike proprietary operating systems, Linux is a combination of open-source programs, including the Linux kernel, GNU tools, desktop managers, and installation and package management systems, plus many other system-, server-, and user-level applications. Anyone can create different combinations of these components, perhaps also change or improve them, and create a *Linux distribution* with unique characteristics. Thus, it is not surprising that many companies and groups all over the world have been distributing somewhat different versions of Linux ready to install on your computer.

Top Linux versions, in terms of user base, include

- Ubuntu—"Ubuntu" means "humanity" in Zulu. Ubuntu Linux started as a version of the popular Debian GNU/Linux. All versions of Ubuntu Linux are free, and there is no charge for mailing a CD to you. Ubuntu supports the GNOME Desktop environment, while another version, *Kubuntu*, uses the KDE Desktop. Ubuntu is easy to install and very user friendly, which has quickly made it the most popular version of Linux. Ubuntu is sponsored by the U.K.-based Canonical Ltd., owned by South African entrepreneur Mark Shuttleworth.

- Red Hat Enterprise Linux—The original *Red Hat Linux* started in 1994 and was discontinued by Red Hat Inc. in 2004. The company now focuses on *Red Hat Enterprise Linux* (RHEL) for business environments and on *Fedora* as a community-supported software project for home, personal, and educational use.

- CentOS—RHEL largely consists of free and open-source software, but the executables are made available only to paying subscribers. CentOS (Community ENTerprise Operating System) is a completely free version of RHEL (minus the Red Hat logos) made available to users as new versions of RHEL are released.

- Fedora—Fedora is a Linux distribution where features and improvements are tested before being included in RHEL/CentOS.

- openSUSE—This is a major retail Linux distribution supported worldwide by Novell. Novell acquired the SuSE Linux (a German translation of the original *Slackware Linux*) in 2004. In the following year, Novel decided to make the SUSE Professional series more open as a community-developed, open-source software and to rename it *openSUSE*.

- Debian—Debian Linux consists entirely of free and open-source software. Its primary form, Debian GNU/Linux, is a popular and influential Linux distribution. Debian is known for an abundance of options. Recent releases include over 26,000 software packages for all major computer architectures. Ubuntu is a derivative of Debian.

There are many other Linux distributions, but all versions are very similar. In fact, Linux is basically a brand of UNIX and can be studied and used as such. This textbook addresses features common to most Linux systems and indicates important differences where appropriate.

The UNIX/Linux Philosophy: *Small Is Beautiful*

The UNIX philosophy influenced not just the original operating system developed by Ken Thompson at Bell Labs (1969), but also the many UNIX clones and UNIX-like systems created afterward. Taken together, these UNIX-like systems are some of the very best operating systems developed to date.

The generally agreed-upon central tenants of the UNIX Philosophy can be listed as

- Keep programs small—Write a program to do one well-defined task; do it efficiently, and do it well.

- Avoid verbosity—Perform no unessential output from any programs; use short names for commands and command options.

- Make programs modular—Build small, independent, and self-sufficient program parts, with each serving a specific function. These program parts can be combined flexibly to form larger programs. This principle is reflected in the small kernel (core of the operating system) cooperating with a large set of small commands which work well together.

- Compose programs through interfaces—Write programs that are easy to interface with other programs. The famous UNIX pipe, which interfaces the output of a program to the input of another, is a primary example of this philosophy.

Keeping program input/output, configuration, and documentation to plain text (character strings) as much as possible makes elements of the operating system easy to interface, read, understand, and improve.

Linux systems have generally adhered to these principles of UNIX, but have also introduced refinements and improvements.

Linux Features

Linux incorporates all the outstanding UNIX core features and adds graphical user interface (GUI), package management, and many useful applications. Important features of Linux include

- *Multi-user and multi-processing*—The ability to allow multiple users to login at the same time and the ability to run many programs concurrently.

- *Graphical user interface*—Offering a *desktop* environment with windows, icons, panels, and menus, making it easy to use point-and-click for operations. Most Linux systems use the *X Window system* and allow the user to choose between two popular desktop environments, *GNOME* and *KDE*.

- *Package management*—A systematic way to find, install, upgrade, configure, and remove the many software packages available. A package contains the executable program and metadata specifying its title, version, purpose, author/vendor, dependencies (on other packages), etc. Packages are made available in *repositories* for downloading. The Red Hat family Linux systems use the **yum** (Yellow dog Updater, Modified) tool and the *rpm* package format, while the Debian varieties use the **apt** (Advanced Packaging Tool) and the *deb* format.

- *Shells*—A *Shell* is a *command-line interface* (CLI) to the operating system. It provides interactive processing and execution of user commands. The standard (default) Shell for Linux is *Bash* (born-again Sh), but you may easily choose to use a different Shell.

- *Hierarchical file system*—The entire file system is tree structured and is anchored at a single directory called the *root*. The root directory contains files and other directories that, in turn, contain more files and directories. Each user has a *home directory* for his/her own files. The file system tree is divided into *volumes*, which can be *mounted* or *dismounted* by attaching them to a node in the file tree. A physical storage device can contain one or several file system volumes.

- *File access control*—Each file in the file system is protected by a sequence of bits whose value is specified by the owner of the file. Access to files is controlled by the operating system. System-wide access is granted to so-called *super users*, usually the system administrators.

- *Compatible file, device, and inter-process I/O*—I/O to physical devices and I/O to a file look the same to a user program. A user can *redirect* a program's I/O so that without changing the program itself, input or output can be directed to a terminal window, file, or even to another program's I/O. The ability to combine and connect existing programs in this *pipeline* fashion provides great power and flexibility.

- *Concurrent processes*—Following UNIX, Linux provides a set of Shell commands and C-language APIs to initiate and manipulate asynchronous concurrent processes. This allows a user to maintain several jobs at once and to switch between them. It is also critical for pipelining several commands (processes) together.

- *Networking and Web Hosting*—As UNIX, Linux systems provide local and wide area networking through *sockets* that support the *Internet Protocol* (IPv4 and IPv6). The `services` icon on the desktop opens a GUI which can be used to easily start/stop different network services. Linux also works well with the *Apache Web Server* to deliver Web pages to the World Wide Web. As a result, Linux is very popular as a network server platform.

- *Utilities*—The Linux architecture encourages building self-contained programs to add new facilities. Linux systems come with many utility programs including text editors, document processors, email servers and clients, Web browsers, raster and vector image editors, scripting languages, language compilers, file manipulation tools, databases, multimedia tools, GUI design and programming tools, software engineering tools, and networking and other system facilities. These utilities usually come in the form of a *Linux package* which can be downloaded, installed, and managed easily with a package manager.

The Linux kernel, the central part of the operating system which provides a programming interface to the hardware, is robust and highly efficient. Figure 1 shows how the Linux kernel relates to various elements on your computer. This textbook covers many of these elements, and this organizational diagram helps to tie them together.

The Linux Environment

Linux is a multi-user, time-sharing system that offers both a GUI (desktop and application windows) as well as a CLI (the Shells). The desktop is the first thing you see after login. The desktop displays one or more *panels* at the top and/or bottom of your screen. A panel provides menus, launchers, and workspace switchers which perform various tasks. Icons on the screen provide access to *Computer*, *Services*, *File System*, and so on.

Application programs fall into two broad categories: GUI applications and CLI applications. A GUI application displays its own graphical window with which the user may interact via the mouse and the keyboard. In contrast, a CLI application must run inside a *terminal window* and interacts with the user only through the keyboard.

Launchers in panels or on the desktop make starting programs easy. However, any program can be invoked by typing a command inside a *terminal*

6

FIGURE 1: Elements of Linux

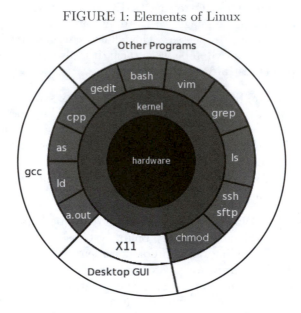

window. You can control and switch among multiple windows on the screen. A command Shell helps you control and manage multiple jobs inside any single terminal window.

The file system contains public files and programs for all users. Each user also has a personal file directory known as the user's *home directory.* Access to files and directories is controlled by the file owner and group designations.

Linux allows a high degree of *customization* on a per-user basis. The Shell, as well as important utilities such as the X Window System and text editors, refers to *initialization* and *configuration* files. You can tailor these files to make the utilities run according to your needs and preferences. You can even choose among different Shells to serve as your CLI. Documentation for Linux and its utilities can be conveniently accessed locally on your computer as well as on the Web.

Learning Linux

Linux systems are used widely in colleges, universities, and companies, both as servers and as workstations. Many users have Linux/Windows dual boot on their personal machines. Knowledge of Linux is important for both learning and employment.

This book covers a set of carefully selected topics that enable you to understand operating system concepts, to use Linux effectively, and to take full advantage of your Linux computer.

The chapters are sequenced in a drill-down progression starting with a *primer* to get you started quickly on Linux with hands-on learning and meaningful tasks.

Next, we present the standard Linux CLI (the Bash Shell) and the interactions with the Linux GUI (the Gnome desktop). Then we discuss how to use commands and filters to build pipelines and how to use *regular expressions* for pattern matching. All this paves the way for writing Bash programs called *Shell scripts*.

Digging deeper, we discuss how to control files and folders and how Linux organizes and manipulates files in a set of *filesystems* that is an important part of the Linux kernel.

Computers are rarely used in isolation, and, like other modern operating systems, Linux relies heavily on networking for many operations. With a good foundation from the earlier chapters, we discuss networking, Web, Internet, and Linux as a platform for Web hosting.

Attention then turns to C-level programming, kernel system calls, processes, and inter-process communication. These topics shed light on the internals of Linux and provide a deeper understanding of concepts and topics covered in earlier chapters. The material should prove especially important for CS/CE majors.

Last, but not least, we cover GUI programming with Ruby/GTK2. Subjects such as widgets, layout, event-driven programming, and object-oriented programming with Ruby can be interesting to many.

Thus, you will find traditional as well as contemporary topics important for the modern Linux environment. The material in this book applies to most popular Linux systems. The knowledge gained will enable you to use any Linux with ease. Major differences among Linux versions are noted where appropriate.

Because Linux is best learned through frequent experimentation and practice, we begin with a *primer* that gets the new user started quickly. We offer examples and practical ways to use Linux throughout the book. Many examples are provided to illustrate concepts and to demonstrate programming techniques. This textbook also contains an *example code package*[3] which provides complete programs ready to download and run on your computer. The material is presented for smooth reading as a textbook, but also for convenient reference later on.

---

[3]See page 405 for downloading instructions.

Chapter 1

A Linux Primer

If you are serious about computing and understanding how an operating system works, Linux is the system of choice. To learn Linux you must use it, and, of course, to use it you must learn it. Such a paradox is rather common—you probably learned to drive a car this way. You just need some basic help and pointers to get started. Here we present an overview of basics. Once you understand the material in this chapter, you will be able to use the operating system to learn more in each successive chapter. At first, you need a learner's permit to drive a car. Consider this chapter your learner's permit for Linux; with a little practice you will be using Linux almost right away.

Learning Linux involves understanding how to use it from the user level and how to program it from the system level. This primer provides basic information and a selection of topics designed to get you started using Linux quickly. As you read this chapter, try the different commands and features as you come to them. In each case, you are given enough information to get you on the system and learning.

1.1 What Is an Operating System?

The operating system controls a computer and makes it usable. It brings life to the innate electronic hardware components and orchestrates all activities on a computer. The same hardware under a different operating system is literally a different computer.

The operating system provides service and control functions to users, programs, files, operators, display monitors, printers, network connections, and everything else on a computer system. A computer operating is one of the most complicated and sophisticated objects humans ever built.

A modern operating system like Linux consists of two parts: a *kernel* and a set of commands and applications. The kernel deals with central functions, including concurrent program execution, memory management, input/output (I/O), file services, and network interface. Commands and applications supply other operations such as Shells, language compilers, text editors, email processors, Web browsers, software package managers, sound and video tools, and so on.

A *Shell* is a special program that is very important in Linux. It is a com-

mand interpreter that allows users to type commands and run programs. We
will have much to say about the Shell.

1.2   Getting Started: Login and Logout

To access your Linux system, you must have a user account, identified by
a *userid* and a *password*, that have been created by a system administrator.
At most installations, your userid will be your last name or your first initials
and last name (often all lowercase).

Your password is a safeguard against unauthorized use of your computer
account. You need to choose a password of at least eight characters (your
local system may enforce other conventions as well, such as a minimum length
or that there be at least one numeral or symbol). Correctly spelled words or
names of relatives are bad choices for passwords. A sequence containing upper
and lower case characters, digits, symbols, and control characters is usually
better. Since you are the only one who knows your password, you must be
careful with it. Forgetting your password means the system administrator
must create a new one for you. Giving your password to the wrong person
could have even more dire consequences; you could be blamed for whatever
damage is caused, intentionally or otherwise, by the other person. The best
rule is do not tell anybody your password, but keep it written down somewhere
safe.

Once you have a userid and password, you can begin your Linux session.
The first step is the login procedure. Login protects the system against unauthorized use and authenticates the identity of the user. You can use Linux
from the *console* or across a network.

FIGURE 1.1: Linux Login Screen

Desktop Login

Find a computer displaying the Linux *desktop login screen* (Figure 1.1). This
can be the *console* where the keyboard, mouse, and display are directly con-

nected to the computer hardware running the Linux system. Or it can be a different computer on the LAN (Local Area Network). Colleges, universities, and companies often run computer labs with Windows or Mac stations that can access Linux servers and display their desktop screens.

In any case, enter your correct password carefully. If you are a new user and, after several careful tries, you are unable to log in, it may be that the system administrator has not yet established your userid on the computer. Wait a reasonable length of time and try again. If you still have a problem, contact your system administrator.

After login, you'll see your *desktop* displayed (Figure 1.2). The desktop enables the use of full-GUI (Graphical User Interface) applications that allow point-and-click operations with the mouse.

FIGURE 1.2: A Typical Desktop

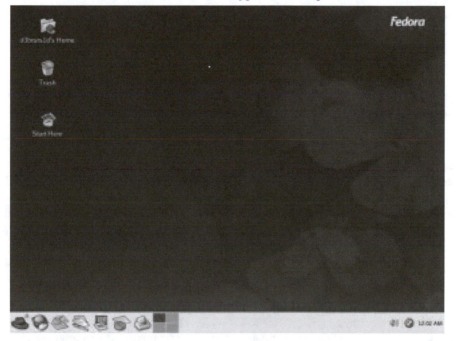

Usually, you will be able to choose between two major desktop environments, *GNOME* or *KDE*, to control and manage your *login session*. From the desktop, you can click on the *start icon*, usually a Linux distribution logo located on the left end of your *desktop Panel* (normally a horizontal bar across the top or bottom of your screen) to pull up the Start menu listing many tasks you can do.

To log out from Linux, you may use the Start menu logout option. Sometimes, a *logout icon* is present on the desktop Panel to make logout easy.

More will be said about desktops in Chapter 3.

Starting a Terminal Window

From the desktop, you can conveniently initiate many operations including starting a terminal window (Figure 1.3) that runs a Shell (Section 1.3). The Shell provides you with a *command-line interface* (CLI) where you can enter commands to perform almost any task on Linux quickly and efficiently.

FIGURE 1.3: A Terminal Emulation Window

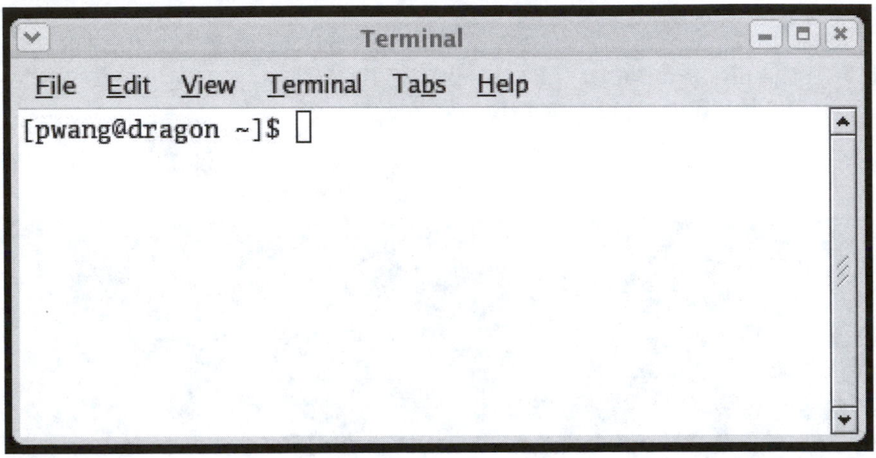

To start a terminal window, go to the `Start` menu and click on the `System tools->Terminal` option or the `Accessories->terminal` option, depending on your Linux and desktop. For the GNOME desktop, the terminal would be **gnome-terminal**, and for KDE it would be **konsole**. A terminal window emulates a character-based computer terminal and allows you to use Linux through a *command interpreter* called the *Shell* (Section 1.3).

As it starts, the Shell also positions you at your *home directory* (Section 1.4), the file folder reserved for you as a user on Linux. The *Shell* indicates its readiness to take your commands by displaying a *prompt* at the beginning of a line.

When you are finished with a terminal window, you may close it by

exit   (exits from Shell and closes the terminal window)
logout   (same as **exit**)

The character CTRL+D (the character **d** typed while holding down the CTRL key) typed alone on a command line often can be used in place of the **exit** command. Exit with CTRL+D is convenient but dangerous, because one typing error can close your terminal window. See Chapter 2 for how to disable exit via CTRL+D.

By the way, we shall use the notation

CTRL+X

to denote a control character, where X is some character. Note also that although the convention is to show an uppercase character, you do not need to hold down SHIFT when typing a control character.

Remember, you need a terminal window to use a Shell, which is your CLI to interact with Linux. Thus, it is important to call up a terminal window quickly and easily. You can create a *keyboard shortcut* to run a terminal window. Go to the `Start` menu

```
Start->System->Preferences->Personal->Keyboard Shortcuts
```

to see all the defined keyboard shortcuts. Find `Run a terminal` and click on that row. The key you type after seeing `New accelerator` will become the keyboard shortcut.

Remote Login

Universities and other institutions often run large Linux servers for users to access through a LAN or even the Internet. You can use TELNET, or more likely SSH (Secure Shell), to access a Linux system from another computer, which can be a PC, another Linux system, or any other platform. Figure 1.4 shows SSH access to a Linux host `monkey.cs.kent.edu` from MS Windows®.

On Linux, the Shell-level command **ssh** provides SSH and is used to access a remote Linux server from a Linux system. For example,

ssh `pwang@monkey.cs.kent.edu`
or
ssh `-X` `pwang@monkey.cs.kent.edu`

networks to the computer `monkey.cs.kent.edu` (the domain name of the computer) and attempts to log in with the userid `pwang`. Remote login normally supports only CLI access. The `-X` (capital `X`) option allows the remote computer to open the graphical display on the local Linux and therefore enables you to also launch remote applications that require a GUI. Running GUI programs remotely involves much heavier network traffic and can be slow.

Without the `-X` option you'll be able to run only command-line applications on the remote computer which is usually the efficient and sensible thing to do. We will return to SSH in Chapter 7 (Section 7.6) where networking is discussed. Download, installation, and usage information for SSH/SFTP can be found appendices on the companion website (`ml.sofpower.com`).

Successful remote login via SSH results in your SSH window being connected to a *login Shell* running on the remote Linux. After login, Linux will record your login in a system log, display a message showing the time and place for your last login, and initiate a Shell to take your commands.

When you see the prompt, you are ready to begin computing. After you are done, you will need to log out from the remote Linux. To log out, first close any programs that you have been running and then issue the Shell-level command **exit** or **logout**. It is a good practice to first close all running

FIGURE 1.4: Login via SSH

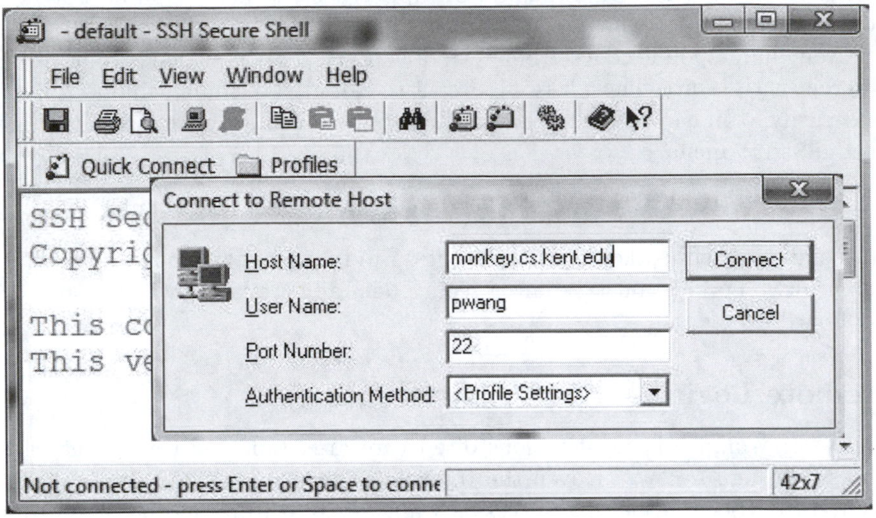

programs manually instead of relying on the logout process to close them for you. It is also good to turn off the power to the display to save energy.

1.3 Understanding the *Shell*

The Shell displays a prompt to signal that it is ready for your next command, which it then interprets and executes. On completion, the Shell resignals readiness by displaying another prompt.

There are several available Shells: the original Shell written by S. R. Bourne known as the *Bourne Shell* or *Sh*, the C-Shell or *Csh* developed at UCB by William Joy, and an enhanced Csh named *Tcsh*. The standard Shell for Linux is the *Bash* (Bourne-Again Sh), which is a powerful and much improved version of Sh. The default Shell on most Linux systems is Bash.

At the Shell prompt, enter the command

echo $0

to display the name of the Shell you are using. Here **echo** displays the value of the Shell variable $0. Don't worry, Chapter 2 explains how this works.

You can change the default Shell with the **chsh** (change Shell) command. For security reasons, usually only approved Shells can be used. In this text we will assume the Bash Shell, although basic features of all Shells are very similar.

Entering Commands

In Linux, you can give commands to the Shell to start application programs, manage files and folders, control multiple jobs (tasks that are running), redirect I/O of programs from/to files, connect one program to another, and perform many other tasks. Virtually anything you want done in Linux can be accomplished by issuing a command to the Shell.

Many different commands are available, but some general rules apply to all of them. One set of rules relates to *command syntax*—the way the Shell expects to see your commands. A command consists of one or more words separated by blanks. A *blank* consists of one or more spaces and/or tabs. The first word is the *command name* (in this book the name of a command will appear in **boldface**); the remaining words of a command line are *arguments* to the command. A command line is terminated by pressing the RETURN (or ENTER) key. This key generates a NEWLINE character, the actual character that terminates a command line. Multiple commands can be typed on the same line if they are separated by a semicolon (;). For example, the command

ls *folder*

lists the names of files in a folder (directory) specified by the argument *folder*. If a directory is not given, **ls** lists the *current working directory* (Section 1.4).

Sometimes one or more *options* is given between the command name and the arguments. For example,

ls -F *folder*

adds the -F (*file type*) option to **ls** telling **ls** to display the name of each file, or each *filename*, with an extra character at the end to indicate its file type: / for a folder, * for an executable, and so on.

At the Shell level, the general form for a command looks like

command-name [*options*] ... [*arg*] ...

The brackets are used to indicate *optional* parts of a command that can be given or omitted. The ellipses (...) are used to indicate possible repetition. These conventions are followed throughout the text. The brackets or ellipses themselves are not to be entered when you give the command.

Command options are usually given as a single letter after a single hyphen (-). For example, the *long listing option* for the **ls** command is -l. Such single-letter options can sometimes be hard to remember and recognize. Many Linux commands also offer full-word options given with two hyphens. For example, the --help option given after most commands will display a concise description of how to use that particular command. Try

ls --help

to see a sample display.

After receiving a command line, the Shell processes the command line as a character string, transforming it in various ways. Then, the transformed command line is executed. After execution is finished, the Shell will display a prompt to let you know that it is ready to receive the next command. Figure 1.5 illustrates the Shell command interpretation loop. *Type ahead* is allowed, which means you can type your next command without waiting for the prompt, and that command will be there when the Shell is ready to receive it.

Trying a Few Commands

When you see the Shell prompt, you are at the Shell level. Now type

echo `Hello Linux`

FIGURE 1.5: Command Interpretation Loop

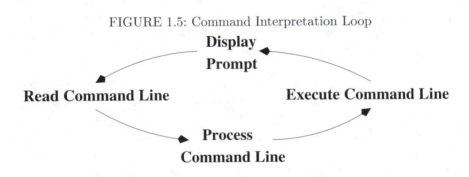

You'll see that the **echo** command displays what you type. Next, enter

echo `-n "Hello Linux "; echo user`

This command line contains two commands separated by the ; command separator. (If you make a mistake typing these commands, glance ahead to the next subheading on correcting typing mistakes.) The option `-n` causes **echo** to omit a NEWLINE character at the end of its output, so the word `user` appears on the same line as `Hello Linux`. Note also the use of quotation marks for the string `Hello Linux` which has a trailing SPACE.

One use of **echo** is to examine the value of a *Shell variable*. For example, if you type

echo `$HOME`

you'll see the value of the Shell variable `HOME` which is the location of your home directory in the file system. Note that the *value* of a Shell variable is obtained by prefixing the variable name with a dollar sign (`$`). More on Shell variables can be found in Chapter 2.

A computer on a network is known as a *host* and is usually identified by a *hostname*. To find out your Linux system's hostname, give the command

hostname

To identify the operating system version running on your computer, enter the command

uname --all

Another command is **who**. Type

who

to list current users signed in on the system. This gives you an idea of how many people are sharing the computing facility.

The **ls** command will not list *hidden files*, any file whose name begins with the period (.) character, unless the **-a** option is given.

ls -a

lists the names of all your files, including the hidden ones. Hidden files are usually standard operating system or application files for configuration or other prescribed purposes and ought not be mixed with other files created by the user.

For the Bash Shell, one standard file is `.bash_profile` in a user's home directory. You can place in this file your personal initialization to be used when **bash** starts as a login Shell.

If you are curious about what's in the file `bash_profile.`, type the command

more .bash_profile

to display its contents. Press SPACE to continue to the next page or q to quit from the **more** display. Don't be discouraged by what you don't understand in this file. When you have progressed further in this book, the contents will become clear.

The Linux system keeps track of the time and date precisely, as you would expect any computer to do. The command

date

displays the current date and time as given by the following typical output showing Eastern Daylight Time

```
Thu Dec  4 16:37:07 EST 2008
```

The Linux system has a dictionary of words for spell checking purposes. The command

spell *file*

will display suspected misspellings for you. Or you can use

aspell -c *file*

to interactively spell check the given *file*. To look for words, you can use

look *prefix*

on most Linux systems, and all words in the dictionary with the given prefix
are displayed.

Another useful command is **passwd**. Type

passwd

to change your password. This command will prompt as follows

```
Changing password for your userid
Old password:
New password:
Retype new password:
```

pausing after each prompt to wait for input. Many Linux installations give out
new userids with a standard password, and the new user is expected to use
the **passwd** command to change to a personal password as soon as possible.

The command **man** consults the on-line manual pages for most commands.
Thus,

man *command*

will display the on-line documentation for the given command. Try

man passwd

just to see what you get. Learn about **man** with

man man

Details on the **man** command can be found in Section 1.11.

The **man** command documents *regular commands* (application programs),
but normally not commands built in to Shells or other application programs.
For Bash you can use

help *builtin_command*

to see a summary of any particular Bash built-in command. Many Linux
systems add a `Bash_Builtins` man page so the **man** command will work for
Bash built-in commands as well.

Correcting Typing Mistakes

As you entered the preceding commands, you may have made at least one keystroke error, or you may wish to reissue a command you have entered previously. Linux Shells provide easy ways to correct typos and to reuse previous commands. Basically, you can use the arrow keys to move the character cursor left and right on a command line and up to a previous command or down to the next command.

The DELETE (BACKSPACE) key deletes the character under (before) the cursor. The ENTER (RET) key issues the command no matter where the cursor is on the line.

The Bash Shell has great support for editing the command line. It actually allows you to pick a text editor to help do the job. We will return to this in Chapter 2, Section 2.3.

Aborting a Command

Apart from correcting typing mistakes, you can also exercise other controls over your interaction with Linux. For instance, you may abort a command before it is finished, or you may wish to halt, resume, and discard output to the terminal window.

Sometimes, you may issue a command and then realize that you have made a mistake. Perhaps you give a command and nothing happens or it displays lots of unwanted information. These are occasions when you want to abort execution of the command.

To abort, simply type the *interrupt character*, which is usually CTRL+C. This interrupts (terminates) execution and returns you to the Shell level. Try the following

1. Type part of a command.

2. Before you terminate the command, press CTRL+C.

It cancels the command and gives you a new prompt.

Exercise A

1. How do you start a terminal window?

2. What command and option should be used to list all the files in your home directory?

3. Set up `ctrl+alt+T` as the keyboard shortcut for running a terminal window.

4. What command is used to change your password? Can you change you password to something like `123`? Why? Make up a longer password and

change your password to it. Why did you have to type your password twice this time?

5. Try input editing with the arrow keys under Bash. After doing a command **ls -l**, press UP-ARROW once and LEFT-ARROW twice. Where is the cursor now? Now, press RIGHT-ARROW once and the cursor should be over the letter l which is the last character on the command line. Can you press RIGHT-ARROW again to move the cursor beyond l? If not, can you find a way? (Hint: Limit yourself to using only the arrow keys.)

6. What is the hostname of your Linux computer? How do you obtain this information?

1.4   Using Files and Directories

FIGURE 1.6: A Sample File Tree

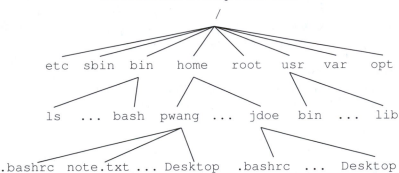

Like other modern operating systems, Linux stores files for users, applications, and the operating system itself on hard disks for ready access. The structure used to store and manage such files is called a *file system*. Files under Linux are organized into a tree structure with a root named by the single character /.

A *regular file* stores a program or data. A *directory* or *folder* contains files and possibly other directories. Internal nodes on the Linux file tree represent directories; leaf nodes represent regular files. This *hierarchical* file structure is widely used by different operating systems. A sample Linux file tree is shown in Figure 1.6.

By clicking on the `computer` icon then the `File System` link, you can launch a visual *file browser* (Figure 1.7) utility which allows you to navigate the file system and perform operations on files and folders. The way to reach the *file browser* may depend on the Linux version you use. While the *file browser* makes moving about the file system more visual, many Linux users still find dealing with files and folders via the Shell command line more efficient.

Current Working Directory and Filenames

FIGURE 1.7: Linux File Browser

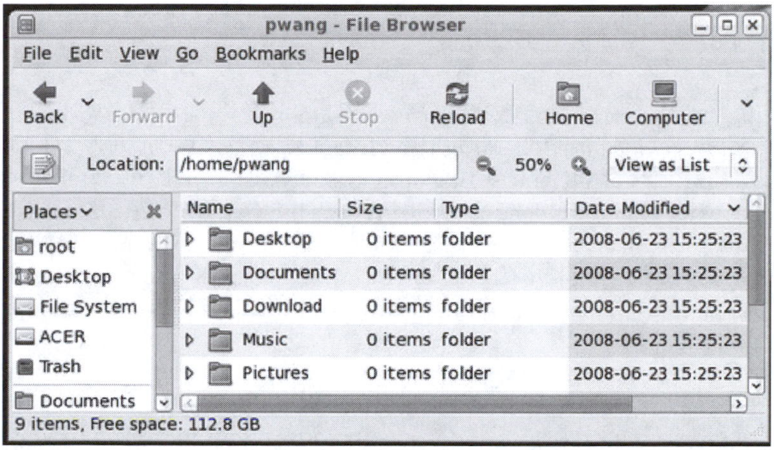

When you get a userid and account on your Linux system, you are given a personal file directory known as your *home directory*. Your home directory will have your userid as its name, and it will usually be a child of a directory called `home`. Your files and folders are kept in your home directory.

To access a file or directory in the file system from the command line, you must call it up by its name, and there are several methods to do this. The most general, and also the most cumbersome, way to specify a *filename* is to list all the nodes in the path from the root to the node of the file or directory you want. This path, which is specified as a character string, is known as the *absolute pathname*, or *full pathname*, of the file. After the initial /, all components in a pathname are separated by the character /. For example, the file `note.txt` in Figure 1.6 has the absolute pathname

```
/home/pwang/note.txt
```

The full pathname is the complete name of a file. As you can imagine, however, this name often can be lengthy. Fortunately, a filename also can be specified relative to the *current working directory* (also known as the *working directory* or *current directory*). Thus, for the file `/home/pwang/note.txt`, if the current working directory is `/home`, then the name `pwang/note.txt` suffices. A *relative pathname* gives the path on the file tree leading from the working directory to the desired file. The third and simplest way to access a file can be used when the working directory is the same as the directory in which the file is stored. In this case, you simply use the filename. Thus, a Linux file has three names

- A full pathname (for example, `/home/pwang/note.txt`)

- A relative pathname (for example, `pwang/note.txt`)

- A (simple) name (for example, `note.txt`)

The ability to use relative pathnames and simple filenames depends on the ability to change your current working directory. If, for example, your working directory is `/tmp` and you wish to access the file `note.txt`, you may specify the absolute pathname

`/home/pwang/note.txt`

or you could change your working directory to `pwang` and simply refer to the file by name, `note.txt`. When you log in, your working directory is automatically set to your home directory. The command

pwd   (print working directory)

displays the absolute pathname of your current working directory. The command

cd *directory*   (change working directory)

changes your working directory to the specified directory (given by a simple name, an absolute pathname, or a relative pathname).

Two *irregular files* are kept in every directory, and they serve as pointers

File .   is a pointer to the directory in which this file resides.
File ..   is a pointer to the *parent* directory of the directory in which this
           file resides.

These pointers provide a standard abbreviation for the current directory and its parent directory, no matter where you are in the file tree. You also can use these pointers as a shorthand when you want to refer to a directory without having to use, or even know, its name. For example, the command

cd .

has no effect, and the command

cd ..

changes to the parent directory of the current directory. For example, if your working directory is `jdoe`, and you want to access the file `sort.c` in the `pwang` directory, you may use `../pwang/sort.c`. Why does this work?

Your home directory already comes with a name, your userid. However, you name your files and subdirectories when you create them. Linux is lenient when it comes to restrictions on filenames. In Linux you may name your file with any string of characters except the character `/`. But, it is advisable to avoid white space characters and any leading hyphen (`-`).

Handling Files and Directories

Generally, there are two kinds of regular files: text and binary. A Linux text file stores characters in ASCII or UNICODE and marks the end of a line with the NEWLINE character.[1] A binary file stores a sequence of bytes. Files may be copied, renamed, moved, and destroyed; similar operations are provided for directories. The command **cp** will copy a file and has the form

cp *source destination*

The file *source* is copied to a file named *destination*. If the destination file does not exist, it will be created; if it already exists, its contents will be overwritten. The **mv** (move) command

mv *oldname newname*

is used to change the file *oldname* to *newname*. No copying of the file content is involved. The new name may be in a different directory—hence the name "move." If *newname* already exists, its original content is lost.

Once a file or subdirectory has outlived its usefulness, you will want to remove it from your files. Linux provides the **rm** command for files and **rmdir** for directories

rm *filename1 filename2 ...*
rmdir *directoryname1 directoryname2 ...*

The argument of **rm** is a list of one or more filenames to be removed. **rmdir** takes as its argument a list of one or more directory names; but note, **rmdir** only will delete an empty directory. Generally, to remove a directory, you must first clean it out using **rm**.

To create a new directory, use the **mkdir** command, which takes as its argument the name of the directory to be created

mkdir *name*

When specifying a file or directory name as an argument for a command, you may use any of the forms outlined. That is, you may use either the full pathname, the relative pathname, or the simple name of a file, whichever you prefer.

Standard Personal Directories

It is easy to change to a home directory, just do

cd   (goes to your home directory)
cd ˜*userid*   (goes to the home directory of *userid*)

---

[1]On Windows or DOS systems, end of line is indicated by RETURN followed by NEWLINE.

In Linux, there are a number of standard folders under each user's home directory, usually including

- `Desktop`—Files in this folder appear as icons on your graphical desktop display, including regular files and *application launchers* (with filename suffix `.desktop`)

- `Documents`—Textual documents such as PDF files and those created using tools such as **openoffice.org**

- `Download`—Files downloaded from the network

- `Music`—Sound and music files

- `Pictures`—Pictures from digital cameras

- `public_html`—Files under this folder are made available to the Web via an HTTP server on your Linux system

- `Videos`—Files from video cameras and recorders

In addition to these, you may consider setting up a `bin/` for your own executables, a `tmp/` for temporary files, a `templates/` for reusable files, a `homework/\` for your classes, and so on.

1.5   Protecting Files: Access Control

Every file has an owner and a group designation. Linux uses a 9-bit code to control access to each file. These bits, called *protection bits*, specify access permission to a file for three classes of users. A user may be a *super user*, the owner of a file, a member in the file's group, or none of the above. There is no restriction on super user access to files.

u   (The owner or creator of the file)
g   (Members in the file's group)
o   (Others)

The first three protection bits pertain to u access, the next three pertain to g access, and the final three pertain to o access. The g type of user will be discussed further in Chapter 8.

Each of the three bits specifying access for a user class has a different meaning. Possible access permissions for a file are

r   (Read permission, first bit set)
w   (Write permission, second bit set)
x   (Execute permission, third bit set)

The Super User

Root refers to a class of super users to whom no file access restrictions apply. The *root* status is gained by logging in under the userid `root` (or some other designated root userid) or through the **su** command. A *super user* has read and write permission on all files in the system regardless of the protection bits. In addition, the super user has execute permission on all files for which anybody has execute permission. Typically, only system administrators and a few other selected users ("gurus" as they're sometimes called) have access to the super user password, which, for obvious reasons, is considered top secret.

Examining the Permission Settings

The nine protection bits can be represented by a 3-digit octal number, which is referred to as the *protection mode* of a file. Only the owner of a file or a super user can set or change a file's protection mode; however, anyone can see it. The **ls -l** listing of a file displays the file type and access permissions. For example,

```
-rw-rw-rw- 1 smith 127 Jan 20 1:24 primer
-rw-r--r-- 1 smith  58 Jan 24 3:04 update
```

is output from **ls -l** for the two files `primer` and `update`. The owner of `primer` is `smith`, followed by the date (January 20) and time (1:24 A.M.) of the last change to the file. The number 127 is the number of characters contained in the file. The *file type*, *access permissions*, and *number of links* precede the file owner's userid (Figure 1.8). The protection setting of the file `primer` gives read and write permission to u, g, and o. The file `update` allows read and write to u, but only read to g and o. Neither file gives execution permissions. There are ten positions in the preceding mode display (of **ls**). The first position specifies the file type; the next three positions specify the r, w, and x permissions of u; and so on (Figure 1.8). Try viewing the access permissions for some real files on your system. Issue the command

```
ls -l /bin
```

to see listings for files in the directory /bin.

FIGURE 1.8: File Attributes

file type	user access	group access	other access	links	userid	size	date	time	file name
↓	↓	↓	↓	↓	↓	↓	↓	↓	↓
-	rw-	r--	r--	1	smith	127	Jan 24	2:04	update

Setting Permissions

A user can specify different kinds of access not just to files, but also to directories. A user needs the x permission to enter a directory, the r permission to list filenames in the directory, and the w permission to create/delete files in the directory.

Usually, a file is created with the default protection

```
-rw-------
```

so only the file owner can read/write the file. To change the protection mode on a file, use the command

chmod *mode filename*

where *mode* can be an octal (base 8) number (for example, 644 for rw-r--r--) to set all 9 bits specifically or can specify modifications to the file's existing permissions, in which case *mode* is given in the form

who op permission op2 permission2 ...

Who represents the user class(es) affected by the change; it may be a combination of the letters u, g, and o, or it may be the letter a for all three. *Op* (operation) represents the change to be made; it can be + to add permission, – to take away permission, and = to reset permission. *Permission* represents the type(s) of permission being assigned or removed; it can be any combination of the letters r, w, and x. For example,

chmod o-w *filename*
chmod a+x *filename*
chmod u-w+x *filename*
chmod a=rw *filename*

The first example denies write permission to others. The second example makes the file executable by all. The third example takes away write and grants execute permission for the owner. The fourth example gives read and write permission (but no execute permission) for all classes of user (regardless of what permissions had been assigned before).

A detailed discussion on the Linux file system can be found in Chapter 6.

Exercise B

1. Go to your home directory and list all files (hidden ones included) together with the permission settings.

2. Using the **ls** command, list your files in time order (most recent first).

3. List the permission settings of your home directory. Use the **chmod** command to make sure to forbid read and write from g and o.

4. Create a folder `public_html` directly under you home directory and make sure you open read and execute permissions on this folder.

5. Connect your digital camera to your Linux box and download pictures. Where are the pictures placed. Can you find them under your `Pictures` folder?

1.6 Text Editing

FIGURE 1.9: Gedit

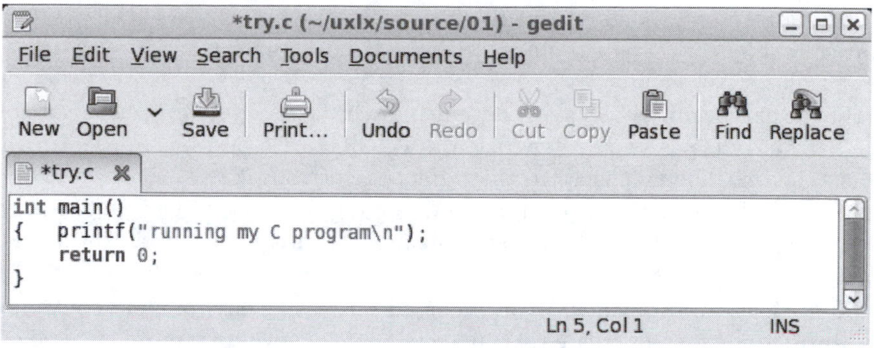

Creating and editing text files is basic to many tasks on the computer. There are many text editors for Linux, including **gedit**, **nano**, **vim/gvim/vi**, and **emacs**.

The editor **gedit** (Figure 1.9) comes with the GNOME desktop. It requires almost no instructions to use. Start it from the `Start` menu `Text Editor` or the command

gedit *file* &

An editor window will display. Then you can type input; move the cursor with the arrow keys or mouse; select text with the mouse; remove text with the DELETE or BACKSPACE key; and find, cut, copy, and paste text with the buttons provided or with the `edit` menu options. It is very intuitive.

The **gedit** is a GUI application. If you want a terminal-window–based editor then consider **nano**, which is very easy to learn but is less powerful or convenient than **vim** or **emacs**. Guides to **vim** and **emacs** can be found in the appendices on the companion website (`ml.sofpower.com`).

Editing power aside, there is something to be said about an editor that is easy and intuitive for simple tasks, especially if you are a beginner. In any case, pick a text editor and learn it well. It can make life on Linux so much easier.

To invoke the editor **nano** for editing *file*, type from the Shell level

FIGURE 1.10: Nano

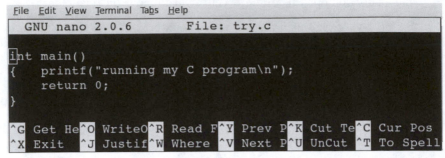

nano *file*   (starts **nano**)
nano -w *file*   (starts **nano** without line wrapping)

If the file exists, **nano** displays it for editing. Otherwise, you are creating a new file by that name. As you enter text, **nano** will start a new line automatically when the text line gets close to the right edge of your editor window. The -w option asks for no such automatic line wrapping. It is also advisable to always use the -z option which allows you to suspend **nano** and get back to the Shell level.

Once inside **nano**, you are working in a text-editing environment controlled by **nano**, and you can create text, make changes, move text about, and so on. Common operations are indicated by the **nano** editor window (Figure 1.10). Here is a list to get you started.

- To save the file, type CTRL+O.

- To quite and terminate **nano**, type CTRL+X. You can then elect whether to save the buffer or cancel to change your mind about quiting.

- To move the cursor, use the arrow keys.

- To cut and paste whole lines, CTRL+K cuts one line at a time and CTRL+U pastes the lines cut.

- To cut and paste selected text, type CTRL+6, move the cursor to highlight selected text, and then use CTRL+K and CTRL+U.

- To look for text in the editing buffer, type CTRL+W (where), the text to find, and ENTER or RETURN.

- To get help on operations, type CTRL+G.

1.7   Getting Hard/Saved Copies

To get a printed copy of a file use

lpr [*options*] *filename*

This command sends *filename* to a printer. Your printing request joins a queue of such requests that are processed in order. Note that only text files (plain text, postscript, or pdf) can be printed this way. Do not send a binary file, such as a compiled program (`.o` file), or a compressed file to a printer this way. The `print` option on the `file` menu of application programs, such as your Web browser, PDF (Portable Document Format) reader, or document editor (**openoffice.org** for example), can also be used.

Often, you can avoid wasting paper by using the `print to file` option. You can easily save the resulting file (mostly in PDF) and share with others by email or SFTP (Secure File Transfer Protocol, Chapter 5, Section 5.20).

1.8   Communicating with Others

As soon as you log in, you can potentially interact with others, whether they are users on the same Linux computer or on other *hosts* (computers) connected by networking. Commands such as **who** (who is logged in) and **finger** help to identify members of your user community; email applications allow the sending and receiving of messages and files; and *instant messaging* (IM) programs enable immediate interaction among on-line users anywhere on the Internet.

Who's Who on the System: `finger`

If you are a new user, you may not know many people on the system, and although the information provided by **who** and **w** is useful, you don't know who these users are. You only know their userids, which may not resemble their actual names even faintly. The command **finger** will give you such data as full name, office, address, and phone number for each user; this is sometimes referred to as the *finger database*, because finger is used to look up information from this database. The general form is

finger *name* ...

This command will display all entries in the finger database that contain a userid and first, middle, or last name matching any of the given arguments. For example, either **finger** `smith` or **finger** `clyde` will result in the entry shown in Figure 1.11.

This multiline output includes a *project* field, which is the first line in the `.project` file in the user's home directory. The *plan* field these two files to supply additional information about themselves for the finger database. The

FIGURE 1.11: A Sample **finger** Output

```
Login name: csmith    In real life: Clyde Smith
(803) 555-5432
Directory:/user/grad/csmith    Shell:/bin/bash
Last login Tue May 27 14:49 on ttyhd
Project: Automation Technology Research
No Plan.
```

no plan line in the example indicates that csmith has no .plan file. On some systems, **finger** gives only a very short summary unless the −l option is given.

Used with an argument, **finger** will access information on any user known to the system, whether that user is logged on or not. If **finger** is used without an argument, an abbreviated finger entry is displayed for each user currently logged in. The **f** command is sometimes available as a shorthand for **finger**.

FIGURE 1.12: Thunderbird Email Program

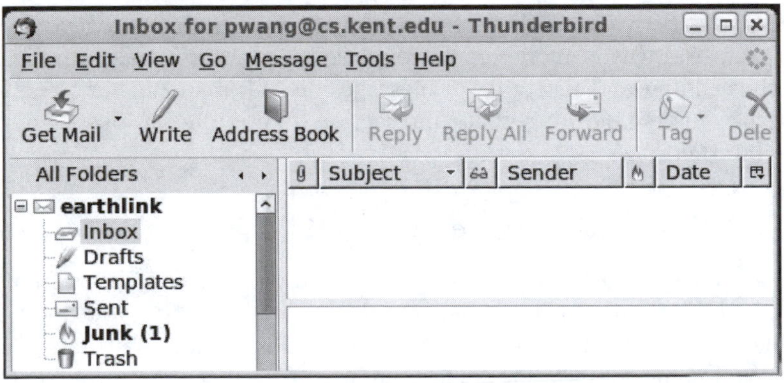

Email

Electronic mail gives you the ability to send and receive messages instantly. A message sent via *email* is delivered immediately and held in a user-specific mailbox for each recipient. You can send email to users on the same computer or on other computers on the Internet.

Many utilities are available on Linux for email, including the popular *Mozilla Thunderbird* (Figure 1.12), *Evolution*, and *Kmail*. These full-GUI email programs are nice when you are at a Linux console. Command-line email programs such as *elm* and *mutt* are useful from a terminal window. Let's explain how to use **mutt**.

mutt *userid@host-address*   (Internet mail)
mutt *userid*   (local mail)

Then just follow instructions and enter the message subject and type/edit your message. **Mutt** lets you edit your message using your favorite text editor. For **mutt** and many other applications that need text editing, set your favorite editor by giving a value to the *environment variable* `EDITOR` (Chapter 2, Section 2.10).

```
EDITOR=vim or EDITOR=emacs
export EDITOR
```

When you finish editing your message, it will be sent out automatically.

```
mutt --help | more
```

displays more information on **mutt** usage. Here, the output is piped via the | notation (Chapter 2, Section 2.5) to the **more** paginator which displays the information one page at a time.

To receive email (to check your mailbox), type **mutt** with no argument and follow instructions. Try to send yourself some email to get familiar with the usage.

FIGURE 1.13: IM with Pidgin

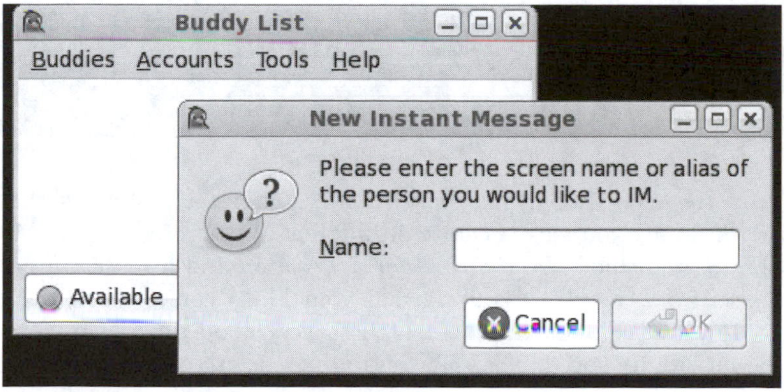

Instant Messaging

Email is fast, but not instant or interactive. On Linux, you can do IM. For example, you can use **pidgin** (Figure 1.13) to IM with friends and business contacts who are online and available. Pidgin is a free IM application capable of working with many different IM services, including AIM by AOL, GTalk by Google, and MSN by Microsoft.

To get started, simply invoke pidgin:

pidgin &

and add your your screen name and password for each IM service you wish to use. Then build up your buddies list and enjoy instant communication whenever you like. The final **&** character causes **pidgin** to run *in the background* so you can use the Shell to perform other tasks.

1.9   Browsing the Web

FIGURE 1.14: Firefox Browser Accessing Linux Documentation

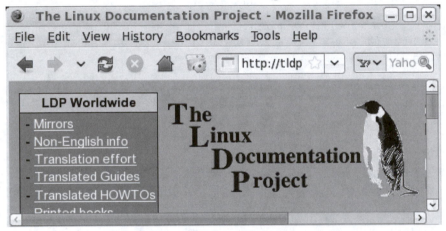

One of the most important tools on any computer system is the Web browser. On Linux you have a choice of different Web browsers. The Mozilla-family browsers include *Mozilla, Netscape Navigator,* and *Firefox. Chrome* is a fast browser from Google. It is likely that your Linux comes with the Mozilla browser. However, Firefox (Figure 1.14) is a very good browser preferred by many because of its speed, efficiency, and adherence to open Web standards. You can easily download and install Firefox from `mozilla.com`. Then you'll have the command **firefox** to use.

You can enter a URL (*Uniform Resource Locator*) in the browser `Location` window to visit a specific Web address. A local file URL, taking the form `file://`*full_pathname* can be used to visit your local file system.

Normally, Web browsers are full-GUI programs used interactively, but Linux also offers a command-line Web browser called *lynx,* a text-only browser that does not display images. However, lynx can be used inside a terminal window to interactively browse the Web using the arrow keys or to download files from the Web.

Exercise C

1. Try the **mutt** email program. Use it to send an email and attach a file.

2. Create a text file using **nano**.

3. Try the **vi** or **emacs** editor. Read the related appendix on the book's website.

4. If possible, set up Thunderbird as your email program and Firefox or Chrome as your Web browser.

5. Download a file using **lynx** from the Web.

1.10 Creating and Running Your Own Program

Skip this section if you have no immediate interest in writing a program in a general programming language. You can always return to this section later. The Linux system offers many languages: C, C++, Java, Fortran 77/95, Python, Ruby, and Perl, just to name a few. You can also write Shell scripts (Chapter 5) to automate frequently used operations.

Linux follows a set of conventions for naming different kinds of files. Table 1.1 illustrates some commonly used filename suffixes. A source code file cannot be executed directly. The program usually must be compiled into machine code before execution can take place. An alternative to compilation is to interpret a high-level language program directly using an *interpreter*. We shall

TABLE 1.1: File Name Suffixes

Suffix	File Type	Suffix	File Type
.html	HTML	.c	C source
.C .cpp	C++ source	.java	Java source
.f77 .f95	Fortran source	.jpg	JPEG image
.pdf	Portable Document Format	.o	Object code
.sh	Sh or Bash script	.bash	Bash script
.tar	Tar archive	.so	Shared library

follow an example of creating and running a simple C-language program. Use your favorite text editor and create a C source file `try.c` (**Ex: ex01/try.c**) with the code

```
#include <stdio.h>

int main()
{   printf("running my C program\n");
    return 0;
}
```

This is a simple source program in C that displays the line "running my first program." The notation \n stands for the NEWLINE character.

Compiling

Before `try.c` can be run, it must be compiled. *Compiling* is the process of translating a program written in a high-level language such as C or Pascal into a low-level language for execution on a particular computer. On many systems the compiler will output a file of *object code*, which must be *linked* (combined with routines supplied by the system library) by a separate program called a linker. Once linkage is complete, the file is considered executable and is ready to be loaded into memory and executed.

Linux-based compilers will handle both the compiling and the linking of a program unless you specifically tell them not to, and their output will be an executable file. Available compilers include produce object files (.o).

gcc   GNU C compiler
g++   GNU C++ compiler
javac   Java compiler
gfortran GNU Fortran 77/95 compiler

Let's try compiling the sample program in the file `try.c`

gcc try.c

This will produce an executable file `a.out` which can be invoked simply by typing it as a command.

a.out

Note that in Linux the command to run a program is simply the pathname of the executable file. Thus,

./a.out   (runs the executable)

At some point, you probably will want to name your executable file something other than `a.out`, especially since `a.out` will be overwritten the next time you invoke a compiler in this working directory. We already know that the **mv** command can be used to rename a file, but the -o option to **gcc** or **g++** can be used to provide a name to use instead of the default `a.out`. For example,

gcc -o mytry try.c

produces the executable `mytry`.

No matter which language program you run, you probably will want a record of your results (to submit as a homework, for example). One way to do this is to use *output redirection*. For example,

./a.out > results

The symbol > tells the Shell to *redirect output* of **a.out** from the terminal screen into a new file named **results**. Thus, you will see no output in the terminal window after the preceding command. Instead, you'll find the output in the file **result**. A full account of Shell I/O redirection can be found in Chapter 2, Section 2.5.

Another way to do this is to use the **script** command

script *record_file*

to record your terminal session into a *record_file*. While **script** is active, all I/O to and from your terminal window is written to the file you specified (or to a file named **typescript** if you entered **script** without an argument). Recording stops when you type CTRL+D at the beginning of a command line. The file then can be viewed later with a text editor or emailed to someone. For example, to run a C program with **script**, the following sequence of commands may be used

script `display_record`
cc `myprogram.c`
a.out
CTRL+D

The **script** command requests that all subsequent I/O be recorded in the file `display_record`. The CTRL+D on the last line stops the recording and gets you out of **script** and back to the Shell level.

An advantage of using script over simply redirecting output is that the file produced by script will contain both input to and output from the program. The file created by redirecting output will contain only output.

Exercise D

1. Type in a simple program (in your favorite programming language) to type out the message: `Linux is nice once you know it`. Compile it and run it. Use **script** to get a display record of the program's execution.

2. Use **more** or **nano** to view the file produced by **script** and then send the file to someone by email.

1.11   Consulting Linux Documentation

The *Linux Documentation Project* website `http://tldp.org` (Figure 1.14) provides comprehensive documentation for almost all aspects of Linux. You'll find FAQs, topic-specific step-by-step instructions called *HOWTOs*, guides, and manual pages for commands. A search feature makes it easy to find what you need.

You can also find documentation provided by your own Linux. The `Help` menu on the tool bar of GUI applications, such as the File Browser, the Terminal Emulation Window, and Pidgin, provides tool-specific documentation.

Command-line programs often provide brief and concise usage information with the `--help` command option. For example, try

`ls --help`

The **man** command displays *manual pages* set in standard UNIX-defined formats. Each manual page provides a concise description of a Linux command. The main body of the manual pages is divided into chapters. Although the exact organization of the chapters may vary with the particular Linux system, the following is a typical organization

1. User-level commands

2. Linux system calls in the C language

3. System library functions for C, Fortran, networking, and other purposes

4. Special files, related device driver functions, and networking support

5. Formats for executable and system database files

6. Miscellaneous useful information

7. Linux maintenance, operation, and management

You can use the command

man man

to see the organization of your manual pages. To display an introduction to chapter *n*, type

man *n* intro

To display the manual for *command_name*, type

man [*n*] *command_name*

where the chapter number *n* is optional.

When an application or utility is added to your Linux by downloading and installing it, a manual page for that new program is often also added automatically. For example, after installing Firefox, you can do **man** `firefox` to see its manual page.

A typical manual page contains the following information: `NAME` (and principal purpose), usage `SYNOPSIS`, `DESCRIPTION`, `OPTIONS`, related `FILES`, and `SEE ALSO` (related commands).

If the manual page is too large to fit on one screen, the program will display one page at a time until the entire entry has been shown. You can type q to quit

man and return to the Shell prompt. This is especially useful if the man page is large and you don't want to see it all. The SYNOPSIS part of the manual page gives a concise description of the command syntax. In the synopsis, certain characters are to be typed literally when using the command; other characters or strings are to be replaced by appropriate words or filenames supplied by the user. Portions enclosed in brackets are optional, and the brackets themselves are not part of the command. Ellipses (...) are used in the synopsis to indicate possible repetitions. Most Linux commands receive *options* that modify the behavior of the command. As mentioned earlier, an option is usually given as a single character preceded by a dash (-), but more verbose options are also possible.

The FILES section of the manual page gives the locations of files related to the particular command. The SEE ALSO section gives related commands that may be of interest. The BUGS section lists some known problems with the command.

The command **man** also can perform a keyword search through the name and purpose part of the manual pages, displaying each line containing any of the given keywords. This is done by using the -k option

man -k *keyword* ...

This feature is useful when you want to do something, but can't remember the appropriate command. For example, to figure out how to copy a file, you could try **man** -k copy. The *keyword* can be any character sequence. So you can find a command if you remember only a part of its name or description.

There are also Web page versions of the Linux man pages (for example, linuxmanpages.com) that can be much easier to use as a reference.

Exercise E

1. How do you ask a command to help you use it?

2. Access the man page for **ls** and read it from beginning to end.

3. Explain how to display the introduction section to the user commands chapter of the Linux man pages.

4. Find and describe a way to do a key-word search of the Linux man pages.

5. Where can you find documentation for a command built in to Bash?

6. Download the most recent Web page version of the Linux man pages from www.tldp.org/manpages/man-html/ to your computer. It can be much easier to use than the regular man pages. Hint: See Chapter 6, Section 6.12.

1.12   Rounding Up Useful Commands

In this chapter, we have run into only a small number of the hundreds of Linux commands. The richness and variety of Linux commands are major strengths of the system. It is doubtful, however, that many users know all of them; you learn the commands that accomplish what you need. This section collects the commands that should be in a new user's basic repertoire.

In Linux, both uppercase and lowercase characters are used, and they are not interchangeable. All system-defined Linux commands are entered in lowercase. Also, there are two kinds of commands: (1) built-in Shell commands that are subroutines in the Shell and (2) regular commands that are initiated as jobs controlled by the Shell. The importance of this distinction will become clear. In the following listing of commands, user-supplied arguments are shown in *italics*. Optional arguments are enclosed in square brackets [ ]. Possibly repeated arguments are indicated by ellipses (...). These conventions will be followed throughout the book. Only the most common usages of these commands are given. The information here is intended to get you started and is by no means complete. Details are provided in later chapters, and you should consult the on-line manual for full descriptions of each of these commands.

File-Related Commands

cat *file* ...	Displays on the terminal the contents of the file(s) specified in the order they are specified.
more *file*	Displays the file on your terminal, pausing after each complete screenful and displaying `--more--`, at which point you can press SPACE for another screenful, RETURN for another line, or **q** to quit (the command **less** is similar).
cp *file1 file2*	Makes a copy of *file1* in *file2*, overwriting it if it exists and creating it otherwise.
cp *file* ... *dir*	Makes a copy of the file(s) using the same name in the given directory.
mv *file1 file2*	Renames *file1* as *file2* overwriting it if it already exists.
mv *file* ... *dir*	Moves the file(s) to the given directory.
rm *file* ...	Deletes the files specified.

Login-Related Commands

ssh [*user@*]*host*	Remote login.
chsh	Changes your default Shell.
passwd	Changes your password.
exit or **logout**	Terminates the Shell and closes the terminal window.
.	A single dot at the beginning of a line quits the Shell, unless disabled.
su [*user*]	Enters a subshell as *user* or `root`.

Directory-Related Commands

cd *dir*	Changes the current working directory to the given directory (a Shell built-in command).
pwd	Prints the absolute pathname of the current working directory.
mkdir *dir*	Creates a new directory.
rmdir *dir*	Deletes the directory which must be empty.
ls [*op*] [*file* ...]	Lists the filenames in the current working directory if no file argument is given. For each *file* given, if *file* is a directory, then the files in that directory are listed; if *file* is a regular file, then that file is listed. The -1 option causes the listing to be in "long form." Other options include: -a, which lists all files, including those files whose names begin with the period character (.); -d, which lists the directory entry itself instead of the files in it; and -F, which indicates executable files with a trailing asterisk (*) and directories with a trailing slash (/). Options for **ls** can be combined, as in -1d.

Text Editors

gedit *file*	Standard GUI editor from GNU.
nano *file*	Simple and easy terminal-window editor.
vi *file*	Full-featured original UNIX/Linux editor.
vim *file*	Vi improved.
gvim *file*	GUI version of Vim.
emacs *file*	Flexible and powerful editor.

Informational Commands

date	Displays the date and time of day.
finger *name*	Consults the user database for anyone with the userid or *name* given.
look *prefix*	Displays all words from the on-line dictionary having the given prefix.
man *cmd*	Consults the on-line manual pages dealing with the specified command.
who	Displays a list of current users on the system. The command **w** is the same, but gives more information on each user.

Compiling and Running Programs

cc *prog*	Invokes the C compiler on the program (name ending in .c), producing an executable file a.out.
g++ *prog*	Invokes the C++ compiler on the program (name ending in .C), producing an executable file a.out.
gfortran *prog*	Compiles the given Fortran 77/95 program.
javac (**java**)	Compiles (runs) Java programs.
a.out	Runs the executable file a.out. In general, the file name of an executable file is also the command to execute that file.

Communications and Web Commands

talk *userid* [*@host*]	Initiates a chat.
mutt *userid@host*	Sends email from command line.
pidgin	Sends and receives instant messages.
thunderbird	Sends and receives email.
firefox	Recommended Web browser.
lynx	Terminal-window–based Web browser.

Printing-Related Commands

lpr *file* or (**lp** *file*)	Prints the file on a printer. Linux puts this printing request in a spooling queue. Your file will be printed when it gets to the front of the queue.
lpq (**lpstat**)	Displays the printer queue and job numbers for your print requests.
lprm *job* or **cancel** *job*	Removes your printing request from the queue.
script *file*	Records terminal I/O in the given file. The **script** command creates a subshell to accomplish this task. To stop recording and exit script, use **exit**.

1.13   Summary

Linux provides both full-GUI applications and command-line programs. The GUI is visual and more intuitive to use, but many basic Linux utilities are more convenient on the command line. A Shell (typically *Bash*) running in a terminal window provides a CLI to enter and execute commands. Learning to use both the GUI and the CLI effectively will make life much easier on Linux. The CLI is especially important for remote access of Linux using SSH.

The desktop main menu leads to many useful operations. Chapter 3 presents the Linux desktop. The Shell executes commands you input from the keyboard and displays results in your terminal window. Typing errors can be corrected through input editing.

Both the system and the users store data in files managed by the Linux file system, which has a tree structure. Each file can be referred to by a full (absolute) pathname, a relative pathname, or a simple filename. Each user has a home directory in which personal files and directories can be kept. Files and directories can be created, moved, copied, listed, and destroyed. Read, write, and execute permissions are used to control file access by u (owner), g (group member), and o (others). The owner can set and change the access permissions of a file.

You can communicate directly with other users by using **talk** to chat directly, by email, and by instant messaging (**pidgin**).

Linux offers several text editors. The full-GUI **gedit** is a good choice. For a terminal window, the simple and easy **nano** is good for beginners and light editing tasks. Serious editing is more efficient with an editor such as **vim**. Editing, compiling, and running of a simple C program have been presented.

Linux offers many facilities and a complete set of manuals. The **man** command can be used to consult the manual pages, and the Linux Documentation Project website provides a variety of comprehensive Linux documentations.

Refer to the final section of this chapter for a list of useful commands for Linux beginners.

Chapter 2

Interactive Use of the Shell

An important purpose of any operating system is to provide users with a convenient interface to manage and achieve tasks on their computers. Linux provides a GUI (graphical user interface) and a CLI (command-line interface) in the form of a *Shell*.

A Shell normally runs inside a terminal window such as **gnome-terminal** or **konsole** (Chapter 1, Section 1.2). It takes input from the user (keyboard) and serves as a *command interpreter* to start applications and to perform all other available operations.

When accessing a Linux system from another host, such as a PC (Windows or Mac) or Linux box, through a remote login program such as SSH (Chapter 1, Section 1.2) or Telnet, the full-GUI of a desktop (Chapter 3) is hard to achieve, and the Shell is usually the only feasible user interface choice.

We already know that Linux offers a number of different Shells including *Sh* (the original Bourne Shell), *Ksh* (the Korn Shell), *Csh* (the Berkeley C Shell), *Tcsh* (TC Shell, an improved C Shell), and *Bash* (the Bourne-Again *Sh*). A user can pick which Shell to use. Although these Shells are comparable, Bash is the standard and preferred Shell on Linux systems.

We will present interactive use of Bash in this chapter. Programming in Bash is presented in Chapter 4.

2.1 Bash

Developed in 1987 for the GNU Project (Free Software Foundation), Bash is a freely available Shell based upon the original Sh (Bourne Shell, 1978). The Bash Shell incorporates features from Sh, Ksh, Csh, and Tcsh; adds new features such as Shell-defined functions; and conforms to the IEEE POSIX (pronounced pahz-icks for Portable Operating System Interface) specification.

Today, Bash is the most popular Shell on Linux systems. Improved versions of Bash have been released regularly. Normally, your default Shell is /bin/bash. If not, you can always set your default Shell to /bin/bash (recommended) with the command

chsh -s /bin/bash

In a Bash Shell, the command

echo $BASH_VERSION

displays its version information. It is a good idea to have your Bash ready for experimentation when reading this chapter.

2.2   Interacting with Bash

Inside a terminal emulator window, Bash serves as your command interpreter and continually executes the *command interpretation cycle*:

1. Displays a prompt

2. Enables the user to type, edit, and enter the next command line

3. Breaks the command line into tokens (words and operators) and performs well-defined *Shell expansions*, transforming the command line

4. Carries out (by calling Shell-level functions) or initiates (by starting external programs) the requested operations

5. Waits for initiated operations to finish

6. Goes back to step 1

The default prompt for Bash is $, but it can be customized to become more useful (Section 2.9).

A command line consists of one or more *words* separated by *white space* or *blanks* (spaces and/or tabs). Pressing the ENTER (RETURN) key completes input typing and sends the Shell to the next step. The ENTER key (generating a NEWLINE character) completes a command line unless preceded by a backslash character (\), in which case the ENTER is *escaped* (Section 2.14) and becomes a blank. The first word in a command is the *command name* and indicates the program to be executed; the other words are *arguments* to the command. There are two types of commands: *Shell built-in commands* and *regular commands*. A built-in command invokes either a routine that is part of the Shell (**cd**, for example) or a *function* or *alias* defined by the user. To execute a built-in command, the Shell simply calls up the appropriate subroutine within itself. A regular command is any other executable program in Linux that is not built into the Shell. These include system commands such as **ls**, **rm**, and **cp**, as well as your own executable programs such as **a.out**.

Each executing program is known as a *process* controlled and managed by the operating system. Your interactive Shell is a process. The Shell *spawns* (initiates) a separate *child process*, known as a *subshell*, to execute a regular command. The distinction between built-in and regular commands is an important one, as you will discover.

A *simple command* is just the command name followed by its arguments. Several commands can be given on a single command line if they are separated by semicolons (;). The Shell will execute the commands sequentially, from left to right. Two commands separated by a vertical bar (|) form a *pipe* (Section 2.5). The *or operator* (||) and the *and operator* (&&) specify conditional execution of commands:

cmd1 || *cmd2*	(executes *cmd2* only if *cmd1* fails)
cmd1 && *cmd2*	(executes *cmd2* only if *cmd1* succeeds)

These are examples of *compound commands* where several simple commands are grouped together to form a single command. In Linux, a command returns an *exit status* of zero when it succeeds and non-zero otherwise.

If you enclose one or more commands inside a pair of parentheses (), the commands will be executed as a group by a subshell.

After issuing a command, it is not necessary to wait for a prompt before typing in additional input. This feature is known as *type ahead*. What you type ahead will be there for the Shell when it is ready to receive the next command.

You also can instruct the Shell not to wait for a command to finish by typing an AMPERSAND (&) at the end of the command. In this case, the Shell immediately returns to process your next command, while the previous command continues to run detached from the Shell. Such *detached* processes continue to execute and are said to be running in the *background*. For example,

firefox &

will start the browser and return you to the Shell level without waiting for **firefox** to finish, which is not likely to be any time soon. Basically, the AMPERSAND instructs the Shell to skip step 5 in the command interpretation cycle. A background process also gives up read access to the keyboard, allowing you to continue interacting with the Shell.

A background process can be reattached to the Shell—that is, brought to the *foreground*—by the command

fg *jobid*

Please refer to Section 2.6 for job IDs and job control.

A foreground program receives input from the keyboard. If we bring the **firefox** job to the foreground, we can type a CTRL+C to abort it, for example.

There can be only one running foreground program at any given time.

2.3 Command-Line Editing and Command Completion

Let's look at typing input to Bash. We have seen in Chapter 1 (Section 1.3) how the arrow keys together with DELETE and BACKSPACE can be used to correct input errors and to reuse previous commands. These and other command-line editing features are provided by the *readline library*.

You, in fact, have a choice of using **vi** or **emacs** (see the appendices at the companion website) for editing the command line with

```
set -o vi
set -o emacs
```

In case of **vi** mode, you would type ESC to get into the **vi** *command mode* and then use **vi** commands to do any editing. When you are done editing the command line, press RETURN (or ENTER) to issue the command.

While entering a command line, Bash helps you complete your typing in various useful ways. Basically, you engage the completion feature by pressing the TAB key. If there is a unique completion, it will be done. If there are multiple ways to complete your typing, a second TAB will reveal the choices.

For example, if you enter un followed by two TABs, a list of choices

unalias	uniq	unlink	unstr
uname	uniqleaf	unopkg	. . .

will be displayed. The technique not only saves typing, but also shows you all the Bash built-in and regular commands with a given prefix, which can be very handy if you forgot the exact command name to use.

Some users prefer getting the choices listed directly with the first TAB by putting

```
set show-all-if-ambiguous on
```

in the readline init file `~/.inputrc`

Standard completions performed are

- *Command name completion*—Completing Shell built-in commands, aliases, functions, as well as regular commands; performed on the first token of the command line

- *Filename completion*—Completing names for files; performed on arguments to a command

- *User name completion*—Completing userids for all users on your Linux system, performed on any word starting with a ~

- *Hostname completion*—Completing domain names; performed on any word starting with @

- *Variable name completion*—Completing names for existing Shell variables; performed on any word staring with $

The `bash-completion` package can be installed (Chapter 8, Section 8.24) to provide many additional useful completions.

```
yum  install bash-completion
sudo apt-get  install bash-completion
```

The `bash-completion` package enables you to TAB-complete common arguments to often-used commands. For example, the argument to the **ssh** command

ssh pwang@mTAB TAB

displays

```
pwang@magicalmoments.info        pwang@mapleglassblock.com
pwang@monkey.cs.kent.edu         pwang@mathedit.org
pwang@monkey.zodiac.cs.kent.edu
```

On top of these completions, you can define your own with the Bash built-in **complete** command which implements a *programmable completion* API. See the **complete** documentation for details.

The *readline escape character* CTRL+V is used to quote the next character and prevent it from being interpreted by readline. Thus, to get a TAB into your input instead of invoking the completion function, you would type CTRL+V followed by TAB. For example, you can define the CTRL+L alias with the following:

alias CTRL+V CTRL+L=clear

2.4   Bash Command Execution

The first word of a command line is the command name. It can invoke a procedure within the Shell (in order): an alias (Section 2.7), a function (Section 2.15), or a built-in command. If not, then the command name invokes a *regular command* implemented by a program independent of the Shell.

In a regular command, the command name indicates an executable file and can be in one of two forms. It can be the absolute or relative pathname of the executable file, or if the executable file is *on the command search path*, the simple filename itself will suffice. The procedure by which the Shell finds the executable file is as follows:

1. If the command name is an absolute or relative pathname, then the name of the executable file has been given explicitly and no search is necessary.

2. If the command name is a simple filename (containing no / character), the executable file is found by searching through an ordered sequence of directories specified by the *command search path*. The *first* file found along this search path is used.

If the executable file cannot be found, or if it is found but the execute permission on the file is not set, then an appropriate error message is displayed. The error message most likely will be `file not found` or `permission denied`.

The Shell environment variable `PATH` (Section 2.10) defines the *command search path*, a list of directories containing executable commands. The Shell looks sequentially through these directories for any command you give on the command line. The `PATH` usually includes the system folders `/bin`, `/sbin`, `/usr/bin`, and `/usr/sbin`, where most system-supplied executable programs can be found. The search path can be modified to include additional directories. For example,

export `PATH=$PATH:/usr/local/bin:$HOME/bin`

adds two directories at the end of `PATH`: a `/local/bin` where you install extra applications to your Linux and a `bin` in your home directory.[1] Now, you can use a simple filename to run a program whose executable file resides in the `$HOME/bin` directory.

Bash uses a hash table to speed up command search and only needs to search through `$PATH` (and update the table) when a command is not found in the table. The built-in **hash** command allows you to display and manipulate this table (see **help** `hash`).

The special period symbol (.) is often placed at the end of the search path to enable you to invoke any command in the current directory with a simple filename.

export `PATH=$PATH:.`

The built-in **export** command tells the Shell to transmit this value to the execution environment (Section 2.10) that will be inherited by subsequent regular commands.

Because of aliasing (Section 2.7), user-defined functions (Section 2.15), and command search, the command actually executed may not be exactly what you intended. To be sure, you can check by issuing

which *command_name*

to display the alias/function or the full pathname of the executable file invoked by the *command_name*. For example,

which `gnome-terminal`

displays

`/usr/bin/gnome-terminal`

Once an executable file has been found, the Shell spawns a child process to run the program taking these three steps:

1. A new (child) process is created that is a copy of the Shell.

2. The child process is overlaid with the executable file. Then the command name together with any arguments are passed to it.

---

[1] The value of the Shell variable `$HOME` is the filename of your home folder.

3. The interactive Shell waits for the child process to terminate before returning to receive the next command, unless the command has been given with a trailing ampersand (&).

4. If the command ends with &, the Shell returns without waiting, and the command is run in the background.

2.5 Bash Input/Output Redirection

Until now, our use of Linux has been limited to issuing commands and observing their output. However, you certainly will want results in a more useful form, either as hard copy or stored in a file. Furthermore, many instances will arise when you want input to come from somewhere other than the keyboard, such as a file, or perhaps even from another command or program running concurrently. Linux provides an elegant solution: *input/output redirection*.

When processing a command line, the Shell arranges any I/O redirections before executing commands contained in the command line.

Standard Input and Output

As an operating system, Linux provides input and output (I/O) services for processes. For each process, a set of *file descriptors* numbered 0, 1, 2, and so on is used for I/O transactions between the process and the operating system. When a process opens a file or a device for I/O, a file descriptor is assigned to the process to identify the *I/O channel* between the process and the open file or device. When a new process is created, its first three file descriptors are automatically assigned default I/O channels.

- File descriptor 0, the *standard input* or simply stdin, is connected to the keyboard for input.

- File descriptor 1, the *standard output* or simply stdout, is connected to the terminal window for output.

- File descriptor 2, the *standard error* or simply stderr, is connected to the terminal window for error output.

Most CLI commands receive input from standard input, produce output to standard output, and send error messages to standard error. The Shell-provided *I/O redirection* can be used to reroute the standard I/O channels.

I/O Redirection

The special characters >, <, and | are used by the Shell to redirect the standard I/O channels of any command invoked through the Shell. Let's look at a simple example. The command line

**ls > **`filelist`

creates in your current directory a file named `filelist` containing the output
of the **ls** command. The symbol **>** instructs the Shell to redirect the `stdout`
of **ls** away from the terminal screen to the file `filelist`. If a file by the same
name already exists, it will be wiped out and replaced by a new file with the
same name, unless you set the `noclobber` option with the Bash built-in **set**
command

set `-o noclobber`	(turns on the `noclobber` option)
set `+o noclobber`	(turns off the `noclobber` option)
set `-o`	(displays all options)

When the `noclobber` option is **on**, redirecting output with **>** to an existing
file will result in an error. This feature protects against accidental loss of a file
through output redirection. If you do mean to wipe out the file, add a vertical
bar (|) after the **>**. For example,

**ls >| **`filelist`

Many users set the `noclobber` variable in their Bash initialization file
`.bash_profile` (see Section 2.13). One exception is that `/dev/null` is a special *data sink*. Output redirected to it disappears without a trace. It is useful
when you wish to discard output from a command.

The symbol **>>** operates much the same as **>**, but it appends to the end of
a file instead of overwriting it. For instance,

**cat **`file1` **>> **`file2`

appends `file1` to the end of `file2`. If `file2` does not exist, it will be created.

So far, we have only talked about redirecting the standard output. But
redirecting the standard error follows the same rules, except you need to use
2> and **2>>** instead to explicitly indicate the file descriptor being redirected.
To redirect both standard output and standard error, use

someCommand **>** *file* **2>&1**   (`stderr` joins `stdout` into *file*)
someCommand **>** *file1* **2>***file2*   (sends to different files)

Let's look at another example.

cat > *file*

After giving this command, what you type on the keyboard (or copy and
paste) is put into *file*. Keyboard input is terminated by CTRL+D given at the
beginning of a line.

Next, let's consider redirection of `stdin`. Using the operator **<**, a command
that takes interactive input from the keyboard can be instructed to take input
from a file instead. For example,

vi *textfile* **<** *cmd-file*

where *cmd-file* contains commands to the **vi** text editor. Let's say *cmd-file* contains

```
dd
ZZ
```

then the first line of *textfile* will be deleted. Many Linux commands take input from a file if the file is given as an argument (**sort** *file*, for example); the usage **sort** < *file* is correct but unnecessary.

Pipes

In addition to being able to redirect I/O to and from files, you also can redirect the output of one program as input to another program. The vertical bar symbol (|) is used to establish a *pipe*, which connects the output of the first command to the input of the second. Thus,

ls -lt | more

pipes the standard output of **ls -lt** to the standard input of **more**. The resulting command is called a *pipeline*. Sometimes, for new users, it is hard to understand the difference between | and >. Just remember that the receiving end of a pipe | is always another program and the receiving end of a > or >> is always a file. You can pipe the standard error together with the standard output using &| instead of |. More elaborate examples of pipelines are described in Chapter 4, Section 4.6.

Table 2.1 summarizes Bash I/O redirection. Optional parts in the notation are enclosed in square brackets.

TABLE 2.1: Bash I/O Redirection

Notation	Effect
cmd >[|] *fileA*	Sends `stdout` to overwrite *fileA*
cmd 2>[|] *fileB*	Sends `stderr` to overwrite *fileB*
cmd &>[|] *file*	Combines `stdout` and `stderr` to overwrite *file*
cmd >> *fileA*	Appends `stdout` to *fileA*
cmd 2>> *fileA*	Appends `stderr` to *fileA*
cmd 2>&1	Joins `stderr` to redirected `stdout`
cmd < *fileC*	Takes `stdin` from *fileC*
cmd1	*cmd2*	Pipes `stdout` to `stdin` of *cmd2*
cmd1 &| *cmd2*	Pipes `stdout` and `stderr` to `stdin` of *cmd2*

2.6 Bash Job Control

On the desktop, we know we can run multiple applications, each in a different window, and we can switch input focus from one window to another.

Within a single terminal window, the Shell also allows you to initiate and control multiple commands (called *jobs*). At any time there is one job that is *in the foreground* and connected to the keyboard input for the terminal window. Other jobs are in the *background*. We already mentioned that if you add a trailing **&** to a Shell-level command, the Shell will run the job in the background. Here is another example.

xclock &   (runs **xclock** in the background)

Then you may start another job, say, for text editing, by the command

nano -z notes.txt

This job is in the foreground, enabling you to control **nano** and perform editing functions using the keyboard. At any time, you can type CTRL+Z to *suspend the foreground job* and get back to the Shell level. If you do that, then you'll see[2]

```
[2]+  Stopped              nano -z notes.txt
```

and a new Shell prompt will appear in your terminal window to provide confirmation that the current job has been suspended and will be in the background waiting to be resumed. Now you can issue any Shell-level command, including one to start another job (which may itself be suspended with CTRL+Z in the same way).

Let's say that you then start a third job,

gimp picture.jpg

to do image processing on a picture and then suspend it also. In this way, it is possible to start then suspend or put in the background quite a few jobs, and it is easy to see how this can become unmanageable quickly. Fortunately, if you issue the Shell built-in command

jobs

you'll see all your jobs displayed

```
[1]   13519 Running          xclock &
[2]- 12656 Stopped           nano -z notes.txt
[3]+ 13520 Stopped           gimp picture.jpg
```

In this case, there are two suspended jobs with job numbers 2 and 3, and one job running in the background with job number 1. The Shell also allows you to resume a suspended job, pull a background job into the foreground, or kill a job entirely.

To identify a job, a *jobid* is used, which can be given in a number of ways: %*job-number*, %*name-prefix*, %+, and %-. For example, the jobids %3, %+, and

---

[2]Note that **nano** ignores CTRL+Z unless given the **-z** option.

%g all refer to same job in the preceding example. The job %+ is always the most recently suspended (the current job), and %- is always the next most recently suspended (the previous job). The %- is useful when you are going back and forth between two jobs. When using the name-prefix form, you need just enough prefix of the command name to disambiguate it from other jobs. For example, %vim, %vi, or %v all refer to job 2.

A job can be resumed (brought to the foreground) by the Shell-level command

fg *jobid*

You can abbreviate the command to just *jobid*. For example, %1 will bring job 1 to the foreground, %+ (or simply **fg** by itself) resumes the current job, and %- resumes the previous job. If no *jobid* is specified, the most recently suspended job will be activated and run in the background.

If a background job produces output to stdout, it will be displayed in the terminal window and interfere with output from any foreground job. Further, if the background job requires input from the terminal, it will stop itself and wait to be brought to the foreground to receive the input it needs. Thus, for jobs to run efficiently in the background, redirecting standard I/O to files usually is essential.

When a background job terminates, the Shell displays a message to notify the user:

[*jobnumber*] Done *command as given*

The message is displayed after normal completion of a background process. The following message is displayed when a background process terminates abnormally:

[*jobnumber*] Exit 1 *command as given*

To switch a suspended job to run in the background, use the command

bg *jobid*

Suspending a job using CTRL+Z is not the same as exiting or terminating it. It is good practice to exit all jobs properly and close all windows before you log out. Each job provides its own way for exiting (quitting); for example, CTRL+X for **nano**, :q! or ZZ for **vim**, q for **mutt**, and **exit** for the Shell.

Sometimes you may need to force a program running in the foreground to terminate. This can be done by typing the *interrupt character*, usually CTRL+C, which aborts the executing job and returns you to the Shell level. If the interrupt character does not stop your program for some reason, your last resort is the **kill** command. Use CTRL+Z to suspend the job and get to the Shell level, then type

kill -9 *jobid*

In this will surely terminate the job. The −9 is optional, but it makes the termination mandatory. A process number also can be used as an argument to the **kill** command. The process number is displayed when a job is put in the background, and it also can be recalled with the command

jobs -1

which gives the process numbers for all jobs. Occasionally, the **kill** command will fail to terminate a job. In this case, the command

kill −9 *jobid*

will execute a mandatory kill on the process. The argument −9 instructs **kill** to send a specific signal to the process, which forces it to terminate. Signals are described further in Chapter 10, Section 10.16. The **kill** command discussed here is built into Bash. There is also a regular command, **/bin/kill**, that can be used. Among other differences, **/bin/kill** allows process specification only by process number. Table 2.2 lists useful job control commands. To sum up,

TABLE 2.2: Job Control Commands

Command	Action
jobs -1	Lists all your jobs. If the −1 option is given, the process number of each job is also listed.
fg *jobid*	Resumes the given job in the foreground.
bg *jobid*	Resumes the given job in the background.
kill [-9] *pid*	Terminates the given job or process specified by a jobid or process number. The −9 option makes the termination mandatory (otherwise, a job may refuse to be killed under certain circumstances).

a job may be in one of three states: running in the foreground, running in the background, or stopped (suspended). No more than one job can run in the foreground at any time, but many jobs can run concurrently in the background. Many also may be stopped. To see the states of the jobs under control of your Shell, use the command **jobs**. Use **fg** along with the jobid to bring a particular job from suspension or from the background into the foreground. Use the suspend character (usually CTRL+Z) to suspend a foreground job. Use the interrupt character (usually CTRL+C) to kill a foreground job. If a job is stopped or running in the background, it can be killed by issuing the command **kill** [-9] *jobid*.

If you give the **exit** (**logout**) command while there still are unfinished jobs, the Shell will remind you of the fact. It is best to terminate all unfinished jobs before exiting the Shell. However, if you insist by issuing an immediate second exit command, the Shell will abort all your unfinished jobs, and your terminal window will close.

2.7 Bash Shell Expansions

Each command line undergoes a number of transformations before it is executed by the Shell. These transformations are called *Shell expansions* and are designed to provide power and convenience to the user. For example, you can use

```
ls -l *html
```

to see a listing of all files with a name that ends with `html`. This works because of *Filename Expansion*. Let's see how these expansions work.

FIGURE 2.1: Bash Expansions

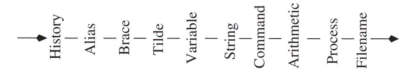

Bash transforms each command by applying the following expansions (Figure 2.1) in sequence:

1. *History expansion*—Allows reuse of parts of previous commands

2. *Alias expansion*—Replaces command aliases by their definitions

3. *Brace expansion*—Treats expressions within curly braces `{}`

4. *Tilde expansion*—Expands a ~ prefixed word to a certain directory name

5. *Variable expansion*—Replaces variables by their values

6. *String expansion*—Interprets standard escape characters, such as `\n` (NEWLINE), `\r` (RETURN), and `\t` (TAB), in strings of the form `$'xyz'`; for example, `$'Name\tAge\r\n'`

7. *Command expansion*—Inserts the output of a command into the command line

8. *Arithmetic expansion*—Includes results of arithmetic expressions in a command (this feature is mostly used in Shell scripts and will be covered in Chapter 5, Section 5.11)

9. *Process expansion*—Specifies output produced by a command to become a filename argument for another command.

10. *Filename expansion*—Adds filenames to the command line by pattern matching

After all transformations, the resulting command line gets executed. You are encouraged to experiment with the expansions as you read their descriptions. The built-in command **echo** which displays the after-expansion state of its arguments can be very useful. By putting the **echo** in front of a command line, the effects of all but alias expansion can be examined.

History Expansion

TABLE 2.3: Common History Expansions

Specification	Meaning
!*n*	The event with sequence number *n*
!-*n*	The *n*th previous event
!!	The last event (same as !-1)
!*prefix*	The most recent event with the specified *prefix*
^*bb*^*gg*	The last event, with the string *bb* replaced by *gg*
!*	All the arguments of the last event
!$	The last argument of the last event
!^	The first argument of the last event
!:*n*	The *n*th argument of the last event
event:s/*xx*/*yy*/	The given history event with the string *xx* replaced by *yy*

The *Bash history* mechanism is very similar to that of the *Csh* Shell. The purpose is to record and allow easy reuse of previous commands. Each command line issued by you, whether successful or not and whether consisting of one or more commands, is kept as an item in the *history list*, which can be displayed using the built-in command **history**. Each item in the history list is known as a *history event*, and each event is identified by a sequence number. The total number of events kept on the history list has a limit (defaults to 500) which is set by

```
HISTSIZE=number
```

Normally, keeping 50 events or so is quite enough. Entering your own **HISTSIZE** setting in the `.bash_profile` file (Section 2.13) makes good sense. We already know from Chapter 1 that you can use the up and down arrow keys to go back and forth on the history list and reuse previous commands. Furthermore, history expansion enables you to substitute history events into the current command line with just a few keystrokes. It also makes modifying and reissuing past commands, or parts of them, easy to do.

History expansion is *keyed* (activated) by the exclamation point character (!), and it works by recalling items from the history list. Items that can be recalled from the list and substituted into the current command include any history event, any word or words of any event, and parts of certain words.

These items also can be modified before their inclusion into the current command. Table 2.3 shows some common history expansions. Table 2.4 contains some applications of history expansion in commands. Each example is de-

TABLE 2.4: History Examples

No.	Last Event	Current Command	Effect
1	**diff** *file1 file2* > *file3*	**nano !$** or **nano !:4**	**nano** *file3*
2	**ls** –l *name*	`^-l^-ld`	**ls** –ld *name*
3	**srot** *file* (srot: not found)	`^ro^or`	**sort** *file*
4	**nano** *file* (file not found)	**cd** *dir*; **!nano** or **!-1**	**cd** *dir*; **nano** *file*
5	**cd** *dir* (no such file or dir)	`cd;!-1`	**cd** ; **cd** *dir*
6	**ls** dir	`^s^s -F`	**ls** –F *dir*

scribed here, and the numbers correspond to the numbers in Table 2.4.

1. Reuse the name *file3*.

2. *Name* turns out to be a directory.

3. Mistyped the command name **sort**.

4. The desired *file* is not in the current directory but in the directory *dir*.

5. The `dir` is not in the current directory but in the home directory.

6. Note that blanks are allowed in the string replacement.

Having seen a number of examples, you are ready to proceed to the general form of a history expansion:

event [: *word designator*] [: *modifier* ...]

The event is given in one of the following ways:

Event number	!12 gives event 12 on the history list.
Relative position	!-2 gives the second most recent event. A special case is !!, which refers to the last event.
Command prefix	!nano gives the most recent event prefix `nano`.
Matching string	!?*string*? gives the most recent event containing *string* anywhere within the event.
^*str1*^*str2*	Repeats the last command, but with *str1* replaced by *str2*.

Following the event are the optional word designators. The purpose of a word designator is to choose certain words from the history event. If no word designators are used, the entire event will be selected. The following word designators can be used:

n	Selects the *n*th word (the command name is word 0); for example, `!-3:2` gets the third word of the event `!-3`
`$`	Selects the last word; `!$` gives the last word of the last event
`^`	Designates the first argument (second word); `!^` gives the first argument of the last event
i–j	Designates words *i* through *j* inclusive *–j* same as 0*-j*
i–	Designates word *i* up to but not including the last word
`*`	Indicates all argument words; `!*` gives all arguments of the previous command
i∗	Same as *i-$*

An optional sequence of modifiers also can be used. One frequent usage is

event:s/*xx*/*yy*/

to substitute the string *xx* by *yy* in *event*. If a word is a long pathname, it is sometimes convenient to use a modifier to extract a portion of it, but most modifiers are seldomly used interactively. Writing programs in the Shell language (Shell procedures) is discussed in Chapter 5, and at that point you will be able to see why modifiers are needed. A number of modifiers are listed in Table 2.5; refer to the Bash manual for a complete list. Once a command line has gone through history expansion, it too becomes part of the history list as the most recent event.

TABLE 2.5: History Modifiers

Modifier	Meaning	Example	Value
`:h`	head	`!$:h`	`/usr/local/kent`
`:t`	tail	`!$:t`	`prog.c`
`:r`	root	`!$:r`	`/usr/local/kent/prog`
`:e`	extension	`!$:e`	`.c`

Note: `!$` is `/usr/local/kent/prog.c`.

The Bash built-in command **fc** (fix command) puts a range of history items into your favorite text editor, allows you to modify any parts at will, and then executes the resulting commands automatically when you exit the editor.

fc *first_event last_event*

Finally, when you are finished interacting with it and exit, Bash saves the command history to the history file specified by the *environment variable* `$HISTFILE`, which defaults to `.bash_history` in your home folder. Next time you start Bash, the saved history will be restored from the history file.

The history file feature can be disabled by

```
export HISTFILE=
```

Alias Expansion

The *alias* feature allows you to define shorthands for often-used commands, making them easier to enter. To create an alias (any single word) and give it a value (a character string), use the Bash built-in command **alias**. The notation

alias *name=value* ...

defines the given name as an alias for the specified string value. Multiple name-value definitions are allowed. The *value* part often requires quotes around it to prevent unwanted Shell expansions (see Section 2.14 for when and how to use quotes). Here are some simple but useful alias definitions.

```
alias dir="ls -l" back='cd $OLDPWD'
alias monkey="ssh -l pwang monkey.cs.kent.edu"
alias append2end="cat >>"
```

With these aliases defined, the command **dir** works because it expands to **ls -l**. The alias **back** works its magic because the Bash variable $OLDPWD always holds onto the previous working directory.

Alias expansion means that if the first word of a simple command is an alias, Bash will replace that first word with the alias value. The first word of the replacement text is again tested for aliases, but a word that is identical to an alias being expanded is not expanded a second time. This means the following is correct and does not result in an infinite loop.

```
alias ls='ls -F'
```

Thus, the **ls** command always is given with the **-F** option, which causes, among other things, directory names to be marked with a trailing **/**, symbolic links to be marked with a trailing **@**, and executable files (files with *execute permission*; see Section 1.5) to be marked with a trailing *****.

To display existing aliases, use

alias   (displays all aliases)
alias *name*   (displays the alias)

To remove alias definitions, use

unalias *name* ...

Brace and Tilde Expansions

Brace expansion provides a shorthand for similar words on the command line. With *brace expansion*, the command line

nano `memo{Sep, Oct, Nov}2011.txt`

becomes

nano `memoSep2011.txt memoOct2011.txt memoNov2011.txt`

and **lpr** `chap{2..5}.pdf` becomes

lpr `chap2.pdf chap3.pdf chap4.pdf chap5.pdf`

The sequence notation (`..`) works for numbers and single letters, for example `{a..z}`.

The character TILDE (`~`) expands to the user's own home directory, `~userid` to the home folder of some other user, `~+` to the current folder, and `~-` to the previous folder.

Thus, the alias **back** earlier can also be defined as

alias `back="cd ~-"`

Variable Expansion

The Shell allows the use of variables. A variable's value is a character string. Some variables are reserved for Shell use. For example, `USER`, `HOME`, `PATH`, and `HISTSIZE` are *Shell variables* having prescribed meaning in Bash (see Section 2.9). In addition, you can also set and use your own *user-defined variables*.

Generally speaking, a variable identifier can be any word whose first character is a letter and the rest consists of letters, digits, and underscore characters. Use

var=*value*   (sets variable value)

to assign a value to a variable. The *value* can be a single word or multiple words in quotes, and no white space is allowed immediately before or after the equal sign (=). After being set, a variable can be used in subsequent commands. For example,

`ldir=/usr/local`

gives the variable `ldir` a string value `/usr/local`. With this variable set, you can input

cd `$ldir`

which is a command with a variable in it. After variable expansion, this command becomes

cd `/usr/local`

As you can see, variable expansion is keyed by the character $. That is, a word that begins with a $ is a variable. If $ is followed by a blank or preceded by a backslash (\), then it stands for itself. The **echo** command can be used to display the value of a variable. For example,

echo $ldir

displays /usr/local. Use **unset** *var* to remove any variable *var*.

The *extent* of a variable name can be delineated by braces ({ and }). For example,

x=abc
echo ${x}de

displays the string abcde, whereas

echo $xde

displays an empty line because the variable $xde has no value.

Variables often have string values. However, they may also have integer values. Inside $((...)), you may perform integer arithmetic operations (including + - * / +% ** ++ --) on variables using C-language syntax. For example,

```
count=7
echo $(( 3*count ))        (displays 21)
echo $(( count%5 ))        (displays 2)
echo $(( count++ ))        (displays 7, sets count to 8)
```

You can display variables (Shell built in and user defined) and function definitions (Section 2.15) with

set   (displays all variables and functions)
declare (displays all variables and functions)
declare (displays all functions)

Command Expansion

Command expansion makes it possible to use the standard output of a command as a string of words in another command. Either $(**command**) or `command` (note the BACKQUOTE) can be used for command expansion. For example,

dir1=$(pwd)   (or dir1=`pwd`)

assigns the output of the **pwd** (print working directory) command to the user variable dir1. Another example,

files=$(ls)

assigns to `files` words produced by the **ls** command, namely, the file names in the current directory. The substitute string of a command expansion also can form part of a single word, as in

`file1=$(pwd)/test.c`

The substitute string is normally broken into separate words at blanks, tabs, and NEWLINES, with null words being discarded.

Process Expansion

Bash extends the ideas of I/O redirection one step further by allowing the notation

<(**command** *args* ...)

to be used where a filename argument is expected for a command. Thus, the notation <(...) produces a temporary file, with the output produced by the command inside, which can be given to another command.

For example,

nano <(ls -l -F)

opens **nano** to view/edit the results produced by the given **ls** command. This ability can be handy sometimes. It is possible to supply multiple files in this way. For example,

diff -u <(ls -F /usr/bin) <(ls -F /usr/bin.old)

displays the differences between the two directory listings.

Filename Expansion

Because command arguments often refer to files, the Shell provides *filename expansion* to make it easier to specify files. When a *filename pattern* or *glob pattern* is used in a command line, the pattern is *expanded* to become all the filenames matching the pattern. A pattern may match simple filenames, in the current working directory, as well as full or relative pathnames. If a pattern does not match any file, then it stands for itself and is not expanded. Glob patterns are specified using the special characters *, ?, and []. The * matches any sequence of zero or more characters. For example,

`ls -l *.c`

produces a listing of all files with a name ending in .c. The *.c is a pattern, and it expands to match all filenames in the current working directory ending with .c. The command

`ls -l ../*.c`

does the same for files in the parent folder. The command

```
ls ~/Pictures/2011*/*.jpg
```

conveniently displays a listing of all pictures, ending in `.jpg`, under folders with a name prefix 2011, in the `~/Pictures` directory.

Filename patterns are matched against existing filenames. Rules for filename patterns are as follows:

`*`	Matches any character string of of length zero or more (the "wildcard").
`?`	Matches any single character.
`[...]`	Matches any one of the characters contained between `[` and `]` (*a range pattern*). For instance, `a[rxz]b` Matches `arb`, `axb`, or `azb`. The pattern `chapter[0-9]` Matches `chapter0`, `chapter1`, and so on.
`[^...]`	Matches any character not in `[` and `]`. The `!` character can be used instead of `^`.
`[:class:]`	Specifies a *class* of characters, in a range pattern. The *class* can be `alnum` (alpha-numeric), `alpha`, `digit`, `lower`, or `upper`.

For example, in the command

```
ls [[:digit:]]*
```

the pattern matches all files whose name starts with a digit.

Filename expansion is also known as *globbing*. Filename expansion can be deactivated with the Bash built-in command

`set -f`	(or `-o noglob`, filename expansion off)
`set +f`	(or `+o noglob`, filename expansion on)

Filename expansion should normally be on when using the Shell interactively.

The character `.` at the beginning of a filename must be matched explicitly unless the `dotglob` option is set.

shopt `-s dotglob`	(enables matching leading dot)
shopt `-u dotglob`	(disables matching leading dot)
shopt	(lists Bash options)

Hence, the command **ls** `*` normally does not list any files whose name begins with a dot. Additionally, the character `/` in a filename must be matched explicitly.

A filename pattern can contain more than one pattern character. When more than one filename is matched, the pattern is expanded into a sorted list of the matched filenames. Matching is case sensitive unless you do **shopt** `-s nocaseglob`. If a pattern matches no filenames (match failure), then it is not expanded (stays unchanged in the command line) unless

shopt `-s failglob`	(match failure causes an error)
shopt `-s nullglob`	(match failure expands to empty string)

2.8   Bash Built-in Commands

We have seen a number of Bash built-in commands. A few more are introduced in this section. To see a list of all Bash built-in commands, you can use the built-in **help**.

help   (lists all built-in commands)
help *commandName*   (describes the given command)
help `help`   (tells you how to use **help**)

Bash maintains a *directory stack* that, by default, contains the current working directory. The built-in **pushd** *dir* changes to the given directory and pushes it onto the stack. The built-in **popd** changes to the top directory on the stack after popping it off the stack. Thus, the sequence

pushd *dir*
popd

brings you back to where you were without changing the directory stack. The built-in **dirs** lists the folders on the stack.

While interactive input usually comes from the keyboard, it is convenient to save and edit commands in a file and then ask the Shell to execute those commands from that file. The Bash built-in command **source** (or simply a dot **.**) can read a file of Bash commands and process them one by one. A file of Shell commands is known as a *Shell script*. Thus, either of

source *script*
. *script*

causes your interactive Bash to read commands from the given *script* as though they were entered from the keyboard individually. Since **source** is a built-in command, the script is not read by a subshell (Section 2.2).

2.9   Shell Variables

Bash uses a number of special variables, with all uppercase names, for specific purposes. Setting special variables controls the way certain Bash operations are carried out. For example, setting the CDPATH to a list of often-used directories enables you to use simple folder names with the **cd** command (**cd** *simpleFolderName*). Bash will then search for the target folder under directories on the CDPATH. Be sure to include the . on the CDPATH. Some variables that affect interactive use of the Shell are listed here. Other special variables affecting the processing of *Shell scripts* are discussed in Chapter 5.

USER	The userid
PWD	The full pathname of the current directory
OLDPWD	The full pathname of the previous directory
HOME	Full pathname of home folder
SHELL	The pathname of the Shell executable
HISTSIZE	Number of history events to keep
HOSTNAME	Name of the Linux host
PATH	Command search path
CDPATH	Search path for **cd**
PS1	Your primary Shell prompt string (see Section 2.11 for customizing your prompt)
PS2	Your secondary Shell prompt tring (default >)

2.10   Environment of a Program

The exact manner in which a program works depends on the *execution environment* within which it is supposed to do the job. For example, the text editor **nano** or **vim** needs to know the capabilities of the terminal emulator it is dealing with, and so does the command **more**. The current working directory is something almost all programs will want to know when they run. For file access permission purposes, any program that accesses files needs to know the userid of the user who invoked it. The execution environment of every process consists of two parts: user defined and system defined. The userid, current working directory, open files, etc. are determined by the system and passed on from your Shell to any invoked application; whereas quantities such as the home directory, the command search path, and the default editor are defined by the user. These are known as *environment variables*. Many applications use certain specific environment variables of their own; for example, DISPLAY for any GUI application, CLASSPATH, and JAVA_HOME for the Java compiler, MOZILLA_HOME for Firefox, and EDITOR for **mutt**.

Command Execution Environment

A principal task of a Shell is to launch applications by interpreting user commands. When Bash launches an application, it creates a *child process* (another running program) and transmits to it an execution environment that includes the following attributes:

- Standard I/O and other open files

- Current working directory

- File creation mask (Section 2.12)

- Environment variables already in the Shell's own execution environment and additional ones defined by the user

A child process (an application, for example) is said to *inherit* its initial environment from its parent process (the Shell, for example). Any changes in the environment of the child process does not affect that of the parent.

Let XYZ be any variable. You can make it part of the Shell's environment by

export XYZ

therefore making it available to any child process the Shell initiates later. If a variable is unset, then it, of course, is also removed from the Shell's environment.

Instead of exporting and then unsetting a variable, you can add variables to the environment on a per-command basis. When you issue any regular command, you can set variables in front of the command name to add them to the *environment passed to the command* without affecting the environment of the Shell itself. For example, if we start a subshell with

YEAR=2011 **bash**

The subshell will have an environment variable YEAR set to the value 2011 while your Shell remains unchanged.

The environment variable TERM records the terminal type. For Linux users, TERM is most likely set to xterm (X Terminal) by a terminal-window program such as **gnome-terminal** (Chapter 3, Section 3.7). The command search path is another environmental parameter whose value is contained in the environment variable PATH. Also, X Windows client programs use the setting of the variable DISPLAY (Chapter 3, Section 3.4). The Bash built-in command **printenv** (or **env**) displays all currently set environment variables and their values. Here are a few more common environment variables.

TERM	Type of terminal
EDITOR	Default text editor
DISPLAY	X server and physical display device designation
MANPATH	Search path for the command **man**

Remember, in Bash any variable can become an environment variable by the **export** command. However, it is good practice to use all uppercase names for environment variables.

2.11   Examples of Bash Usage

By studying examples, you can gain a deeper understanding of how the Shell works and how the various expansions can be used. Almost all examples given here are of practical value, and you may consider adopting any or all of them for your own use.

Customized Prompt

The Shell displays a prompt when it is ready for your next command. For GNU Linux, the default Bash prompt is `PS1='\s-\v\$'`, meaning *-Shell_base_name-version$*. For example,

`-bash-3.2$`

The trailing `$` is automatically replaced by `#` if the user is `root`.

Many users choose to customize the prompt to display more information. A good example is

`PS1='\u@\h:\W{\!}\$'`

which specifies *userid@hostname:current_folder{history_number}$* and produces, for example, the prompt

`pwang@acerwang:ch03{361}$`

You may also set the special variable `PROMPT_COMMAND` to any command to be executed before displaying each prompt. See the Bash documentation for more information on setting the prompt.

Removing Files Safely

Deleting files accidentally is not unusual. This is especially true with the powerful and terse notation of the Shell. It is entirely possible to mistype the command

rm *.o   (deletes all files with the `.o` suffix)
as
rm * .o   (deletes all files and the file `.o`)

by accidentally typing an extra SPACE in front of the `.o`.

It is recommended that you define an alias

alias rm="rm -i"

The `-i` option requires *interactive confirmation* before deleting any file. Consider placing this alias in your `.bash_profile` (Section 2.13).

Some users prefer an even safer alternative, moving unwanted files to a *trash folder* rather than actually deleting them. You should already have a `Trash` folder in your home directory or you can create one with

mkdir ~/Trash

Now, define a function **rm** (Section 2.15) that uses **mv** to move any given files to `~/Trash` (see Exercise 20).

Copy, Paste, and I/O Redirection

You can combine copy-and-paste (using the mouse, see Chapter 3, Section 3.7) with I/O redirection to make certain operations easier. For example, you can mark and copy display text containing information you wish to save and enter it directly into a file. Just type

cat > notes.txt

and paste the marked line followed by CTRL+D on a new line (to signal end of input to **cat**). To mail some screen output to another user, simply do

cat | mail *userid* -s *subject*

and then paste the material.

Setting Up Your Personal Web Folder

Often, the Linux system at school or the office will also serve the Web. If so, the Linux system often also supports per-user Web pages. This means you can set up a public_html folder in your home directory in the following way:

cd	(goes to home directory)
chmod a+x .	(allows Web server access)
mkdir public_html	(creates new folder)
chmod a+x public_html	(allows Web server access)

Now you may create Web pages (*filename*.html) in your pulic_html and make each one Web readable:

chmod a+r public_html/*filename*.html

You can then access them over the Web with the Web address

http://*hostname*/~your_userid/*filename*.html

2.12   Default File Permissions

File protection was described in Chapter 1, Section 1.5. When you create a new file, Linux gives the file a default protection mode. Often, this default setting denies write permission to g and o and grants all other permissions. The default file protection setting is kept in a system quantity known as *umask*. The Shell built-in command **umask** displays the umask value as an octal number. The umask bit pattern specifies which access permissions to deny (Section 10.4). The positions of the 1 bits indicate the denied permissions. For example, the umask value 0022 (octal 022) has a bit pattern 000010010, and it specifies denial of write permissions for g and o. The Shell built-in command **umask** also sets the umask value. For example,

umask 0077

sets the umask to deny all permissions for g and o. If you find yourself using **chmod** go-rwx a lot (Section 3.5), you might want to consider putting **umask** 0077 into your `.bash_profile` and `.bashrc` files (Section 2.13).

2.13   Shell Startup and Initialization

As mentioned, the Shell itself is a user program. The term *user program* refers to programs not built into the Linux operating system kernel. Examples of kernel routines are file system routines, memory management programs, process management programs, and networking support. The commands **ls**, **nano**, **mail**, and **cat**, as well as Shells **bash**, **csh**, and so on, are user programs. In fact, all Linux commands are user programs.

The login Shell is selectable on a per-user basis and is specified in the user's *password file entry* in the password file `/etc/passwd`. This file contains a one-line entry for each authorized user on the system. Each **passwd** entry consists of the following fields:

- Login name (contains no uppercase letters)

- Encrypted password or x

- Numerical userid

- Numerical groupid

- User's real name, office, extension, and home phone

- User's home directory

- Program to use as the Shell

The fields are separated by colons (`:`). For example, a **passwd** entry may look like the following:

```
pwang:x:500:500::/home/pwang:/bin/bash
```

The x password indicates that a *shadow password file* is used to better protect and manage user passwords. The `/bin/bash` at the end specifies the user's *login Shell.*

Immediately after a login window starts, the user's *login Shell* is invoked (Chapter 3, Section 3.7). The login Shell specified in the passwd entry can be changed using the command **chsh** (change Shell). For example,

```
chsh  -s /bin/bash
```

will change your login Shell to `/bin/bash`. At the Shell level, the command

echo $0

displays the name of your current Shell.

When a Shell starts, it first executes commands in Shell initialization files, allowing a Linux installation and individual users to customize the Shell to suit their purposes. Exactly which initialization file Bash loads depends on how it is invoked.

- Login Bash—If **bash** is invoked via a login window or given the option -l or --login, then it is a login Shell. As a login Shell, Bash first loads the system-wide initialization file /etc/profile which defines environment variables such as PATH, USER, HOSTNAME, and TERM. Then it loads a per-user initialization file which is the first of .bash_profile, .bash_login, and .profile found in the user's home directory. The per-user clean-up file .bash_logout is executed when a login Bash exits.

- Non-login interactive Bash—When Bash is run from the command line, it is an interactive Shell (with standard I/O connected to the terminal window) but not a login Shell. Such a Bash loads the system-wide /etc/bash.bashrc first and then loads the per-user ~/.bashrc.

- Non-interactive Bash—Bash started to run a command (**bash** -c *cmd*) or a script (Chapter 5) is non-interactive. Such a Bash does not load any init files by default. It will load a file specified by the environment variable BASH_ENV.

There are some differences among Linux distributions on Shell initialization files. For example, CentOS/Fedora/Red Hat also provides the system-wide /etc/bashrc file for users to load if desired with a conditional expression:

```
if [ -f /etc/bashrc ]; then
    . /etc/bashrc
fi
```

Note that the . command is the same as **source**. Writing Bash programs is the topic of Chapter 5.

Among other things, the /etc/bashrc usually sets the umask to a default value (Section 2.12). It is a good idea to include /etc/bashrc if your system provides one. Here is a sample .bashrc file.

```
# Source system definitions
if [ -f /etc/bashrc ]; then
        . /etc/bashrc
fi
set -o noclobber
umask 0007
```

FIGURE 2.2: A Sample `.bash_profile`

```
if [ -f ~/.bashrc ]; then
    . ~/.bashrc                ## Loads my .bashrc
fi
umask 0007; set -o vi          ## vi-style input editing
set -o noclobber;  alias rm="rm -i"
## defines environment variables
 PS1="\u@\h:\W{\!}\\$"          ## primary prompt
 BASH_ENV="~/.bashrc"; SHELL=bash; USERNAME=pwang
 IGNOREEOF=3; HISTSIZE=50;  EDITOR=/bin/vi
 UNAME="'/bin/uname -s -r'"  ## system name string
 DOCUMENT_ROOT="/var/www/html"
 MOZILLA_HOME=/usr/local/firefox
 JAVA_HOME=/usr/java/latest
 . ~/.bashPATH                  ## source my PATH setting
```

The `.bashrc` is usually included in the `.bash_profile`, which adds other settings important for interactive use of the Shell. Figure 2.2 shows a sample `.bash_profile` (**Ex: ex02/bash_profile**). A non-interactive Bash is a subshell, and the execution of any Bash script (Chapter 5) involves a subshell Bash. Therefore, the setting for aliases, functions, and `PATH` used for Shell procedures ought to be placed in `.bashrc` instead of in `.bash_profile`.

2.14 Shell Special Characters and Quoting

The Shell uses many special characters in establishing the command language syntax and as keys for the various expansions provided. Some often-seen special characters are listed in Table 2.6.

Special characters help achieve many Shell functionalities. However, because the Shell interprets a special character differently from a regular character, it is impossible for a special character to stand for itself unless additional arrangements are made. For example, if there is a file named `f&g.c`, how can you refer to it in a Shell command? The solution to this problem is the use of more special characters, known as *quote characters*. If you are getting the impression that there are many special characters in Linux, you are absolutely right. In fact, any character on the keyboard that is not alphabetic or numeric is probably special in some way. Notable exceptions are the period (`.`) and the underscore (`_`).

TABLE 2.6: Bash Special Characters

Characters	Use	Characters	Use
>, <, &, |	I/0 redirection	|, &	Pipe
$, :, =, [], -	Variable expansion	!, ^, /, :	History expansion
[], *, ?, ~, {}	Filename expansion	&, ;	Cmd termination
`, $()	Cmd expansion	(), {}	Cmd grouping
NEWLINE	Cmd line termination	blank	Word separation
\, ", '	Quoting	CTRL+V	Literal next
TAB, BS	Cmd line editing	DEL, arrows	Cmd line editing
$(( ))	Arithmetic expr.	CTRL+C,DEL	Interrupt, abort
<( )	Process expansion	.. in {}	Sequence notation

Quoting in Bash

Bash provides the backslash (\) escape character, single quotes (' . . .'), double quotes (" . . ."), and ANSI-C quotes ($' . . .').

The character \ quotes or escapes the next character. For example,

nano f\&g.c

and

grep US\$ report.*

The characters **&** and **$** lose their special meaning when preceded by \. Instead, they stand for the literal characters themselves. If a space or tab is preceded by a \, then it becomes part of a word (that is, it loses its special meaning to delineate words). If the NEWLINE character is preceded by a \, it is equivalent to a blank. Thus, using a \ at the end of a line continues the Shell command to the next line. To get the \ character without escaping the next character, use \\.

Whereas the \ escapes the next character, a pair of single quotation marks (') quotes the entire string of characters enclosed.

echo 'a+b >= c*d'

When enclosed by single quotation marks, all characters are escaped. The quoted string forms all or part of a word. In the preceding example, the quoted string forms one word with the spaces included. The command

cat /user/pwang/'my>=.c'

is used to type out a C program in the file **/user/pwang/my>=.c**. In this example, the quoted string forms part of a word. To include a single quotation mark in a string, the \ is used, as in

echo It\'s a good day

The following rules summarize quotation with single quotation marks:

1. All quoted characters, including \, are taken literally. Thus, escaping the single quote with backslash within a single-quoted string does not work.

2. The quoted string forms part or all of one word.

Sometimes it is desirable to allow certain expansions within a quoted string. Quoting with double quotation marks (") serves this purpose. A pair of double quotation marks functions the same as a pair of single quotation marks with three differences:

- First, variable and history expansions are performed within double quotation marks; that is, variable expansion keyed by the $ sign and history expansions keyed by the ! sign work within double quotation marks. For example,

```
echo "your host name is $HOST"
echo "Last command is !-1"
```

work as expected.

- Second, command expansions are allowed inside double quotation marks and are treated slightly differently from normal command expansions. Normally, the output of a command expansion, via $(...) or `...` (Section 2.7), is broken into separate words at blanks, tabs, and NEW-LINEs, with null words being discarded; this text then replaces the original backquoted string. However, when command expansion is within double quotation marks, only NEWLINEs force new words; blanks and tabs are preserved. The single, final NEWLINE in command expansion does not force a new word in any situation. For example,

```
date=`date`
```

and

```
datestring="`date`"
```

are different in that $date includes multiple words, but $datestring is one word.

- Third, escaping " with backslash within a double-quoted string works. Actually, within a double-quoted string, the backslash (\) escapes only $, ‘, ", \, or NEWLINE. Within a double-quoted string, the combination \! escapes history expansion, but the backslash is not removed from the resulting string.

Now, we still need an easy way to include hard-to-keyboard characters in strings. This is where the ANSI-C quotes are useful. A string in the form $' *str* ' allows you to use ANSI-C escape characters in *str*. For example, you can use \b (BACKSPACE), \f (FORMFEED), \n (NEWLINE), and so on. For example,

alias $'\f'=clear

defines a convenient alias, allowing you to clear your terminal screen by typing CTRL+L as a command.

2.15   Simple Functions

You can take a hard-to-enter command or a sequence of commands for a certain task and build a function to make the task easy. To define a function, use the syntax

```
function fnName () {
    command 1;
    command 2;
        ...
    command n;
}
```

A command in a function can be a Shell built-in command, a regular command, or a call to another function. Aliases don't work inside a function. Each command in the function definition must be terminated by a semicolon.

Once defined, you can use the function name as a command name and also pass the function arguments. For example,

```
function office ()
{  /usr/bin/openoffice.org-2.0 $1;  }
```

defines the function office. You can then invoke **openoffice.org** on a document with the command

office note.doc

The special variable $1 in the function definition refers to the first argument in the function call. In general, the *positional parameters* $1, $2, ... are used to access arguments passed in a function call.

In fact, the keyword function is not necessary if the () are there. For example,

```
dir ()
{
    ls -1F --color=auto --color=always "$@" | less -r
}
```

gives you a DOS-like **dir** command.[3] The special variable $@ refers to all the arguments in the function call.

A function is normally not inherited by child Shells unless it is exported with **export -f** *functionName*.

You can remove a function with

unset -f *functionName*

and display all functions with

declare -f

There is no built-in command to display a specific function, but the following function will do the job

```
function which ()
{  (alias; declare -f) | \
   /usr/bin/which --tty-only -i \
   --read-functions $@;
}
```

The pair of parentheses around (`alias; declare -f`) groups commands just like {}, except it calls for a subshell to execute the commands. The `stdout` of that subshell is fed to the `/usr/bin/which` command.

With this function defined, the command **which** *fname* will now display any alias or function definition for the given *fname*. If there is no such function or alias, it will also look for *fname* on `$PATH`. The special variable $@ evaluates to all the arguments passed to the function. Also note we used `/usr/bin/which` instead of just `which` because it is not our intention to call the function recursively.

Here is the display produced by **which which**.

```
which ()
{  ( alias;
     declare -f ) | /usr/bin/which --tty-only -i \
     --read-functions --show-tilde --show-dot $@;
}
```

More will be said about functions in Chapter 5, Section 5.18.

2.16   For More Information

You can use the Bash command

help | more

---

[3]Note that Linux already has a regular command **dir** for listing directories.

to get a listing of built-in commands and how to get more details on them.

The Bash man page

man bash

is a good reference on the Bourne-Again Sh Shell.

The *Bash Manual* from GNU can be found at

www.gnu.org/software/bash/manual.

2.17   Summary

Running in a terminal window, the Bash Shell provides a CLI to your Linux
system. You interact with the Shell via the input-processing-execution-prompt
cycle. The command line goes through a well-defined set of *expansions* before
getting executed. A Shell built-in command is carried out by the Shell itself. A
non-built-in or regular command involves locating an executable program in
the file system, running it in a child process, and passing to it any command-
line arguments and any environment values, including exported Shell variables
and functions.

A command name can be either a simple name or a pathname. In the former case, the command may invoke an alias or a function defined in the Shell.
Otherwise, the command is found by search through the *command search
path*—a list of directories given by the environment variable PATH.

I/O redirection enables you to direct the stdin, stdout, and stderr of
commands to/from files and other commands (forming pipes). Job control
makes it possible to start multiple tasks, suspend them, put them in the
background, or bring any to the foreground to reassert terminal control.

Entering of input is helped by input editing, TAB-completion, history substitution, and filename expansion.

Bash loads initialization files at start-up time. It is important to keep your
favorite settings in the appropriate init files .bashrc and .bash_profile.

The Shell uses many special characters, such as *, =, (), [], blanks, ;, and
so on. Quoting with single and double quotes and character escaping with
\ are necessary to counter the effects of such characters. This is especially
important to remember when issuing commands that require the use of such
characters.

Bash also supports function definition. A function becomes a new built-
in command. A function can take arguments and access them as positional
parameters. If you like Shell aliases, you'll love functions. More about functions
can be found in Chapter 5.

2.18 Exercises

1. The command **cd** is built into the Shell. Why can't it be implemented as a regular command?

2. Consider the special directory symbol **.** and its inclusion on the command search path (`$PATH`). What difference does it make if you do or do not include **.**? If you do include **.**, where should it be placed relative to other directory names on the search path? Why?

3. You have written a program that takes input from `stdin` and writes it to `stdout`. How could you run this program if you wanted input to come from a file named **in** and output to be stored at the end of a file named **out** and any error to `stderr` be recorded in a file named **errlog**?

4. What if you wish to have `stdout` and `stderr` sent to the same file?

5. John wanted to append the file **fb** to the end of the file **fa**, so he typed

   cat fa fb >| fa

   What really happened here? How would you do it?

6. John then wanted to send a line-numbered listing of file **fa** to the printer. He typed

   cat −n fa > lpr

   but no printout appeared. Why? What happened here?

7. John made a typo

   srot file1 file2

   Specify two ways using the Shell history mechanism to correct **srot** to **sort** and reissue the command.

8. How does one set the editor used in Bash command-line editing? Show the code.

9. Name at least two commands that are built in to Bash but also are regular Linux commands.

10. Give a command to edit, using **nano**, every file in the current directory whose filename ends in **.txt** that contains the string **Linux**. (Hint: consider the −l option of **grep**.)

11. What is a foreground job, background job, and suspended job? How does one display a list of all jobs, or switch from one job to another?

12. How do you exit from your interactive Shell? Specify at least three ways.

13. What happens if you exit from your Shell and there are unfinished jobs?

14. Explain the difference between these two commands:

    ```
    ls chap[0-9]
    ls chap{0..9}
    ```

15. What is *command expansion* in Bash? Give the two notations used for command expansion.

16. What is *string expansion* in Bash? Explain and give two examples.

17. Consider the two Bash initialization files: `.bashrc` and `.bash_profile`. What initialization commands should be kept in which? Why?

18. What is the syntax for function definition in Bash? After defining a function, can you undefine it? How?

19. In bash, what are *positional parameters* of a function? How do you export a function into the environment? What good does it do?

20. Write a Bash function **rm** to move its argument files to the ˜/Trash folder. (Hint: Use **mv** -i.)

21. Find the Linux version running on your computer. (Hint: The **uname** command.)

Chapter 3

Desktops, Windows, and Applications

In the beginning, Linux/UNIX systems were used exclusively through the command-line interface (CLI) (Chapter 2). A *graphical user interface* (GUI) employs a pixel-based graphical display and a pointing device such as a mouse, in addition to the keyboard, to interact with the user. The first effective GUI on an affordable personal computer[1] was introduced by the Apple Lisa in the early 1980s.

Linux offers several *desktop environments* (or simply desktop) for GUI, providing workspaces, windows, panels, icons, and menus, as well as copy-and-paste, drag-and-drop operations. Today, it is hard to imagine computer users doing without a desktop GUI. Nevertheless, when you become more of a Linux expert, you may find the CLI more convenient in many situations. The right approach is to combine GUI and CLI to get the best of both worlds.

Many Linux application programs came from UNIX and were written before the graphical display became standard. Others chose not to use any GUI. These *command-line applications* tend to be more efficient (less taxing on the computer, easier to combine with other applications, and simple to access across a network), but can be harder for novice users. GUI applications are generally more intuitive to learn and use interactively, but they can be harder to control or run within other programs.

We will discuss the Linux GUI in this chapter.

3.1 Desktop Overview: GNOME and KDE

After login at the console, the first thing you see is the desktop from which you can launch applications, manage files, control your Linux system, and perform many other tasks. A desktop provides a GUI to make operating your computer more intuitive through a *desktop metaphor* by simulating physical objects. Overlapping windows can be moved and shuffled like pieces of paper. Buttons (icons) can be pushed (clicked) to initiate actions.

Unlike Microsoft Windows or the MAC OS, Linux offers a good number of alternative desktops with a high degree of user customization. Included with most Linux distributions are the most popular desktops: GNOME and KDE. The KDE (K Desktop Environment) is derived from the CDE (Com-

---

[1]Cost about $10K in 1983.

mon Desktop Environment) developed by the X/Open Company (a joint effort by HP, IBM, and Sun Microsystems). GNOME is part of the GNU Project. Both are built on top of the *X Windows System* and offer a complete desktop GUI together with a set of essential applications including a clock/calendar, sound volume control, email client, Web/file browser, instant messenger, image displayer, media player, address book, PDF reader, photo manager, preference/configuration editor, and more.

A good understanding of the desktop and how to use it effectively can make life on Linux much easier. It is perhaps safe to assume that you already have good working experience with MS Windows® or MAC OS. A Linux desktop is often more flexible, but works in very similar ways. Here, we will focus on the GNOME desktop. The KDE follows the same principles.

3.2 GNOME Desktop Components

Even though a user need not invoke it, the actual command to run the GNOME desktop is **gnome-session**, which serves as your Linux *session manager* as well as GUI. The GNOME desktop displays the following components:

- *Root Window*—After login, the entire graphical display screen is covered by the desktop which is the *root window* of GNOME. It is the space where all other GUI objects (desktop components and application windows) are placed and manipulated.

FIGURE 3.1: A Typical Panel

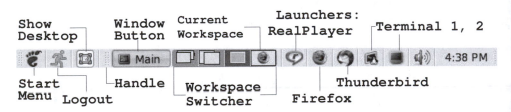

- *Control Panel*—The GNOME desktop displays a *Task Bar*, known as a *Control Panel* (or simply *Panel*), usually in the form of a horizontal bar along the top or the bottom edge of the root window. The Panel displays icons called *launchers* to invoke specific applications such as Firefox, RealPlayer, and a terminal emulator, as well as *applets* (small applications) such as a clock or an audio volume control. See Figure 3.1 for a typical Panel. To add objects to the Panel, right click on any empty space in your Panel and select the `Add to Panel` option to display a dialog window (Figure 3.2).

FIGURE 3.2: Add to Panel Dialog

Add to Panel

Find an item to add to "Bottom Panel":

Custom Application Launcher
Create a new launcher

Application Launcher...
Copy a launcher from the applications menu

Battery Charge Monitor
Monitor a laptop's remaining power

Brightness Applet
Adjusts Laptop panel brightness

Character Palette
Insert characters

Clock
Get the current time and date

Command Line
Mini-Commander

Help   Back   Add   Close

- *Desktop Objects*—In the vast space left open by the Panel on the root window, you can place objects such as files, folders, and application launchers for easy access. Figure 3.3 shows a sample collection of desktop objects. Pre-installed desktop objects usually include the following:

  - The *Computer* icon gives you access to the file systems on the hard disk, CDs/DVDs drives, and removable media such as flash drives. *Nautilus* from GNOME is the file manager/browser used.

  - The *Username*'s *Home* accesses your *home folder* where all of your personal files are kept. You can also open this folder from the Start menu Places.

FIGURE 3.3: Desktop Objects

– *Trash* is a link to the .`Trash` folder in your home directory. Use it to hold files and folders you wish to discard. You can recover items moved to *Trash* and empty trash to finally get rid of what is in *Trash.*

– The *Printing* object allows you to add, configure, and otherwise control printers connected to your Linux system.

– The *Services* object is used to control and configure network services on your Linux.

– A *device icon* object appears when you insert a CD/DVD or plug in a flash drive, a music player, or a digital camera.

Click a desktop object to open the associated program or folder. Drag (depress the left mouse button without releasing) an object to move it anywhere you like on the desktop. Right click an object to select possible operations on it.

• *Application Program Windows*—Displayed in the root window as child windows. Multiple child windows can overlap. You can change the input focus from one window to another, as well as move, resize, maximize, minimize, unmaximize, or close each window as you like.

Often, parts of the desktop can be obscured by application windows. There are two quick ways to reveal the desktop by minimizing all windows: click on the *Show Desktop* icon on the Panel or press CTRL+ALT+D. Either action will also restore your windows to their previous state. If it is more convenient, you can switch to a less crowded *workspace* to see the desktop.

• *Start Menu*—The `Start` menu (Figure 3.4) is exposed by clicking the *start icon* (often in the form of a logo for GNOME, Red Hat, Fedora, or Ubuntu depending on your Linux version) placed at the end of the Panel. The keyboard shortcut for the `Start` menu is usually ALT+F1. From the `Start` menu, you can perform almost all operations and can access files, the network, installed applications, commands, and preference options The Panel may display additional menus.

• *Workspace Switcher*—Enables you to work with multiple *workspaces* and to switch your screen display from one workspace to another from the Panel. A *workspace* is essentially a duplicate root window to provide more space for placing additional application windows. With several workspaces, you can spread out your application windows for easier use. The *workspace switcher* (Figure 3.1) is an applet you may add to the Panel and configure to provide several workspaces. Your root window can display one workspace at a time. The workspace visible is your *current worksapce.* Click on the switcher to change the current workspace.

FIGURE 3.4: The **Start** Menu

- *Window List*—Displays a button on the Panel (Figure 3.1) for each window in a particular workspace. Clicking on a window list button minimizes and restores the window. A different set of window list buttons is displayed when you switch workspace.

- *Notification Area*—Part of the Panel and displays icons from various applications to indicate activity in the particular applications; for example, a system update available icon, an incoming email icon, a Skype notification icon, a pidgin IM notification icon, and so on. Clicking on a notification icon usually displays the application window, and right clicking on the icon reveals a menu of operations. The notification area is usually at the right end of the Panel next to the clock applet.

3.3   Working with the GNOME Desktop

One of the best ways to get familiar with the GNOME, or any other, desktop is to learn how to achieve specific tasks with it. Here are some tasks important for a new Linux user.

Time and Date

If you see time or date displayed on your Panel, the *clock applet* has already been added. Otherwise, add it from the **Add to Panel** dialog. Select **clock** and click the **Add** button (Figure 3.2). You'll see the time/date display added to the Panel. Once there, right click on the time display to set preferences and to adjust the time or date (Figure 3.5). You can also copy the time or date (into an internal `clipboard`) for pasting to another application. Using the same procedure, you can add other applets, drawers, menus, and any application launchers to your Panel. You may drag an icon to move it to a different position in the Panel. Right click an icon to remove it from the Panel, edit its properties, or lock it in position.

FIGURE 3.5: Time Adjustment

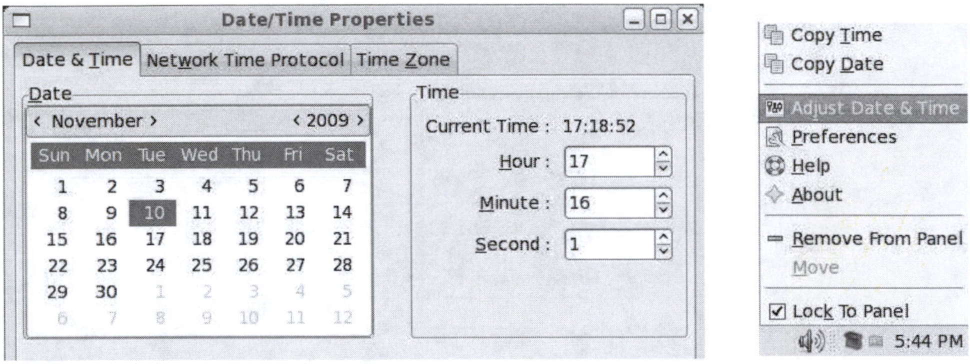

Launching Application Programs

Perhaps the single most important purpose of a desktop is to help you run and manage application programs. Linux offers a large number of applications. You'll find many of them from the `Start` menu organized into several general groups such as *accessories, office, graphics, Internet, programming, multimedia, games* and so on. In addition, there are many character-terminal–oriented applications that you can invoke directly from the command line in a terminal window.

In fact, you have multiple ways to start applications:

- Single clicking a Panel launcher icon (You may add a launcher for any application you like.)

- Single or double clicking a desktop object, depending on your preference setting

- Selecting an application from the `Start` menu or a submenu thereof

- Adding a launcher to the Panel

- Issuing a command from the command widget on the Panel

- Issuing a command in a terminal window

Most users place additional launchers for a Web browser (say, Firefox), an email agent (say, Thunderbird), and a media player (say, RealPlayer) on the Panel.

To initiate a graphical application, say **gedit**, from the command line without the Shell waiting for it to finish or the Shell job control mechanism getting involved, use

(**gedit** *filename* &)

This way, a subshell puts the graphical application in the background, disassociates it from the terminal window, and gives you back your command prompt.

Desktop Appearance

You can customize the look and feel of your desktop display. Either right click on an empty spot on your desktop working area and choose `change desktop background` or go to

`Start->System->Preferences->Look and Feel->Appearance`

to expose the appearance dialog (Figure 3.6). From this dialog you can choose a scheme, fonts, button style, and a background image for your desktop. From

`Start->System->Preferences->Look and Feel->ScreenSaver`

you can chose different screen savers. When your computer goes idle for a prescribed time, your session will be locked and the screen saver images will be displayed. You'll need your password to resume your session. To easily activate screen-locking at any time manually, add the `lock screen` icon to your Panel.

Using and Managing Desktop Objects

For the average user, the most useful object on your desktop is perhaps the `Home` folder. Click on it and you'll invoke the GNOME file browser **Nautilus** (Section 3.5) to display files and folders in your own home directory. Each user's home directory often contains these standard folders: `Documents`, `Download`, `Music`, `Pictures`, `Videos`, `Desktop`, and the hidden `.Trash`. If your Linux serves the Web, each user may also have a folder `public_html` where per-user Web pages reside. Other files and folders can be set up as you work on your Linux computer.

From the `Start` menu you can select any items to place on the desktop. As you gain more experience with Linux, you will undoubtedly place more objects on your desktop. You can drag and reposition these icons on the desktop and rearrange them in ways you see fit. desktop objects and other files can be moved to `Trash` and then eventually discarded when you empty trash (from `File` menu of `Trash`). It is also possible to retrieve items from `Trash`.

FIGURE 3.6: Desktop Appearance

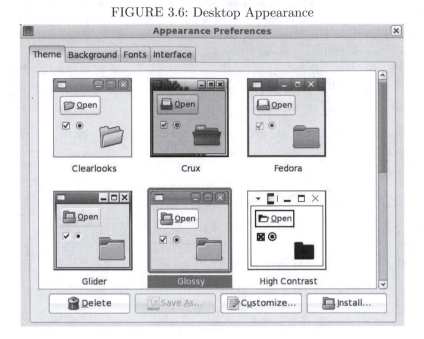

Right click a desktop object to change its name, move it to `Trash`, add an *emblem* (a small indicator icon), or otherwise manage it.

Multiple Workspaces

The desktop workspace can get crowed with multiple application windows quickly when you work on your Linux. Fortunately, you can set up more than one workspace and switch from one to another to work on different tasks. For example, you may have one workspace for Web browsing, another for email and instant messaging, yet another for text editing and word processing, and so on.

To enable multiple workspaces, simply add the `Workspace Switcher` to your Panel. This will give you four workspaces by default. Right click on the switcher to customize it to your liking (Figure 3.7).

The current workspace is highlighted in the switcher. Switch from one workspace to another by clicking on the switcher. As you go to a different workspace, the `window list` buttons change in the Panel to show windows in the new current workspace. Click on a `window list` button to show/hide the corresponding window.

FIGURE 3.7: Workspace Switcher and Preferences

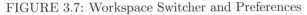

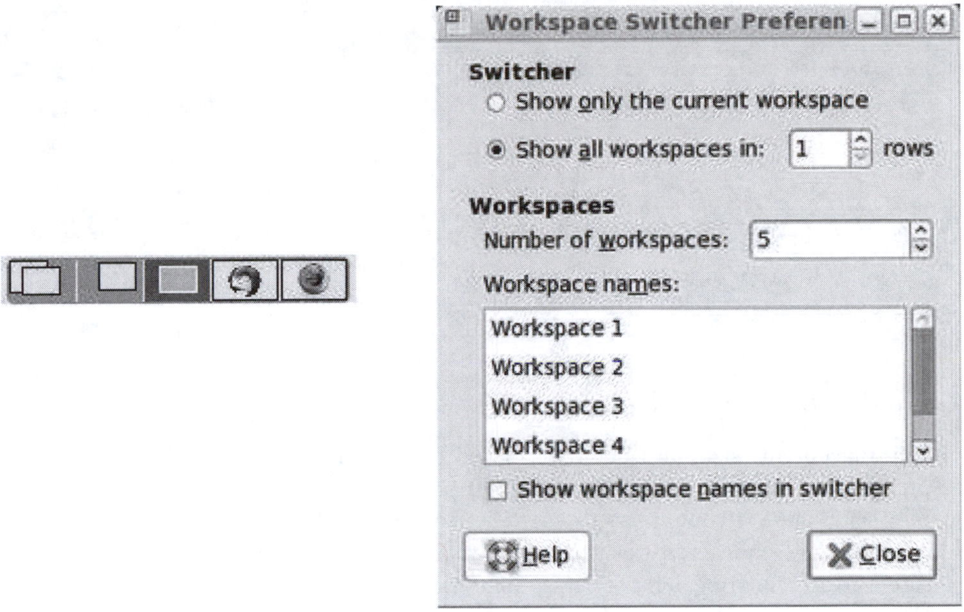

Sessions

A Linux *session* starts after you log in and ends when you log out. At any given time, a session consists of the current state of all the application programs that you are running. Your desktop (GNOME, KDE, or some other system) is your session manager. Often, Linux lets you choose which desktop to use when you log in. After login, your session manager (**gonme-session** in this case) will automatically start designated applications after displaying the desktop, keep track of your currently running applications, and remember any programs not closed before logout so that they can be restored the next time you log in (the *restore application feature*).

Go to `Preferences->Personal->Sessions` to customize your session manager (Figure 3.8 left). It allows you to add/remove applications that automatically start at login, to view a list of currently running applications under session management, and to enable/disable the restore application feature.

To leave the computer for a moment or two, you can prevent unauthorized use by locking your session with the `lock screen` icon or the `Start->Lock Screen` option.[2] The screensaver displays and you'll need your password to unlock the screen when you return.

---

[2]For security reasons, screen locking is not available for the `root` user.

FIGURE 3.8: Session Preferences and Panel Customization

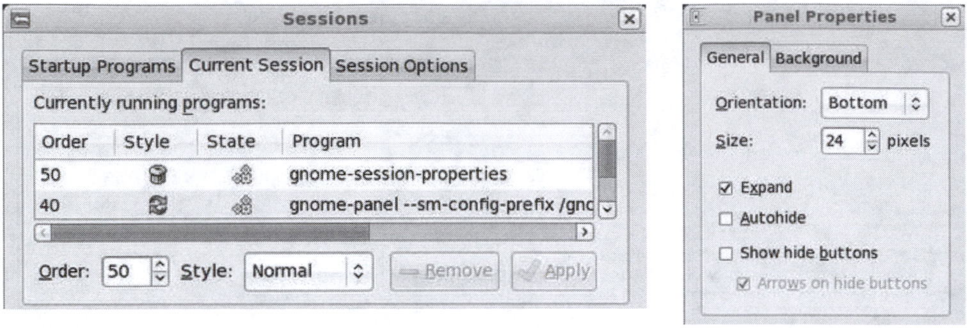

Customizing the Panel

The placement, size, and behavior of the Panel can be set according to your preferences. Simply right click on any unoccupied spot on the Panel and display the `Panel Properties` dialog (Figure 3.8 right).

You may choose to position the Panel on one of the four sides of your screen: top, bottom, left, or right. The `Expand` option controls whether the Panel spans the entire width/height of the screen or not. If `Autohide` is selected, the Panel will hide itself automatically and will reveal itself when the mouse is moved to its edge of the screen.

3.4   Windows

The X Window System

In Linux/Unix, graphical applications use the *X Window System* (originally developed at the Massachusetts Institute of Technology) to create GUIs. Windowing software such as X enables pixel-based graphical displays and the use of windows, menus, and the mouse to interact with application programs. The X Window System works as a *GUI server* (the *X server*) that enables client programs (*X clients*) to interact with users through GUIs. X clients can connect to the local X server running on the same host or a remote X server running on a host across the network. Furthermore, the X server can handle multiple *stations*, each potentially with multiple displays. (Two or three 20-inch LCD displays for your Linux desktop computer, anyone?)

For Linux, the X server is basic and is started within the boot sequence. If the X server is not running, no GUI programs will work. Figure 3.9 shows the X Window System architecture.

When an X client starts, it needs to connect to an X server running on the local host or a remote computer. The X server is always specified by the `display` option. For example,

FIGURE 3.9: X Window System Architecture

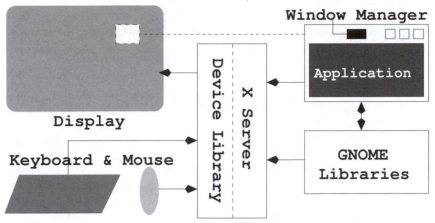

xclock -display *hostname:s.m*

says the **xclock** display will be rendered on *hostname*, station number *s*, and monitor number *m*. A *station* on a computer consists of a keyboard, a pointing device (mouse), and one or more graphical display monitors. A computer may have one or more stations, each with one or more monitors.

If the X server is local (on the same host as the client), the hostname part can be omitted. For a single-display computer, the monitor-station designation would be :0.0 and can usually be omitted also.

The Shell environment variable **DISPLAY** specifies the *default X server* for any client program started without an explicit **-display** option. Try the command

echo $DISPLAY

to see the value. Most likely, it will be the string :0.0.

Window Manager

You control windows displayed on your desktop through a *window manager*. The window manager is the piece of software that controls the display windows in the X Window System environment. The opening, closing, size, placement, borders, and decorations of any window are managed by the window manager. The X Window System calls for the window manager to be a client rather than a built-in program. In this way X can work with many different kinds of window managers. One of the original window managers is **twm**.

GNOME works with the window manager to display application windows on your screen (Figure 3.9). The default window manager for GNOME is *Metacity*, but any *GNOME-compliant* window manager can be used instead. Examples of such window managers include *Enlightenment, Icewm,*

Window Maker, *FVWM*, and *AfterStep*. Let's see how to control windows under GNOME with Metacity.

A GNOME window under Metacity consists of a thin rectangular frame enclosing a *title bar* and a display area for the application program using the window. An application may use one or more windows. Figure 3.10 shows such a window for the **xclock** application.

FIGURE 3.10: GNOME Window under Metacity

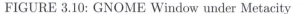

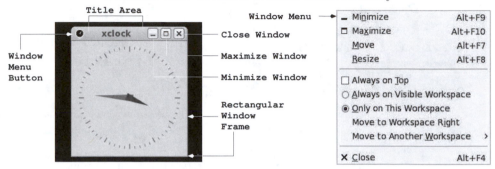

You can easily minimize, maximize/unmaximize, or close a window using the usual buttons on the title bar. If you close all the windows of an application, then the application will be closed. Clicking the top-left button (often an icon of the application), right clicking on the title area, or right clicking on a window button in the Panel displays the *Window Menu* from which you can control the window. A window can be moved by holding onto the title area and dragging it and can be resized by holding onto and dragging a side or a corner of the frame.

A window is in the workspace where it is created but can be moved to another workspace any time you wish. You can also make a window (an **xclock** window, for example) visible in every workspace.

Windows in a workspace may overlap. Clicking on a window or its Panel button shifts *input focus* to it and brings it to the top.

In addition to regular windows, an application will sometimes display a *dialog window*. Such popup windows are used to display alerts, to solicit user confirmation, or to obtain user input. For example, an application may ask if a user really wants to quit before closing. A dialog window can be *modal* or *transient*. A *modal* dialog will not allow you to interact with the main application window until the dialog is closed. Thus, you must deal with a modal dialog and close it before you can resume working with the application.

Window Information

Under X Windows, all windows form a containment hierarchy with the *root window* (desktop) sitting at the top. Each window has a unique *window ID*.

The *root* window's ID is **root**. The command **xwininfo** displays the window ID and many other items of information about any window. Run **xwininfo** first, then click on any target window to get the information display. Here is a sample on the Firefox window.

```
xwininfo: Window id: 0x12c4116 "Homepage - Mozilla Firefox"

  Absolute upper-left X:  195
  Absolute upper-left Y:  95
  Relative upper-left X:  4
  Relative upper-left Y:  24
  Width: 1321
  Height: 789
  Depth: 24
  Visual Class: TrueColor
  Border width: 0
  ...
  Corners:  +195+95  -164+95  -164-166  +195-166
  -geometry 1321x789+191+71
```

Note that the window ID is a hex number 0x12c4116.

Now let's take a look at some useful GUI applications on Linux.

3.5   Nautilus: the GNOME File Manager

An important aspect of any operating system is the ability to store and manage files. The Linux file system has a hierarchical structure. Either a regular file, a directory (folder), or a hardware device (special file) is considered a *file* in the Linux file system. A directory (folder) is a file that records the names and attributes of files it contains. Any of the contained files can, in turn, be folders as well.

The Linux file system is all encompassing because almost everything in Linux has a representation in the file system. Chapter 6 discusses the Linux file system in detail.

Nautilus (Figure 3.11) from GNOME is a powerful GUI application that helps you navigate the Linux file system and manage your files. In fact, the desktop launchers `Computer`, `Home`, and even `Trash` simply invoke the **nautilus** command on different *URI*s (Universal Resource Identifier). For example,

The Computer icon	**nautilus** `computer:///`
The Home icon	**nautilus** `$HOME`
The Trash icon	**nautilus** `$HOME/.Trash`
The Network icon	**nautilus** `network:///`

Nautilus enables you to interactively navigate the file system, manage files and folders, access special places on your computer, use optical drives, and reach available networking places. See the network part of Nautilus in Chapter 7, Section 7.4.

Navigating the File Tree

You browse the file system by following the folder-subfolder path until you find your target file(s). Thus, at any given time you are located at a *current directory*. The contents of the current directory can be displayed in a list view or an icon view, and you can switch between them easily. The icon view is recommended for easier visual interactions.

Double click a folder to open it and see the files in it, and click on the **up** button to move up to the parent folder of the current folder. The **Location** box shows the *pathname* leading from the root directory / to the current directory. Normally, any file whose name begins with a period (.) is hidden. Select **View->Show Hidden Files** to reveal them.

FIGURE 3.11: Nautilus File Manager

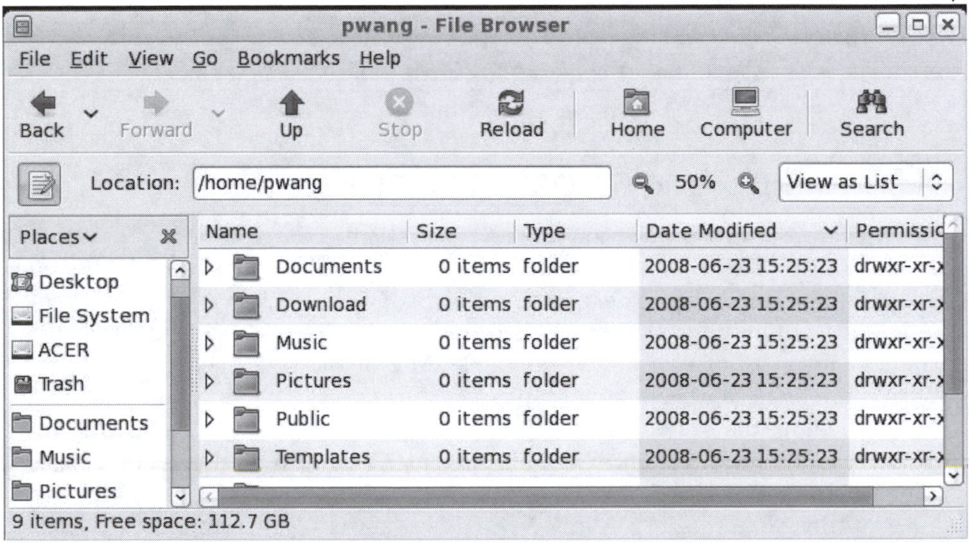

Opening a File or Folder

Double click a folder to open it and display its contents. Double click an ordinary file to open it with the default application, for example, PDF files with **evince**, a .txt files with **gedit**, or .html files with your preferred Web browser. Right click an ordinary file to open it with any application you choose and that application will be remembered as a possibility to run that particular type of file. By right clicking, you can also elect to remove a file/folder to **Trash** or to change its properties, including access permissions and the open-with application (see Section 3.5).

It is also possible to drag a file (a PDF file, for example) and drop it on a desktop object (the Adobe Acroread, for example) to open the file using that particular application.

Some users find double clicking difficult. In that case, you may go to `Edit->Preferences->Behavior` and select `single click open` instead. This means folders will open and files will execute by a single click. Hence, you need a different way (a CTRL click) to select a file/folder without opening it. Since icons displayed on your desktop are actually files/folders in your `Desktop` folder, the single or double click to open behavior also applies to them.

Finding Files

By clicking on the `Search` button, you change the `Location` box into a `Search` box. Type a string of characters in the name or contents of the file(s) you wish to find and press ENTER. The search results will be displayed. If too many files are found, you can narrow your search by file type and by location conditions and use the + and the − buttons to add/remove such conditions. Click the `Reload` button to see new search results (Figure 3.12).

FIGURE 3.12: File Search in Nautilus

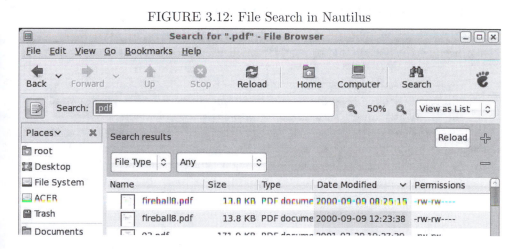

Managing Files and Folders

From the file display area, you select one or more files to manage. Click (or CTRL click) a file or folder to select it. Use CTRL click to select additional items. The selected items are highlighted. In icon view you may drag a rectangle around a group of icons to select them.

From the `Edit` menu you can easily select all displayed items or use the `Select Pattern` option to give a string pattern to match the filenames to be selected. Example patterns are `*.html` (all names ending in `.html`), where the character `*` matches any string of zero or more characters. See Section 2.7 for more about such patterns.

Making a new selection cancels the previous selection. If you CTRL click on a highlighted item or click on an empty spot in the file display area, the selection is also canceled.

After making a selection, you can perform operations on the selected files.

- Open the `Edit` menu to select operations on the selected items.

- Drag and drop the selected items onto a folder (or the desktop which is a folder anyway) to move them there.

- Grab the selection, then hold down ALT, and drag to a new folder and release the mouse. Then select the operation you wish, including *move here*, *copy here*, or *link here*. A link is a shortcut or a pointer to an actual file (Section 6.2).

- Right click on your selection to see the available operations. These include *moving to trash*, *open with*, *copying*, *send to*, and changing file properties (name, permissions, list of applications to open files of this type, and so on).

Access Control for Files and Folders

On Linux a file is owned by the user who created it. The file owner can set the *permissions* for the file to control which users will have what accesses to it.

Users are also organized into groups. A user can belong to multiple groups. The file `/etc/groups` lists groups and group members. The file owner can also set the *group* attribute to any group to which the owner belongs.

As a file owner, you can set the *read* (`r`), *write* (`w`), and *execute* (`x`) permissions for three disjoint set of users: the file owner (`u`), other users in the file group (`g`), and all others (`o`). Each permission is independent of other permissions. For example, a file may have no permissions for `u`, but `r` and `w` for `o`. It may not make much practical sense, but it does drive home the point of the `u`, `g`, and `o` permissions being independent. The meaning of each permission is as follows:

- `r`—Permission to open a file or to inspect contents of a folder

- `w`—Permission to modify a file or to create or delete files in a folder

- `x`—Permission to run a file as a program or to enter a folder to reach files contained in it

You set/change file permissions by right clicking the file and selecting the `properties` dialog (Figure 3.13 left).

The *root users* are system managers and have all permissions to all files in the file system, regardless of any file permission settings.

FIGURE 3.13: File Permissions and Associated Applications

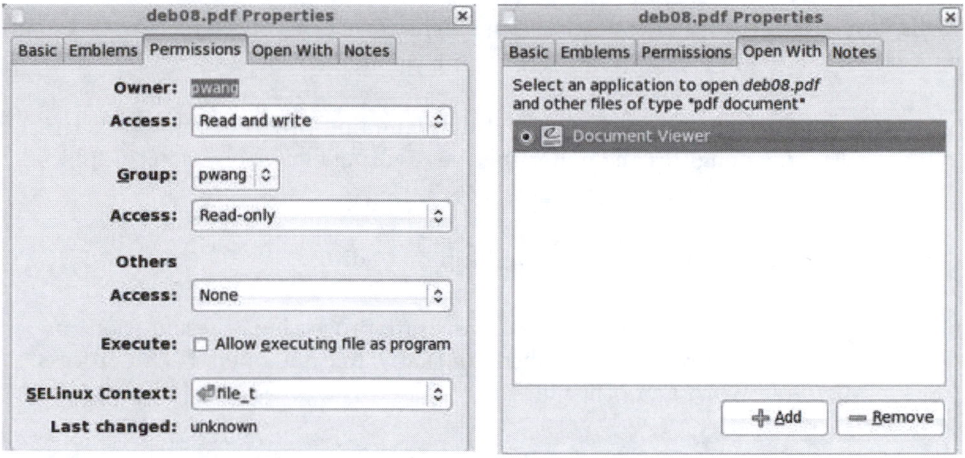

Writing CDs or DVDs

To create a data CD, click the `CD/DVD Creator` option on the `go` menu or simply insert a blank disc into the optical drive. This leads to a special folder (`burn:///`). Drag files and folders you wish to write to the disc into the `burn` folder. When ready, click on the `Write to Disc` button.

To copy a CD/DVD, simply insert the disc in the optical drive, right click on the resulting desktop icon (or find it from the `Computer` icon), and choose `Copy Disc`.

3.6   Graphical Applications

An application is *graphical* if it employs a GUI for interacting with the user. For example, **xclock** is a graphical analogue clock (Figure 3.10). Some graphical applications allow you to specify the size and location of the window by the `geometry` command option:

`-geometry` *string*

The geometry *string* is given in the form $c \times l \times y$, where c and l give the number of characters and lines for the window and (x, y) gives the window position on the screen. For example, `-geometry 80x42+100+8` says the window is 80 columns and 42 lines with its upper left corner located 100 pixels from the left edge and 8 pixels from the top edge of the screen. The coordinate `-1-4` locates the lower right window corner measured from the right and the bottom edges in the obvious way. Also `-1+1` locates the upper right corner and so on. Therefore, the command `xclock -geometry -0+0`

displays an xclock window on the upper right corner of your screen.
Here is a list of some other useful graphical applications.

- The **xfig** command—A powerful diagramming program for interactive authoring of figures and exporting the resulting figure to various graphics formats.

- The **gedit** command—GNU desktop text editor.

- **openoffice.org**—A comprehensive word processing, presentation authoring, and spreadsheet application that is free and open. It can process Microsoft Word and other files.

- The **evince** command—A document viewer that supports multiple document formats, including PDF, Postscript, tiff, and dvi.

- The **eog** command—The *Eye Of Gnome* image viewer to display most image formats, including BMP, GIF, JPEG, PNG, SVG, and TIFF.

- The **gimp** command—The *GNU Image Manipulation Program* is a full-function raster image authoring and editing tool similar to Adobe Photoshop.

- ImageMagick—A set of tools for editing and converting raster and other image formats. The **display** command in this tool set provides a GUI for many operations on images.

- Screen Capture—The **import** command allows you to capture any visible window (by window ID) or rectangular area (by mouse operation) and save the image in a designated format. The **ksnapshot** program presents a visual interface for screen capture and also allows a delay before the actual capture.

- The **gcalctool** command—A basic desktop calculator. Some may prefer the command-line tool **bc** which takes infix notation input.

- Skype—A popular Internet telephony application for making phone calls to and receiving phone calls from other Skype users online (free service), as well as regular landline or cell phones (paid service).

- RealPlayer—A popular and free audio and video media player from RealNetworks.

FIGURE 3.14: Starting a Terminal

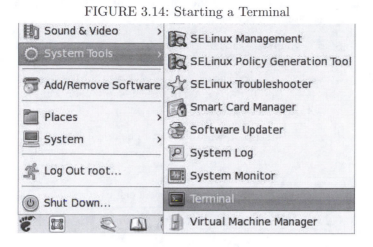

3.7   The GNOME Terminal

Because Linux is based on UNIX and offers many programs that run under character-based terminals, character terminal emulation programs are provided to run such applications. Linux commands requiring a terminal window include **bash** (the default Linux Shell), **vim** (text editor), **ssh** (secure remote login), **sftp** (secure file transfer), and many other command-line applications such as **ls** (directory listing), **rm**, **mv**, **more**, **man** (displaying manual pages), and so on.

The classic **xterm** terminal emulator is available on most Linux distributions, but most users prefer to use the **gnome-terminal**.

Starting a GNOME Terminal

A *GNOME Terminal* emulates a character-based computer terminal and allows you to run a Shell or command-line applications. Here is how it works. A GNOME Terminal emulates the *xterm* from the X Consortium which can, in turn, emulate the various DEC (Digital Equipment Corp.) terminals such as VT220 and VT320.

From the `Start` menu select `System Tools->Terminal` (Figure 3.14) to get a GNOME Terminal which will run Bash by default. Without customization, the terminal attributes are specified by a *default profile*. You can customize the size, font, colors, and other attributes for the terminal window from the `Edit->Current Profile` menu option.

By creating different terminal window profiles and giving them names, you can use them for different instances of GNOME Terminals you run. Let's assume that you have saved your favorite window profile under the name `main`.

Because easy access to a terminal can be such a convenience, we recommend that you add a terminal launcher on the Panel to start your customized terminal window. Follow these steps:

1. Go to `Start->System Tools->Terminal` and right click on `Terminal` to select `Add to Panel`, or you can right click on any empty space on the Panel and select `Add to Panel` to find a launcher under `System Tools`. After this step, you'll have a terminal icon on your Panel which will launch a default terminal window.

2. To customize this terminal window, right click on its Panel icon to expose the `Properties` dialog (Figure 3.15) to customize this launcher. Now you

FIGURE 3.15: Terminal Launcher Properties

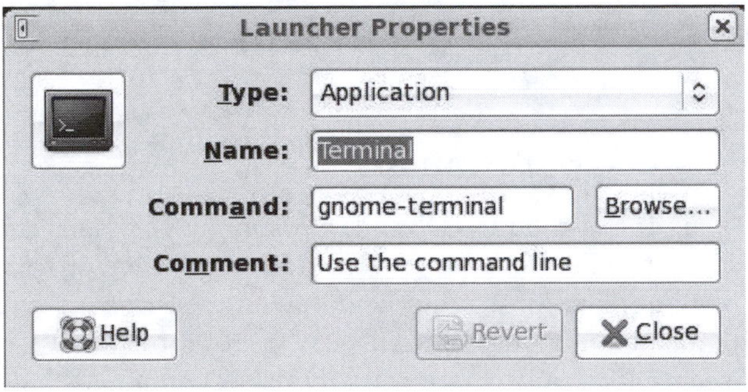

can modify the `command` entry to something suitable. For example,

gnome-terminal `--geometry=80x30+130+200 \`
                    `--window-with-profile=main`

This command invokes the program **gnome-terminal** to display an 80-character by 30-line terminal window whose upper left corner is positioned 130 pixels from the left edge and 200 pixels from the top edge of the screen. The `80x30+130+200` notation is known as the *geometry* of the window (Section 3.6). The window preferences come from the profile `main`.

3. If you wish, you may also change the Panel icon for the launcher. When you have several terminal launchers with different profiles (Figure 3.1), placing distinct icons on the Panel makes good sense.

Terminal Window and the Shell

When you start a terminal window, your designated Shell (**bash** by default) will be the application running in the window. The Shell can run in the window as a regular Shell or a *login Shell*. The GNOME Terminal allows you to make this choice as part of the window profile setting. The difference is that a regular Shell reads only the *Shell initialization* file, whereas a login Shell will also read the *Shell login initialization file* (Section 2.13).

In some situations, you may want to start a terminal window to run something other than a Shell. For example,

gnome-terminal -e "ssh -X pwang@monkey.cs.kent.edu"

gives an **ssh** command to run in the terminal window instead of the Shell. The result is a GNOME Terminal connected to the remote host **pwang monkey.cs.kent.edu** for the user **pwang** to log in.

The terminal window closes when the application, whether a Shell or some other program, terminates.

Select, Copy, and Paste

It is generally a good idea to use what is already displayed on the screen and avoid typing the information again to save time and preserve accuracy. With the GNOME Terminal, and other text-displaying windows such as a Web browser window or a text editor window, you can select, copy, and paste text with ease.

- Select—Press the left mouse button, click a character, double click a word, or triple click a line and then drag the mouse over a section of text to highlight and select the desired text.

- Copy—Simply selecting the text copies it into a *clipboard*. You can also right click the highlighted text (or use the **Edit->Copy** menu option) to explicitly copy the text into a *copy buffer*. Any previous content in the clipboard or copy buffer is lost.

- Paste—Go to the target application, position the input cursor where you wish, and then click the middle mouse button to paste from the clipboard. Or use the **Edit->Copy** option of the target application window to paste from the copy buffer.

A GNOME Terminal remembers displayed text lines (500 by default). Use the *scroll bar* to go back and forth on the text.

Web and Email Links

The GNOME Terminal recognizes Web and email addresses. For example, it recognizes **http://www.kent.edu** and **pwang@cs.kent.edu**.

Move your mouse cursor over such an address and it will be automatically underlined, signaling that the GNOME Terminal has recognized the address. Right click and select the **Open Link (Send Email To)** option to launch a Web browser (an email application) directly. This feature is very convenient.

The application launched in response to such usage is defined by your *Preferred Applications* setting under the **Start** menu.

3.8   Saving Energy

We all need to make an effort to save energy in big and small ways. GNOME provides power management that you can control. From the `Start` menu, use `Preferences->System->Power Management` or simply run the command

gnome-power-preferences

to set your power preferences. If you place a launcher for it on the desktop, it can be even easier.

If your Linux is running as a network server and needs to be up all the time, then you don't ever want to put your computer to sleep. If it is not running as a network server, set it to put the display and computer to sleep after a reasonable period of inactivity. This can save a lot of energy and your equipment in the long run.

When you leave your computer for the day, you can set the computer to sleep and turn off the display by its physical power switch. It is a good work habit to have.

If you will leave the computer for days, turning it off completely is more than reasonable.

3.9   Accessing Help and Documentation

From the `Start` menu select `System->Help` to access the full manual for your Linux system. You'll find sections on the GNOME desktop, different types of applications, as well as system tools. You can also browse the man pages and other information.

You can also reach item-specific help/documentation. For example, right clicking the Panel or an applet in the Panel also reveals a `Help` option that leads to the part of GNOME documentation specific for that particular item.

In general, the `Help` button on the menu bar of any application program will lead to documentation and user guides for that particular application.

3.10   Summary

GNOME and KDE are the two most widely used GUI environments for Linux. They both rely on the X Window System for graphical display and windowing support. Knowledge and skillful use of the GUI can make life on Linux easier and you more productive.

As a modern GUI system, GNOME provides a graphical point-and-click interface to your Linux computer. The GNOME session manager (**gnome-session**) displays and controls the *desktop* which is the *root window* containing launchers, windows, and the control Panel. Multiple workspaces make working with many windows much easier. *Metacity* is the window manager for the GNOME desktop.

The `Start` button on the Panel exposes an extensive menu for many operations, including logout/shutdown, adding/removing programs, setting preferences, starting applications, administering the system, and so on. Other objects on the Panel provide for workspace switching, minimized window parking, quick launchers for important applications, a clock/calendar, and a notification area. The Panel is also easily customizable.

In addition to the desktop, GNOME also manages your login session. It automatically starts designated applications when you log in, keeps track of your currently running applications, and remembers any programs not closed before logout so they can be restored next time you log in.

Another advantage of the desktop environment is easy launching of applications. In addition to issuing Shell commands, you can also start applications by clicking a Panel icon, a launcher, or an option from the `Start` menu.

GNOME comes with many useful GUI applications. The **gnome-terminal** is important because it provides a terminal window for the Shell and is your ticket to the command line. Cut-and-Paste of displayed text is supported with the clipboard and the copy buffer. Automatic recognition of Web and email addresses enables you to use them directly. The **gnome-terminal** can be customized and your settings can be saved in profiles for reuse. A comfortable terminal window can make life on Linux much easier.

The GNOME Nautilus file browser provides a visual environment to navigate the file tree and to manage files and folders as well as their attributes. Other useful GUI applications include image processing, document preparation/viewing, audio-video playing, and creating CDs and DVDs.

3.11   Exercises

1. How do you move your Panel to the top or bottom of the root window?

2. How do you make your Panel span the entire width of the root window or be centered at the top/bottom?

3. Do you have a logout icon on the Panel? If not, describe how to place one on it.

4. Do you have a workspace switcher on your Panel? If not, describe how to create one.

5. Is it possible to add/remove workspaces? How?

6. How does one place an analogue clock on the desktop?

7. Describe how to place a power management launcher on the desktop.

8. What is **eog**? Place a launcher for it on the desktop.

9. What is **evince**? Place a launcher for it on the desktop.

10. Setting up a terminal window correctly can make a big difference in how comfortable you will be using the Linux command-line interface. Consider the command

    gnome-terminal --geometry=80x30+350+150
    　　　　　　　　--window-with-profile=main

    and explain its meaning. Create a named profile of color, font, and other preferences for yourself. Make yourself a panel launcher to start your customized gnome-terminal.

11. On Linux, which GUI application can be used for MS Word and Excel documents? Install it on your computer and place an icon on the desktop.

12. Find out how to use **xfig**.

13. Find out how to use the commands **gimp** and **display**.

14. Burn a data CD and describe your procedure.

Chapter 4

Filters and Regular Expressions

One of the strengths of Linux is the richness of its command set with nearly 700 different commands. This richness is further enhanced by the ease with which new commands can be crafted by combining existing ones.

Effective use of Linux involves knowing existing commands, learning how to combine them into new commands, and selecting the right commands to apply. Throughout this book, we will introduce many useful Linux commands and demonstrate how they can be put to good use individually and in combination.

Many commands are *filters*. A filter usually performs a simple and well-defined transformation of its input and follows certain conventions to make it easy to connect to other programs. Filters can be strung together using pipes (Chapter 2, Section 2.5) to become *pipelines* that can perform complex functions on data. Many useful filters are presented in this chapter. Examples show how to build pipelines in practice.

For instance, the command **sort** is a filter that orders its input lines. The command **tr** translates specific characters in the input into other characters. You can combine these two filters with others to create and maintain a simple database of addresses.

Utilizing and processing human readable textual data have been an emphasis of Linux. Within textual data, we often need to identify the exact places where transformations or manipulations must take place. *Regular expressions* provide standard ways to specify *patterns* in textual data. It is important to become familiar with regular expressions because they occur frequently and are basic to programming. We explain the regular expression notations and how they are used in applications such as **grep**, **sed/vi**, and **awk**.

4.1 Commands and Filters

Simply put, a filter is any command that produces output by transforming its input by following a set of well-defined conventions. The conventions make filters easy to combine with other programs in a pipeline (Figure 4.1).

A filter is distinguished from other commands by the following characteristics:

1. A filter takes input from the *standard input* (stdin). Thus, when we invoke a filter, it does not need a file argument.

FIGURE 4.1: A Filter

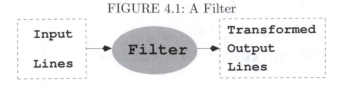

2. A filter sends its results to the *standard output* (`stdout`). Therefore, it does not need an output file argument.

3. A filter performs a well-defined transformation on the input and produces the output with no header, trailer, label, or other formatting.

4. A filter does not attempt to interpret its input data in any way. Thus, it never treats its input as instructions or commands.

5. With few exceptions, a filter does not interact with the user for additional parameters other than those supplied on the command line.

6. Any error or diagnostic output produced by a filter is sent to the *standard error output* (`stderr`). Hence, error messages are never mixed with results produced.

These characteristics make a filter easy to fit into a pipeline. The overall purpose is to make a program produce output that can be fed into another program as input and that can be processed directly. Typically, such input contains lines of text with no decorative labels, comments, or extra formatting. A separate line is used for each data entry. For example, if the data entries are words, then the input should be one word per line. For more complicated data entries (for example, those produced by `ls -l`), the line may consist of several fields separated by spaces, tabs, or colons (for example, `/etc/passwd`).

Many Linux commands are filters that can also work on files. The convention is *If filenames are supplied as arguments, a command can use them for input/output. Otherwise, if no files are given, the command acts as a filter.*

The *process expansion* (Chapter 2, Section 2.7) feature of Bash makes it possible to treat output from filters as input files to other commands.

Let's look at some filters and then show how to build pipelines with them.

Leading and Trailing Lines: `head` and `tail`

The commands **head** and **tail** are available for displaying the leading and trailing lines of a file, respectively. The command

head [-*k*] [*file* ...]

outputs the first *k* (default 10) lines of each given *file* to the standard output. If no file argument is given, the standard input is used. The **head** command is a quick way to examine the first few lines of a file, which are often all that is needed.

The command **tail** is the opposite, displaying the last part of a file on the screen:

tail [*starting-point*] [*file* ...]

outputs the last part (from *starting-point* to the end or, by default, the last 10 lines) of each given *file*. If no file is specified, the standard input is used. The starting point is specified as

+*k*   (line *k* from the beginning)
−*k*   (line *k* from the end)

If the integer *k* is followed immediately by the characters b or c, **tail** will count blocks or characters, respectively, instead of lines. The −f option instructs **tail** to continue, even after the end of the file has been displayed, repeatedly probing the file in case more lines are appended. This option provides a way of monitoring a file as it is being written by another program.

In pipelines, **head** and **tail** are useful for selecting some lines from the input and excluding others. The **more** (**less**) command can be used at the end of a pipeline to manage long output.

Character Translation: tr

The command **tr** copies standard input to standard output, substituting or deleting specified characters. For example,

tr A-Z a-z < *file1* > *file2*

creates *file2* as a copy of *file1*, with all uppercase letters translated to the corresponding lowercase ones. Another example is

tr TAB % < *file1* > *file2*

where TAB must be escaped by CTRL | V when typing this command. This method allows you to see each TAB in *file1* as a % character in *file2* (assuming *file1* does not contain any % characters). Generally,

tr *string1* *string2*

translates *string1* characters to the corresponding *string2* characters, assuming the two strings are of the same length. If *string2* is shorter, it is treated as if it were padded with enough repetitions of its last character to make it the same length as *string1*. A range of characters can be given, as in x-y. A character also can be given by its ASCII code in octal (for example, \040 for SPACE, \011 for TAB, and \012 for NEWLINE). For example, to replace a string of blanks with a NEWLINE, use

```
tr -s '\040\011' '\012'
```

The −s (squeeze) option shortens all strings of consecutive repeated characters in *string1* to just one character. The −c (complement) option is used to specify *string1* by naming characters *not* in it. Thus,

```
tr  -cs 0-9A-Za-z '\012'
```

creates a list of all words (one per line) in the input. In this example, *string1* is all characters except numerals and letters.

When the option −d (delete) is given, characters in *string1* are deleted from the output, and there is no need for *string2*. For example, to rid the input of all CR characters, we can use

```
tr -d "\015" < file
```

Tab Expansion

Tabs often need to be expanded into an equivalent number of spaces or vice versa. However, this transformation is not performed by **tr** because each TAB must be replaced by just enough spaces to move the output column position to the next *tab stop*. Tab expansion and its inverse transformation are slightly more complicated than simple character-for-character replacement. The filters

expand   (substitutes spaces for tabs)
unexpand   (substitutes tabs for spaces)

are used for these purposes. For example,

```
expand 6 < file
```

replaces each TAB in *file* by spaces, assuming that TAB stops are 6 (default 8) spaces apart.

Folding Text Lines

It is sometimes necessary to make sure lines of text are within a certain length for easy display, viewing, or printing. The **fold** filter breaks up long lines by inserting a NEWLINE character where necessary.

```
fold < file
```

The default is to limit lines to a length of 80 characters. Useful options include

−c *n*   (sets width to *n* columns)
−s   (breaks lines only at spaces)

For example,

```
fold -w 72 -s report > new_report
```

creates *new_report* as a version of *report* with all lines folded at spaces to within 72 characters.

Sorting Text Lines

Data often are sorted in some kind of order for easy access and manipulation. You may want to alphabetize a list of names and addresses, combine several such lists into one, look an entry up in a list, or compare two lists already in order.

The **sort** command takes input lines and writes them to the standard output in sorted order. The units being sorted are entire lines. Each line may contain one or more fields, which are separated by one or more blanks (spaces or tabs). For example, a file called students (**Ex: ex04/students**) may contain the following lines:

```
F. Smith 21 3.75 Physics
J. Wang 23 2.00 Accounting
R. Baker 20 3.20 Chemical Engineering
S. Doe 24 3.20 Business
P. Wang 22 4.00 Computer Science
```

The first line contains five fields (separated by white space); the third line contains six fields. The **sort** command allows you to use field positions to specify *sort keys* for ordering the lines. A sort key is defined by a starting and an ending field position in a line. The sort keys in different lines are compared to order the lines.

Thus, if you specify the sort key as the second field to sort the file students, then the lines will be ordered by last name using, by default, the ASCII collating sequence. In the absence of any specification, the sort key is the entire line. Multiple sort keys are given in order of importance. In comparing any two lines, **sort** uses the next sort key only if all previous sort keys are found to be equal.

The command has the general form

sort [*options*] [--**key**=*key* ...] [*file* ...]

All lines in the given files are sorted together. A file named "-" is the standard input. If no file is given, **sort** uses the standard input. It writes to the standard output by default. Keys are given in order of significance. A key is given by two field positions:

begin[, *end*]

which specify a sort key consisting of all characters between the *begin* and *end* positions (field separators excluded). When omitted, *end* becomes the end of line. Each position has the form

f [.c]

where f is a field number, and the optional c is a character number. For example, the position 2.3 indicates the third character of the second field. If

TABLE 4.1: Sort Keys

Specification	Key
2,3.0	Second field
4	Fourth field to end of line
2.3,4.7	Third character of second field to seventh character of fourth field, inclusive

omitted, c is 1. Thus, the position 3 is the same as 3.1. Table 4.1 provides some examples of sort key specifications.

Therefore, the command

```
sort --key=2,3.0 students
```

sorts the file **students** by last name. In this and many other cases, the ending field can be omitted without affecting the search.

Sort keys are compared using ASCII ordering, unless one of several options is used. A few important options are listed here:

f   Treats all uppercase letters as lowercase letters
n   Sorts by increasing magnitude using a leading numerical string in the sort key where the numerical string may have leading blanks and/or a sign followed by zero or more digits, with an optional decimal point
r   Reverses the sense of comparisons and sorts the lines in reverse order

These option characters can be given globally, affecting all sort keys, or immediately after a key specification to affect only that sort key. Note some examples:

```
ls -l | sort -n --key=5,6.0        (sort by increasing byte count)
ls -l | sort --key=5,6.0nr         (sort by decreasing byte count)
```

For multiple sort keys, consider

```
sort --key=4,4.4nr --key=5 students
```

which sorts by grade point average (4th field), highest first, and break ties with the second key, the department name (field 5 to end of line). See **man sort** for more information.

4.2   The grep Command

The **grep** command is a filter that provides the ability to search and identify files containing specific text patterns or to find all lines in given files that contain a certain pattern. The command has many possible applications. You may search for a name, a subject, or a phrase. You may search for something contained in a file whose filename you have forgotten, or you can extract text lines from files that pertain to a particular subject. The **grep** filter is often useful in pipelines. For example,

look men | **grep** gitis

is a cute way to find the word "meningitis."

The name **grep** comes from *generalized regular expressions* which are exactly what **grep** uses to specify search patterns. The general form of the **grep** command is

grep [*options*] [*patterns*] [*files*]

It searches for the given regular expression *patterns* (Section 4.3), using a fairly efficient matching algorithm, in the given files and outputs to stdout the matching lines and/or file names. Making it flexible, many options control how exactly **grep** works . A **grep** command searches the specified *files* or

TABLE 4.2: Options of the **grep** Command

Option	Description
-E	Enables matching of *extended regular expression* patterns (same as the **egrep** command)
-F	Uses a fast algorithm for matching fixed-string patterns (same as the **fgrep** command)
-c	Displays only a count of the matching lines
-f *file*	Takes patterns from *file*, one per line
-i	Ignores the case of letters
-l	Lists only names of files with matching content
-n	Adds a line number to each output line
-s	Displays nothing except errors (silent mode) and returns exit status 1 if no match
-v	Displays all non-matching lines
-w	Matches whole words only
-x	Displays whole-line matches

standard input for lines that match the given patterns. A line matches a pattern if it contains the *pattern*. Each matched line is copied to the standard output unless specified otherwise by an option (Table 4.2). The output lines are prefixed with a filename if multiple files are given as arguments. Generally speaking, the **grep** command is used either to obtain lines containing a specific pattern or to obtain the names of files with such lines.

For example, let's say you have a file of phone numbers and addresses. Each line in the file contains the name of the person, a phone number, and an address. Let's name this file contacts (**Ex:** ex04/contacts). A few typical entries follow:

```
(330) 555-1242 Bob Smith C.S. Dept. Union College. Stow OH 44224
(415) 555-7865 John Goldsmith P.O. Box 21951 Palo Alto CA 94303
(415) 555-3217 Bert Lin 248 Hedge Rd Menlo Park CA 94025
(617) 555-4326 Ira Goodman 77 Mass. Ave. Cambridge MA 02139
```

Consider the command

grep -F *string* contacts

or equivalently

fgrep *string* contacts

If *string* is a name, then any line containing the given name is displayed. If *string* is an area code, then all entries with the same area code are displayed. If *string* is a zip code, then all lines with the same zip code are displayed. Also,

fgrep -v MA contacts

displays all addresses except those in MA.

Here is an application dealing with multiple files. Let's say you have a directory named letters that you use to file away electronic mail for safekeeping and later reference. Suppose you need to find a letter in this directory, but you don't remember the letter's filename. All you recall is that the letter deals with the subject "salary". To find the letter, use

cd letters
fgrep -i -l salary *

The command searches all (non-hidden) files under the current directory for lines containing the string **salary** (ignoring case differences) and displays only the name of any file with matching lines. The Shell variable **$?** records the *exit status* of a command (Chapter 5, Section 5.7). The **grep** command returns *exit status* 0 if any matches are found, 1 if none, and 2 if error.

4.3   Regular Expressions

In the **grep** command and many other text processing situations, the need to find a string of characters matching a particular *pattern* arises. For example, testing if a file name ends in .**pdf**, checking if a particular user input represents a number with an optional leading sign, or making sure that a line of text has no trailing white spaces. In order to define patterns to match, we need a notation to specify patterns for programs. A *regular expression* is a pattern matching notation widely used and understood by programmers and programs.

The simplest regular expression is a fixed string such as **Ubuntu** or **CentOS**.

Such a regular expression matches a fixed character string. However, regular expressions are much more flexible and allow you to match strings without knowing their exact spelling.

In Linux, the applications **grep**, **vi/vim**, **sed**, **egrep**, and **awk**, among others, use largely the same regular expressions. Table 4.3 gives the basics for regular expression notations that most programs understand. The **grep** command accepts many additional pattern notations (see Section 4.4 and the **grep** man page).

TABLE 4.3: Basic Regular Expressions

Pattern	Meaning
x	A character x with no special meaning matches itself.
$\backslash x$	Any x, quoted by $\backslash$, matches itself (exceptions: NEWLINE, <, >).
^	The character ^ matches the beginning of a line.
$	The character $ matches the end of a line.
.	The character . matches any single character.
[*string*]	A string of characters enclosed by square brackets matches any single character in *string*.
[*x-y*]	The pattern matches any single character from x to y.
[^*string*]	The pattern matches any single character not in *string*.
pattern*	It matches *pattern* zero or more times.
$re_1 re_2$	Two concatenated *re*'s mean a match of the first followed by a match of the second.
$\backslash$<	The notation matches the beginning of a word.
$\backslash$>	The notation matches the end of a word.

Consider editing, with **vim**, a recipe that contains many steps labeled sequentially by `Step 1`, `Step 2`, and so on. In revising the recipe, you need to add a few steps and renumber the labels. A search pattern can be specified by the regular expression

```
Step [1-9]
```

where the notation `[1-9]` matches any single character 1-9.

In the **vim** editor (see appendices), you can search with the command

```
/Step [1-9]
```

and make the appropriate modification to the number. After that, you can repeat the search using the **vim** search repeat command **n**, change another number, search, and so on until all the changes have been made.

Let's put the regular expression notations to use and look at some specific patterns.

`[A-Z]`	(any capitalized character)
`[0-9]*`	(a sequence of zero or more digits)
`\<[A-Z]`	(any word that begins with a capital)
`^##*`	(one or more #'s starting in column one)
`;;*$`	(one or more ;s at the end of a line)
`ing\>`	(any word ending in `ing`)
`\<[A-Z][a-z]*\>`	(any capitalized word)

In a regular expression, the * character indicates an occurrence of *zero or more times* of the previous character/pattern. In Table 4.3, we see regular expression special characters: `[`, `]`, `*`, `^`, and `$`, each having a prescribed meaning as a pattern specifier.

Quoting in Search Patterns

The use of special characters in any searching scheme inevitably leads to the question of how to search for a pattern that contains a special character. Let's say that you are editing a report and you want to search for `[9]`, which is a bibliographical reference used in the report. Because the regular expression `[9]` matches the single character 9, you need to *quote* the `[` and `]` so that they represent themselves rather than pattern specifiers. The solution, ironically, is to introduce yet another special character, the backslash (`\`), to serve as a *quote character* that prevents the immediate next character from being treated as a pattern specifier and forcing it to stand for itself. Thus, the pattern `\[9\]` matches `[9]`, and the pattern `\[[1-9]\]` matches the strings `[1]` through `[9]`. To match any such bibliographical reference, use the pattern `\[[1-9][0-9]*\]`. Here are some more pattern examples:

`\.\.\.`	(matches ..., namely three dots)
`\/\*`	(matches /*)
`\\`	(matches \)
`[0-9A-z]`	(matches any of the indicated characters)

Quoting a character that does not need quoting usually causes no harm.

4.4   Patterns for `grep`

Most of the *basic regular expression* patterns listed in Table 4.3 work in programs accepting regular expression patterns. The **grep** command also accepts *extended regular expressions* available via the `-E` option or through the **egrep** command. Extended regular expressions add notations described in Table 4.4 to the basic regular expressions.

In Table 4.4 *re* denotes any regular expression. The precedence of operators used for extended regular expressions is `()`, `[]`, `"`, `+`, `?`, concatenation, and `|`. Care should be taken when entering patterns on the command line because many pattern characters are also special Shell characters. It is safest to always

TABLE 4.4: Extended Regular Expressions

Pattern	Description
\w	Matches an alpha-numerical char, same as [0-9A-Za-z].
\W	Matches a non-alpha-numerical char, same as [^0-9A-Za-z].
re+	Matches *re* repeated one or more times.
re?	Matches *re* zero or one time.
re{*n*}	Matches *re* repeated *n* times.
re{*n*,}	Matches *re* *n* or more times.
re{*n, m*}	Matches *re* *n* to *m* times.
re_1|re_2	Matches either re_1 or re_2.
(*re*)	Matches *re*. Parentheses delineate patterns. For example, (cb)+ matches cbcb, but cb+ does not.

enclose the entire pattern in a pair of single quotation marks. Here are some more examples:

grep '\-s'	(matches -s; the \ prevents -s from becoming a command option)
grep -i '^linux'	(matches linux at the front of a line, ignoring case)
grep 'ch[0-9]*'	(matches ch followed by any number of digits)
egrep \.html?\>	(matches a word ending in .htm or .html)
egrep '\<\w+\.docx?'	(matches any word followed by .doc or .docx)

The **grep** commands are often used in a pipeline with other commands to filter output. More examples of **grep** within pipelines are discussed in Section 4.6.

Information on regular expressions presented here forms a basis for learning more elaborate regular expressions in languages such as Perl, Ruby, Javascript, and Java.

4.5   A Stream Editor: sed

The **sed** program is a filter that uses line-editing commands to transform input lines, from stdin or a file, and produces the desired output lines (Figure 4.2). **Sed** is a non-interactive, line-oriented editor. It applies prescribed

FIGURE 4.2: The Stream Editor **sed**

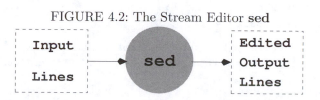

editing actions to lines matching given basic regular expression patterns.

In practice, **sed** is used for such chores as deleting particular lines, double spacing a program listing, and modifying all occurrences of some pattern in one or more text files.

In fact, **sed** and **grep** can perform many of the same functions. However, **sed** is more powerful because it supplies text editing capabilities. The **sed** program buffers one input line at a time, repeating the following steps until there are no more input lines. Figure 4.3 shows the **sed** processing cycle.

1. If there are no more input lines, terminate. Otherwise, read the next input line into the buffer, replacing its old content, and increment the line count (initially 0) by 1.

2. Apply all given editing actions to the buffer.

3. Write the buffer out to the standard output.

4. Go to step 1.

FIGURE 4.3: The Editing Cycle of **sed**

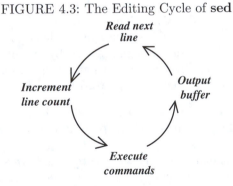

Each editing action may be applicable to all lines or to just a few. Therefore, it is possible for some lines to pass through **sed** unchanged; at the same time, others can be modified or deleted entirely. Frequently, **sed** is used in the simple form

sed *script* [*file*] ...

where *script* specifies one or more editing actions separated by semicolons. For example,

```
sed 's/Web site/website/' chapter1
sed 's/Web site/website/g' chapter1
```

The first command reads the input file **chapter1**, substitutes (the **s** action) any first occurrence of **Web site** in each line with the string **website**,[1] and

---

[1]The AP (Associated Press) style book recently made the change.

outputs all lines, changed or not, to the standard output. If any line contains multiple instances of `Web site`, only the first instance in the line will be replaced. To replace all occurrences, use the second command where the `g` (global modifier) does the trick.

If no file is specified, **sed** edits lines from `stdin`. The single quotation marks around *script* prevent the Shell from interpreting any special characters in the script. The command

sed 's/Red Hat/Fedora/g ; s/ubuntu/Ubuntu/g' chapter1

applies two string replacement actions to each line of **chapter1**. The option `-f` *scriptfile_file* indicates a file containing the desired editing script. If a script file **double** contains the two lines

```
s/$/\
/
```

then

sed -f double *file*

adds an empty line after each line in *file*, producing a double-spaced output. As in **grep**, the pattern `$` means the end of a line.

Each editing action can also be specified to act on a range of lines. Here is the general form:

[*address1*] [, *address2*] *action* [*args*]

where the addresses specify the range of input lines to apply the given *action*. An address can be a line number or a pattern.

- No *address*—The given *action* applies to every line.

- One *address*—The *action* applies to every line matching that address.

- Two *addresses*—The *action* is applied repeatedly to the next set of lines beginning with a line that matches *address1*, up to and including the first line that matches *address2* (but not *address1*).

For example,

sed '/^$/d' *file*

applies the action **d** (delete line) to each line matching the single address `/^$/`, an address obtained by searching for the next empty line. The output will be the same as *file*, but with all empty lines deleted. Another version of this example deletes all blank lines

sed '/^[⊘▷]*$/d' *file*

We use the symbols ⊘ and ▷ to stand for a SPACE and a TAB, respectively. The address matches a line containing zero or more spaces and tabs and nothing else.

Let's look at an example involving a two-address action. Say that in your HTML files tables are sandwiched between two lines

```
<table ... >
```

and

```
</table>
```

Suppose you wish to remove all tables from a given HTML document (**Ex:** ex04/remove_table). You may use

sed '/<table .*>/,/<\/table>/d' try.html > notables.html

The delete line action **d** is applied to all table lines.

A useful **sed** option is **-n**, which skips step 3 of the **sed** cycle. Hence, the command

sed **-n** '/*pattern*/p'

with the output-line action **p**, is equivalent to **grep** '*pattern*', and

sed **-n** '12,20p' *file*

outputs only lines 12–20 of the given *file*.

Hence, if you wish to extract all the tables from a given HTML document (**Ex:** ex04/extract_table). You may use

sed **-n** '/<table .*>/,/<\/table>/p' try.html > tables

to output lines between the beginning and the end of each table using the **p** action and the **-n** option. Alternatively, you can use

sed '/<table .*>/,/<\/table>/!d' *try*.html > tables

The exclamation point (!) reverses the sense of the specified addresses; it applies the specified action to every line except the lines matching the given address.

Also, the **y** action

y/*string1*/*string2*/

when given two equal-length character strings, performs character translations. Thus,

sed 'y/abc/ABC/' *file*

functions the same as

tr abc ABC *file*

Simple scripts are easy to give on the **sed** command line. More complicated scripts should be placed in files and applied with the **-f** option. Storing scripts in files makes them easily reusable.

The **sed** command offers a number of other features and options. Please refer to the **sed** man pages for additional information.

4.6   Building Pipelines

We have discussed a good number of filters and seen some pipelines already. Let's now see a few more examples.

Here is a pipeline to look up the correct spellings of words:

look *prefix* | **fgrep** *string*

All words in the dictionary `/usr/dict/words` with the specified *prefix* are produced by **look** and fed to **fgrep**, which selects only those words that contain the given *string*. For example,

look dis | **fgrep** sion

gives the following output:

```
discussion
dispersion
dissension
```

Another example is a pipeline that saves to a file those commands that you have given to your Shell. The Bash command **history** displays a numbered list of your most recent commands. To enter the last eight commands into a file, you can use the following pipeline:

history | **tail** -8 | **sed** 's/.*▷//' > *file*

where the **sed** command removes the leading sequence numbers (▷ is a tab).

A third example collects a list of directory names from the current working directory:

ls -l | **grep** ^d | **sed** 's/^d.*⊘//'

Here the **sed** editing command deletes a *maximal* (longest) string starting with the letter d at the beginning of a line and ending with a space (⊘) for each line. Another way to accomplish the same task is

ls -F | **grep** '/$' | **sed** 's/\/$//'

A final example has to do with maintaining an address list. Let's assume you have a file of addresses, `myaddr`, in human-readable form. Its entries are multiline addresses, and a single empty line follows each entry. A typical address entry would look like the following (**Ex: ex04/myaddr**):

```
Dr. John F. Doe
Great Eastern Co.
40 North Rd.
Cambridge, MA 02139
```

This form is easy for a user to read, but hard to maintain using filters. However, you can transform this address file with the following pipeline (**Ex: ex04/toaddr**):

```
sed 's/^$/@/' myaddr | tr '\012@' ':\012' \
        | sed 's/^://;s/:$//' | sort -u -t: --key=1,2 >| addr
```

The first **sed** substitutes the character @ for each empty line. The **tr** command translates every NEWLINE character into a colon and every @ into a NEWLINE. At this point, each address entry is on a separate line with a colon separating the fields within each address. The second **sed** removes any colon at the beginning or the end of a line. The final **sort** command orders the address entries using the first field and removes any duplicate entries.

Address Processing

Now your address file **addr** is sorted and contains one address per line in the following form:

```
Dr. John F. Doe:Eastern Co.:40 North Rd.:Cambridge, MA 02139
```

You can extract an address by using (**Ex: ex04/useaddr**)

```
grep 'John F. Doe' addr | tr ':' '\012'
```

You can delete any address by using

```
sed '/John F. Doe/d' addr > temp.file
mv temp.file addr
```

You can insert one or more addresses by using

```
sort -u -t: -key=1,2 addr - > temp.file
```

which allows you to type in the entries from the standard input. You may insert another address file, **addr2**, by using

```
sort  -mu -t: --key=1,2 addr addr2 > temp.file
mv temp.file addr
```

In the preceding example, the first field contains the title and name of a person. The sorted address file is not in alphabetical order with respect to names, unless everyone has the same title. To avoid this problem, you may want to modify the record format to (**Ex: ex04/newaddr**)

```
Doe:John:F:Dr.:Eastern Co.:40 North Rd.:Cambridge, MA 02139
```

and sort the address file using the first, second, and third fields as keys. Then the following can be used to display an entry (**Ex: ex04/usenewaddr**):

```
look 'Doe' newaddr|
awk -F: '{print $4, $2, $3".", $1;
        print $5; print $6; print $7}'
```

For large files, the **look** command, which uses a binary search method for a line prefix, is much faster than the **fgrep** command. We will explain **awk** next.

4.7 Pattern Processing: awk

The **awk** program is a powerful yet simple filter. It processes its input one line at a time, applying user-specified **awk** *pattern actions* to each line. The **awk** program is similar to, but more powerful than, **sed**. The **awk** mechanisms are based more on the C programming language than on a text editor, allowing for variables, arrays, conditionals, expressions, iteration controls, formatted output, and so on. The **awk** program can perform operations not possible with **sed**, such as joining adjacent lines and comparing parts of different lines.

The general form of the **awk** command is

awk [-F*c*] *script* [*file*] . . .

The -F option specifies a character *c* to be the *field separator* (default white space). The argument *script* is an **awk** script given on the command line or in a file with the -f *filename* convention. The files are are processed in the order given. If no files are given, standard input is used. If a dash (-) is given as a file name, it is taken to mean standard input.

The **awk** processing cycle is as follows:

1. If there are no more input lines, terminate. Otherwise, read the next input line.

2. Apply all **awk** pattern commands sequentially as specified in *script* to the current line.

3. Go to step (1).

Note that unlike **sed**, **awk** does not write lines to the standard output automatically.

An **awk** script consists of one or more pattern actions given on different lines or separated by semicolons. Each pattern action takes the form

pattern {*action*}

If the current line matches the *pattern*, the *action* is taken. A missing pattern matches every line, and a missing action outputs the line. Thus,

```
ls -l | awk '/Linux/'
```

is the same as

```
ls -l | sed -n '/Linux/p'
```

Pattern and *action* are described more fully in the following subsections.

The concept of a field here is the same as that used for **sort**: **awk** delineates each of its input lines into fields separated by white space or by a field separator character specified with the −F option. In an **awk** action, the fields are denoted $1, $2, and so on. The entire line is denoted by $0.

While it is impossible to rearrange the order of fields using **sed**, it is easy with **awk**. For instance, the output of **ls −l** has eight fields:

```
-rw-rw---- 2 jsmith 512 Apr 23 21:44 report.tex
-rw-rw---- 1 jsmith 79 Feb 9 15:13 Makefile
-rw-rw---- 2 jsmith 1024 Feb 25 00:13 pipe.c
```

When the preceding lines are piped through **awk**,

```
ls -l | awk '{print $8,$4,$5,$6}'
```

The following output is produced:

```
report.tex    512 Apr 23
Makefile       79 Feb 9
pipe.c       1024 Feb 25
```

Note that, in this example, the pattern action contains no pattern. Also, actions are always enclosed between braces ({ and }).

awk **Patterns**

As with **sed**, the *pattern* determines whether or not **awk** takes an action on the current line. In fact, a **sed** address, specified with one or two match expressions, also will work as an **awk** pattern. If you are familiar with **sed**, you already know many useful patterns. For instance,

```
/^first/        (first at the beginning of a line)
/last$/         (last at the end of a line)
/^$/            (an empty line)
/[⊘▷][⊘▷]*/     (a line with a string of one or more blanks)
/begin/,/end/   (all lines between a begin match and an end match)
```

are valid patterns in both **sed** and **awk**.

In **awk**, a pattern is an arbitrary Boolean expression involving *regular expressions* and *relational expressions*. Boolean expressions are formed with && (and), || (or), ! (not), and parentheses. A regular expression in **awk** must begin and end with a slash (/) and otherwise is defined the same as that for **egrep** (Table 4.4). Relational expressions are formed using C-like operators >, >= , <, <=, == (equal), and != (not equal). In addition, a relational expression can be:

expression ~ *re*
expression !~ *re*

where ~ means "contains" and !~ means "does not contain." For example, the pattern

$1 ~ /GNU/ && $2 ~ /Linux/

is true if the first field contains the string GNU and the second field contains the string Linux.

A pattern may contain two patterns separated by a comma, in which case the action is applied to all lines beginning with a line matching the first pattern up to and including the line matching the second pattern (the same as in **sed**). Thus,

awk 'NR==14,NR==30' *file*

outputs lines 14-30 of *file*, because **awk** keeps a running line count in the built-in variable NR. Other useful built-in variables are listed in Table 4.5.

The special patterns BEGIN and END in

BEGIN {*action*}
END {*action*}

specify actions executed before the first input line and after the last input line, respectively. They are used for initialization and postprocessing when needed.

awk Actions

Now let's turn to the question of how *actions* are specified. An *action* contains a sequence of statements given on different lines or separated by semicolons. Possible statements are as follows:

TABLE 4.5: Built-in `awk` Variables

Variable	Meaning
NF	Total number of fields on current line
NR	Sequence number of current line
FS	Input field separator character (default blanks)
RS	Input record separator (default NEWLINE)
OFS	Output field separator string (default SPACE)
ORS	Output record separator string (default NEWLINE)
OFMT	Output format for numbers (default %g as in `printf`)

Assignment:	*var* = *expression*
Output:	`print` *expression* [, *expression*] ...
	`printf(` ... `)` (as in C)
Flow control	`if (` *conditional* `)` *statement* [`else` *statement*]
(as in C):	`for(`*expression; conditional; expression*`)` *statement*
	`while (`*conditional*`)` *statement*, `break`, `continue`
Additional flow	`next` (skip remaining commands, start next **awk** cycle)
control:	`exit` (exit **awk**)

In the preceding definitions, a *statement* can be a compound statement in the form

{*statement, statement,* ... }

The output statements use the standard output. However, they can be followed by > "*filename*" to redirect the output into a file.

`awk` Expressions

Expressions in **awk** statements can be constants, variables, arrays, fields, or any combinations of these using the following C operators:

+, -, *, /, %, ++, --, +=, -=, *=, %=

Numerical constants in **awk** statements are the same as in C. String constants are placed in double quotation marks ("*string*") and variables are initialized to the null string. An array element is denoted as a[*i*], where *i* can be an integer or any string. A blank between two expressions concatenates them into a string. Thus, for example,

awk '{print $2 ":" $1}' *file*

outputs *field2:field1* of each line from the given file. Built-in functions (Table 4.6 lists a few) can also be used in expressions. In **awk**, conditional expressions use C notation and may involve **awk**-defined relational expressions. In Table 4.6, *e* is an expression, *c* is a character, *s* is a string, and *i* and *j* are integers.

Index Preparation: An Example

The **awk** pattern processing program is powerful and involved. The best way to learn it is through use and experimentation. In this section, we present an example of **awk** usage to prepare an index for a document (**Ex: ex04/index.awk**). Suppose you have several index files, each containing entries such as (**Ex: ex04/index.data**)

```
bash:99
regular expression:155
bash:123
pipe:101
gnome:163
socket:415
pipe:23
```

where each line has two fields: an index item and a page number separated by the . Your goal is to produce an overall index file in alphabetical order with lines such as (**Ex: ex04/index.file**)

```
bash 99,123
gnome 163
pipe 23,101
regular expression 155
socket 415
```

The first step is to order the entries alphabetically and by page number, which can be done with

```
sort -t: --key=1,2.0f --key=2n index.data >| index.tmp
```

in which the following sort keys are used:

```
1,2.0f        (first key, field one ignoring case)
2n            (secnd key, field two with numerical comparison)
```

TABLE 4.6: Built-in **awk** Functions

Function	Meaning
int(e), length(s)	Integer (floor), length of string s
gsub(re, s, t)	Replaces matches of re in t with s
index($s1,s2$)	Position of string $s2$ in $s1$, zero if $s2$ not in $s1$
sprintf(...)	Format conversion, same as in the C language
substr(s,i,j)	Substring of s of length j from position i
split(s,a,c)	Cuts s into substrings a[1] to a[i] at char c; returns i
getline()	Inputs next line, returns 0 on end of file, otherwise 1

It then remains to collect repeated index items to form lines with multiple page numbers. Since repeated items will be on consecutive lines, the **awk** script `index.awk` (Figure 4.4) can be used. To apply the script use

```
awk -f index.awk index.tmp >| index.file
```

There are four pattern commands in `index.awk`. The first command sets the variable i (used for initialization) to zero. The second command compares $1 with the variable pre, which stands for the previous index item and is initially null. If $1 is equal to **pre** (field one is the same as the previous index item), then output the page number ($2), preceded by a comma. If $1 is not equal to **pre** (a new index item), then output NEWLINE, $1, SPACE, and $2 except for the very first line where the leading NEWLINE is not needed. The conditional output is performed in the **if** of the third command which also records the index item ($1) in the variable pre. At the end of the input file, a final NEWLINE is output.

FIGURE 4.4: Program `index.awk` for Index Processing

```
BEGIN {  i = 0; }

$1 == pre {  printf(",%s", $2); }

$1 != pre {  if (i > 0)
                { printf("\n%s %s",$1,$2); }
             else
                { printf("%s %s",$1,$2); i = 1; }
             pre = $1;
          }

END {  printf("\n"); }
```

4.8   For More Information

See the man pages for the commands and filters covered in this chapter. For filters accepting regular expressions, their man pages will specify exactly what patterns are recognized.

The current **awk** program on most Linux is usually **gawk** from GNU where more information can be obtained.

Both **sed** and **awk** are facilities to extract data from a file and to transform the information to a different form. On most Linux systems, there are also more general scripting languages such as Perl (Practical Extraction and Report Language, Chapter 5, Section 5.23) and Ruby (Chapter 12, Section 12.3). Both run under Linux, Unix, Windows, and Macs. See their manual pages for more information.

4.9 Summary

Filters produce output by performing a simple, well-defined transformation on their input and follow a set of well-defined conventions so they can become stages in pipelines that combine them to perform many varied tasks. Filters and pipelines are concrete artifacts of the UNIX/Linux philosophy.

Linux filters range from simple character substitutions (**tr** and **expand**) to finding string patterns in text lines (the grep commands), to ordering text lines, and to complicated stream editing (**sed**) and pattern processing (**awk**). How these commands work individually and in pipelines for realistic applications, such as creating, maintaining, and accessing an address database, have been discussed.

Regular expressions are well-established notations for character string pattern matching. They are used, in very similar ways, in many different programs such as **grep**, **egrep**, **sed/vim**, and **awk**. In Chapter 5, you'll see that the Bash Shell also understands regular expressions. It is important to become familiar with regular expression concepts.

Table 4.7 summarizes the commands described in this chapter.

TABLE 4.7: Commands Summary

Command	Description	Command	Description
awk	Pattern processing	fold	Line wrapping
expand	TAB-to-SPACE conversion	unexpand	blank-to-TAB conversion
fgrep/grep	Fixed/basic re matching	egrep	Extended **grep**
head	Beginning of file	tail	End of file
look	Dictionary search	sed	Stream editing
sort	Line ordering in files	tr	Character translation

4.10 Exercises

1. Consider how **expand** works. Write an algorithm for figuring out how many spaces should be generated for a TAB.

2. Write a pipeline, using **ls** and **head**, to list the ten most recent files in the current directory.

3. How can you use **grep** to locate all lines in a file that do not contain the pattern -`option`?

4. What is a Glob pattern? What is a regular expression pattern? What is the difference? Give a pattern in each case to match a string ending in `.html`.

5. Specify a regular expression to match (a) any word ending in `.html`; (b) any image name ending in `.jpg`, `.png`, or `.gif`; (c) any empty line (line with no characters in it whatsoever); (d) any blank line (line that is empty or contains only white space characters); and (d) any number.

6. Explain the following regular expressions: (a) `^a+$`, (b) `http[s]*:\/\/` and (c) `[^@]+@gmail\.com`.

7. Consider the following **sed** command:

   `sed -n '/begin/,/end/p' `*file*

   Discuss its effect if *file* contains many lines with `begin` and or `end` in them.

8. Consider building pipelines to manage an address file. Suppose you wish to have an address, an email, and a phone nubmer on each address line. How would you design the record format? Write a pipeline to extract a desired email or phone number from the address file.

9. Following the previous exercise, write a pipeline to add/change a phone number or email to an existing address entry.

10. Specify an **sed** command to replace any set of consecutive empty lines in a file with just one empty line. An empty line is one with nothing in it, not even blank characters.

11. Write an **awk** program to transform the file `myaddr` to `addr` so that a single **awk** command can perform a task equivalent to that defined by the pipeline (involving **sed**, **tr**, and **sort**) in Section 4.6.

12. *Rot13* is a method to encode ASCII text files: each letter in the alphabet `A` through `z` is replaced by another 13 positions away (`A` by `N` and `n` by `A`, for example). Write a **tr** command to perform this encoding/decoding.

13. The `y` function of **sed** can perform most of the same translations as **tr**. Is there anything **tr** can do that **sed** cannot? If so, discuss.

14. Take the index generation example from Section 4.7 and improve it so that it can also handle index data enteries such as

Bash:78-82

Chapter 5

Writing Shell Scripts

The Shell is more than just an interactive command interpreter. It also defines a simple programming language. A program written in this language is known as a *Shell procedure* or *Shell script*, which, in its simplest form, is just a sequence of commands in a file. The file, when executed, performs the tasks as if each command in the script had been entered and executed individually, but without all the typing. Shell scripts can save you a lot of time if you find yourself repeating a sequence of commands over and over. The Shell language also provides variables, control structures such as if-then-else, looping, function definition, and means for input and output. If a particular task can be achieved by combining existing commands, then consider writing a Shell script to do the job.

As with other Linux commands, a Shell script can be invoked through your interactive Shell and can receive arguments supplied on the command line. Sometimes, scripts written by individual users also can be of general use. Such scripts can be installed in a system directory accessible to all users.

This chapter covers Shell script writing and techniques for effective Shell-level programming. We will focus on Bash scripts because Bash is currently the most widely used and most advanced Shell. Csh and Sh scripts follow many similar rules.

The presentations in this chapter are oriented toward script writing. However, most constructs discussed here can be used interactively as well. Some topics (for example, command grouping) are as relevant to interactive use as to script writing.

5.1 Invoking Shell Scripts

As mentioned, a Shell script is a program written in the Shell language. The program consists of variables, control-flow constructs, commands, and comments. The Shell script is kept in a text file whose file name is said to be the name of the script.

There are two ways to invoke a Shell script: by *explicit interpretation* and by *implicit interpretation*. In explicit interpretation, the command

bash *file* [arg ...]   (for Bash script)
csh *file* [arg ...]   (for Csh script)

sh *file* [*arg* . . .]   (for Sh script)

invokes a specific Shell to interpret the script contained in *file*, passing to the
script any arguments specified.

In implicit interpretation, the script file containing the script is first made
readable and *executable* with the **chmod** command to turn on the appropriate
protection bits (Chapter 3, Section 3.5). Then the script can be invoked in the
same way as any other command: by giving the script name on the command
line followed by any arguments.

In either explicit or implicit interpretation of a Shell script, *two Shells* are
involved: (1) the interactive Shell (usually the login Shell) that interacts with
the user and processes the user's commands and (2) the invoked Shell that
actually interprets the script. The invoked Shell is a process spawned by the
interactive Shell. Since the spawned process is also a Shell, it is referred to as
a *subshell*. The effect of this can be illustrated by the following experiment.

First create a file named **try** that contains the simple script

```
cd /usr/lib
pwd
```

To run this script, type

bash `try`

The script called **try** displays the string `/usr/lib`, which is the output of the
pwd contained in the script. However, once it is finished, if you type **pwd**
in your interactive Shell, your old working directory will appear. Obviously,
the **cd** command executed in the script has not affected the current working
directory of your interactive Shell. This is because **cd** is executed by a subshell.
To execute the commands in **try** with the interactive Shell, use instead

source `try`

5.2   A First Shell Script

Now let's consider a simple Bash script. The purpose of this script is to
consult a list of email addresses that are kept in a file named `myContactList`
(**Ex:** ex05/myContactList) in a user's home directory. Each line in the contact list gives the name of a person, email address, phone number, and perhaps
some other information.

The script (**Ex:** ex05/contact_one.sh) is

```
#!/bin/bash
## consults myContactList
grep -i "$1" ~/myContactList
```

We will use the suffix `.sh` for Bash scripts as a naming convention. The first line is special. In Linux, the proper way to begin an *executable text file* is `#!`, followed by the full pathname of an executable file together with any arguments to it. This specifies the command to invoke an interpreter for the remainder of the script. Make sure `#!` are the very first two characters in the file, with no empty line, white space, or any other character before them.

The first line of `contact.sh` indicates a Bash script. Similarly, the line `#!/bin/csh` begins a Csh script, and the line `#!/bin/sh` begins an Sh script.

The second line is a comment. In Shell scripts, the part of any line from the first `#` to the end of line is ignored by the Shell.

The symbol `$1` is called a *positional parameter*. The value of the positional parameter `$n` is the *n*th command-line argument. Thus, if the first argument is `smith`, then `$1` has that value, and the script is equivalent to

grep -i smith `~/myContactList`

Recall that `~` expands to your home directory. Now you should issue the command

chmod +rx `contact.sh`

to make `contact.sh` readable and executable. Now the command

contact.sh `smith`

runs the **contact.sh** script (in the current directory). The preceding command assumes that the special period symbol (`.`) is included in your command search path (Section 2.4). Otherwise, you need to use

./contact.sh `smith`

If the **contact.sh** script is put in a directory whose name is on the command search path, then

contact.sh `smith`

will work no matter what your current directory is, without having to specify the **contact.sh** command with a pathname.

Usually, you would create a directory `bin` or `cmd` in your home directory to hold all scripts and other executable commands written or obtained by you. By including the line

`PATH=$PATH:$HOME/cmd:.`

in your `.bash_profile`, you can invoke executable files in your own `cmd` directory just like other Linux commands.

5.3 Shell Script Execution

A Shell script consists of a sequence of Shell built-in commands and regular Linux commands separated by NEWLINE or semicolon (;) characters. Comments are introduced by #, as previously mentioned. Commands in a script are executed in sequence. If the execution of a command results in an error, script execution will be aborted if the offending command is a Shell built-in. Otherwise, for a regular command, the default action is to skip the offending command and continue with the next command in the script.

In describing the Shell languages, the term *commandlist* means a sequence of zero or more commands separated by NEWLINE or semicolon (;) characters. The term *wordlist* refers to zero or more blank separated words.

5.4 Positional Parameters

In Shell scripts, the variables $0, $1, $2, and so on are known as *positional parameters*. The variable $0 refers to the first token of the command line which invoked the script. Thus, $0 may have the value `contact.sh` or `./contact.sh` depending on the command given. The variables $1, $2, and so on refer to the command-line arguments.

When a Bash script is invoked, the special variable $0 is set to the command name. The positional parameters $1, $2, etc. are set to the command-line arguments (use ${n} for n bigger than 9); $* (and $@) is set to the list of arguments given on the command line; and $# is set to the number of positional parameters. The Bash script (**Ex:** ex05/myecho.sh)

```
#!/bin/bash
## echoing Shell parameters

echo '$0 = ' $0
echo '$1 = ' $1
echo '$2 = ' $2
echo '$3 = ' $3
echo '$# = ' $#
echo '$* = ' $*
echo '$@ = ' $@
```

displays these parameter values. For example, the command

myecho.sh A B C D

produces the output

```
$0 =   ./myecho.sh
$1 =   A
$2 =   B
```

```
$3 =   C
$# =   4
$* =   A B C D
$@ =   A B C D
```

Try it yourself.

5.5   The `for` Command

The **for** command is used to execute a set of commands repeatedly. The general form is

for *var* **in** *wordlist*
> **do** *commandlist*
done

The line breaks are needed unless you use the ; command separator as in

for *var* **in** *wordlist* ;   **do** *commandlist* ;   **done**

The *commandlist* is executed once for each word in *wordlist* as, each time through, the control variable *var* takes the next word for its value. As an example, let's rewrite the `contact_one.sh` script given in Section 5.2 as (**Ex: ex05/contacts.sh**):

```
#!/bin/bash
## consult my contacts for args given

for x in "$@"                          ## (0)
    do grep -i "$x" ~/myContactList
done
```

Bash has two built-in variables, $* and $@, referring to the arguments given on the command line. Each is a list of words from the command-line arguments. Consider a command with three arguments:

somecmd a b "c d"

The $* and $@ in this case will both be the list a b c d with four words. The quotation "$*" makes it one word, whereas the quotation "$@" makes it three words a b and "c d". It is important to understand this difference. It turns out also that line 0 can be written simply as **for** x, which means x will take on successive command-line arguments. Now **contact.sh** can be used on one or more names, as in

contact.sh "John Smith" "Paul Wang"

The **for** command can be used to go through each file in a directory. Try the following script:

```
#!/bin/bash
## example to go through all files in the current directory

for file in *
     do echo $file
done
```

Execute this script, and you'll see the filenames in the current directory displayed. Since the filename expansion does not match any filename that begins with a period (.), those filenames will not be displayed. To get *all* files, use

```
for file in .* *
   do echo $file
done
```

Bash supports another form of **for** loop that is similar to that of the C language.

```
#!/bin/bash

for (( i = 0 ; i < 9 ; i++ ))
   do echo $i
done
```

The iteration control involves numerical expressions (Section 5.11). Such loops are useful for indexing through arrays (Section 5.14) and, of course, for numerical computations.

5.6   The if Command

The **if** construct provides for conditional execution of commands. The simple form of **if** is

if *test-expr*
> **then**
>> *commandlist₁*
> **else**
>> *commandlist₂*

fi

If the given *test-expr* is true, then *commandlist₁* is executed; otherwise, *commandlist₂* is executed. The **else** part may be omitted.

For example, the test expression [[-f *file*]], known as an *extended conditional* (Section 5.7), tests if *file* exists and is a regular file. We can improve the `contact.sh` as follows (**Ex: ex05/contact_check.sh**).

```
#!/bin/bash
## consult my contacts for args given

if [[ -f ~/myContactList ]]                      ## (A)
    then
        for x
            do grep -i $x ~/myContactList
        done
    else
        echo "File ~/myContactList not found."
fi
```

In a test expression, the SPACE after [[and the SPACE before]] are part of the conditional notation (line A).

Within the **if** statement, the **elif** construct can be used. The general form is

if *test-expr*$_1$
    then
            commandlist$_1$
elif *test-expr*$_2$
    then
            commandlist$_2$
else
            commandlist$_3$
fi

If *test-expr*$_1$ is true, *commandlist*$_1$ is executed. If *test-expr*$_1$ is not true, and if *test-expr*$_2$ is true, then *commandlist*$_2$ is executed. There can be any number of **elif** constructs. If all test expressions are false, then the **else** part is executed.

Often, it is important for any program to check the arguments it receives, and a Shell script is no exception. Here is some typical argument check code (**Ex: ex05/argCheck.sh**).

```
#!/bin/bash
## check and set command-line arguments
if [[ $# > 2 || $# < 1 ]]                         ## (1)
    then
        echo usage: "$0 [ from-file ] to-file"    ## (2)
        exit 1;                                   ## (3)
elif [[ $# == 2 ]]                                ## (4)
    then
        from="$1"
        to="$2"
else                                              ## (5)
        to="$1"
fi
```

The expression $\$\# > 2$ checks if the number of arguments is greater than 2. The || is *logical or*, whereas < is *less than*. This script expects one or two arguments. If the number of arguments is incorrect, it displays an error message (line 2) and terminates the script with an abnormal *exit status* 1 (line 3). If we have two arguments (line 4), we can set the variables `from` and `to`. Otherwise, we have only one argument and it becomes the value of `to`. Argument checking is critical at the beginning of every program.

FIGURE 5.1: The `cmdsearch` Script

```
#!/bin/bash
## Finds a given command on the search path.
## The pathname found or a failure message is displayed.

cmd="$1"  ## the command to find              ## (a)
path=$(echo $PATH | tr ":" " ")              ## (b)
for dir in $path                             ## (c)
   do
      if [[ -x "$dir/$cmd" ]]                 ## (d)
         then
               echo "FOUND: $dir/$cmd"
               exit 0
      fi
   done
echo "$cmd not on $PATH"                      ## (e)
```

Now let's look at a complete script using **for** and **if** constructs. The script (Figure 5.1) locates a command on the command search path ($PATH) and displays its full pathname (**Ex:** ex05/cmdsearch.sh). The first (and lone) argument is the target command name (line **a**). On line **b**, each : in $PATH is replaced by a SPACE with the **tr** command (Chapter 4, Section 4.1), and the resulting multiword string is assigned to a variable `path` via command expansion (Chapter 2, Section 2.7). For each $dir on $path (line **c**), we see if $cmd is found (line **d**). The conditional expression [[-x *file*]] is true if *file* exists and is executable (see Section 5.13 for more on file queries). If the program ever reaches line **e**, then the target command is not found.

Here are some sample uses of **cmdsearch**.

cmdsearch gnome-terminal
cmdsearch vim
cmdsearch gcc

5.7 Test Expressions and Exit Status

Exit Status

In Linux, a command indicates whether it has succeeded by providing an integer *exit status* to its invoking environment. A zero exit status means okay, and non-zero means error.

The Shell, being a command interpreter, is a primary invoking environment for commands. After executing a command, the exit status is available in the special Shell variable $?.

In a Shell script, use the built-in command **exit** *n* to terminate execution and return *n* as the exit status.

Test Expressions

Test expressions are used in **if** as well as other constructs (**while**, **until**, etc.) to produce true/false values by testing given conditions.

The truth value of a Bash test expression is really determined by its *exit status*. A test expression is true if it returns a zero exit status; otherwise, it is false. Now let's take a look at the different forms of test expressions.

A *test-expr* can be a list of one or more of these expressions:

- A regular or built-in command (Section 5.7)

- An extended conditional expression [[ ]]

- A numerical expression (( )), with 0 being false and non-zero being true (Section 5.11)

- (*test-expr*), using () for precedence grouping

- ! *test-expr* "logical not" of *test-expr*

- *test-expr*$_1$ && *test-expr*$_2$ "logical and" of the two expressions

- *test-expr*$_1$ || *test-expr*$_2$ "logical or" of the two expressions

Here is an example that uses **grep** as a test expression (**Ex:** ex05/condEdit.sh).

```
#!/bin/bash

for file in *              ## for each file in current folder
  do  if grep -q "$1" $file ## if pattern $1 is in $file
        then nano $file     ## invoke nano on $file
      fi
done
```

An *extended conditional* is enclosed by [[SPACE on the left and SPACE]] on the right.[1] Table 5.1 lists test expressions for strings. Within the [[conditional, Glob patterns (Chapter 2, Section 2.7) are allowed on the right-hand sides of == and !=. Furthermore, extended regular expressions (Chapter 4, Section 4.4) following =~ are supported.

TABLE 5.1: Bash String Conditions

Condition	True if:
[[*var*]]	*var* is defined and not null
[[-z *str*]]	*str* is zero length
[[str_1==str_2]]	str_1 and str_2 are equal
[[str_1!=str_2]]	str_1 and str_2 are unequal
[[str_1<str_2]]	str_1 is lexicographically before str_2
[[str_1>str_2]]	str_1 is lexicographically after str_2
[[*str*==*pattern*]]	*str* matches the Glob *pattern*
[[*str*!=*pattern*]]	*str* does not match the Glob *pattern*
[[*str*=~*pattern*]]	*str* matches the **egrep** *pattern*

The extended conditionals also support numerical tests

[[arg_1 *rop* arg_2]]

to compare two integers arg_1 and arg_2 with a relational operator *rop* which can be ==, !=, <, >, -le, or -ge.[2] Often, programmers prefer to use numerical tests provided by (( )) (Section 5.11) instead.

Please refer to Table 5.2 for file-related test expressions and to Section 5.11 for numerical test expressions. Inside [[ ]] you can also use the logical operators ! (not), || (or), and && (and) on test expressions.

5.8   The shift Command

The Bash built-in command

shift

left shifts $2 to $1, $3 to $2, etc. In general,

shift *n*

shifts $n to $1, $n+1 to $2, etc.

The **shift** command is often useful after the first few positional parameters have been processed and you want to use a loop such as (**Ex: ex05/shift.sh**)

---

[1]The earlier Bash construct [ ] can still be used but is superseded by the [[ ]].

[2]Unfortunately, inside [[ ]] the usual <= and >= are not recognized.

for var
  do echo $var
done

to go over the rest of the positional parameters.

5.9 The case Command

While the **if-elif-else** command enables logical branching in general, the **case** command provides branching based on simple pattern matching. The general form of **case** is

case (*str*) **in**
    pattern₁)
            commandlist₁
        ;;
    pattern₂)
            commandlist₂
        ;;
    ...
esac

The given expression *str* is successively matched against the **case** *patterns*. Each **case** pattern can be one or a list of Glob patterns (Section 2.7) separated by | and terminated by). Only the list of commands for the first match will be executed. Nothing is executed if there is no match.

For example, the string ab.c matches the **case** pattern *.c or the pattern a*c. As an example, a script for appending either the standard input or a file

FIGURE 5.2: The **append** Script

```
#!/bin/bash
## append.sh
## appends $1 to $2 or standard input to $1

case $# in
1)    cat >> "$1"
      ;;
2)    cat "$1" >> "$2"
      ;;
*)    echo "usage: $0 [ fromfile ] tofile"
esac
```

to the end of another file is shown in Figure 5.2 (**Ex:** ex05/`append.sh`). The command

append.sh *file1 file2*

appends *file1* to the end of *file2*. The command

append.sh *file*
first line
second line
third line
^D

appends the three lines to the end of *file*. Note the catch-all pattern * as the last case clause to process any unmatched cases.

5.10   The while and until Commands

In addition to the **for** command, the **while** and **until** commands control iteration with an arbitrary condition. The general form of the **while** command is

while *test-expr*
do
            commandlist
done

The *test-expr* is evaluated. If it is true, then *commandlist* is executed, and *test-expr* is retested. The iteration continues until the *test-expr* tests false. For an infinite loop, use the Bash built-in command : (yes, the character COLON) as the *test-expr*. The : command does nothing other than expand any arguments and give a 0 exit status.

In the following script (**Ex:** ex05/`myfortune.sh`), we continue to display a fortune message until the user wishes to stop.

```
#!/bin/bash
## displays fortune until the user quits

go="yes"

while [[ "$go" == "yes" ]]                    ##    (i)
do
    /usr/bin/fortune                          ##   (ii)
    echo -n "**** More fortune? [yes/no]:"    ##  (iii)
    read go                                   ##   (iv)
done
```

The **while** condition is checked (line i). If true, the **fortune** command[3] is invoked (line ii), and a prompt is displayed (line iii) to see if the user wishes to continue. The **-n** option tells **echo** not to output the usual line break after the message. The user input is read into the variable go (line iv), whose value is tested again.

If we replace the **while** test expression with the pattern condition

```
[[ "$go" == y* ]]
```

then the user may enter anything starting with y to continue.

The **until** loop is the same as the **while** loop, except the iteration stops as soon as the **until** condition is met.

5.11 Numerical Expressions

Since Shell variables are string-valued. We need to use the *arithmetic expansion notation*

```
$(( arith-expr ))
```

to perform integer arithmetic computation. The Shell built-in command **let** can also be used to perform arithmetic operations.

let *arith-expr₁* *arith-expr₂* ...

Here are some examples (**Ex:** ex05/arith.sh).

```
#!/bin/bash

a=2
echo $(( a + 3 ))                    ## displays 5
let b=2*++a
echo $b                             ## displays 6
echo $((a * b))                     ## displays 18
let c=-8
echo $(( c > 0 ? c : -c ))          ## displays 8
```

To compare numbers in numerical conditionals use, for example,

if (($a > $b))   (if a is greater than b)

The Bash command

help let

---

[3]If your Linux does not include the **fortune** command, you can get it by installing the fortune-mod package (Section 8.24).

displays a full list of operators available for the numerical expressions for **let** or inside (( )).

Here is a loop that displays the command-line arguments in reverse order (**Ex: ex05/echoback.sh**).

```
#!/bin/bash

output=""
until (( $# == 0 ))
  do output="$1 $output"
     shift
done
echo $output
```

5.12 The break and continue Commands

The **break** command is used inside the iteration control structures **for**, **while**, and **until**. When **break** is executed, control shifts to the first line after the end of the nearest enclosing iteration. This command provides a means to "break out" of an iteration loop before its normal completion.

The **continue** command is used in the same manner as the **break** command, except it transfers control to the beginning of the next iteration instead of breaking out of the loop entirely. The example script **clean.sh** (see Section 5.20) involves some typical applications of **break** and **continue**.

Within nested loops, **break** or **continue** can take an optional integer argument (1, 2, 3, etc.) to break or continue out of the nth level of nested loops.

5.13 File Queries

To make file and directory access and manipulation easy, Bash also provides a set of conditions to query status information for files. File queries are in the form $-x$ *file*, where x is a single character. Common file queries are listed in Table 5.2. To get a complete listing use **help test**.

If the file does not exist or if it is inaccessible, all queries return false. For example, the following code fragment is valid:

```
if [[ -e $file && -f $file && -w $file ]]
then
        cat $1  >>  $file
else
        echo "access problem for $file"
fi
```

TABLE 5.2: Bash File Queries

Expr	True if *file*:	Expr	True if *file*:
-r *file*	Is readable by the user	-w *file*	Is writable by the user
-x *file*	Is executable by the user	-e *file*	Exists
-o *file*	Is owned by the user	-s *file*	has non-zero size
-f *file*	Is an ordinary file	-d *file*	Is a directory

In the file system, an ordinary file is one that stores application data and not one that serves filesystem functions such as a directory (folder) or link (shortcut). See Section 6.2 for more information on Linux file types.

5.14   Variables

There are different kinds of variables:

1. Positional parameters ($1, $2, ...) and special variables ($0, $#, ...).

2. Environment variables such as **DISPLAY** and **SHELL**

3. Ordinary variables and arrays of your own choosing

To assign value to a variable

var=*value*

Shell expansions and evaluations are performed on *value*, and the result is assigned to the given variable. If *value* is omitted, then the variable has value null. Variable attributes can be declared:

declare -i   *var*$_1$ *var*$_2$ ...   (holding integer values)
declare -r   *var*$_1$ *var*$_2$ ...   (read-only)
declare -a   *arr*$_1$ *arr*$_2$ ...   (arrays)
declare -x   *var*$_1$ *var*$_2$ ...   (exported to the environment)

To remove a variable use **unset** *var*. The special operator += performs addition on integer variables and concatenation on string variables. For example (**Ex:** ex05/varusage.sh),

```
#!/bin/bash

declare -i a b;
a=10; b=5
b+=$a;                   ## b is 15
declare -r b;
b=0                      ## error, b is read-only
```

```
unset b                       ## error, b is read-only
name="John"; last="Doe"
echo ${#name}                 ## length of $name is 4
name+=$last                   ## name is JohnDoe
```

5.15   Arrays

To declare an array variable use

declare -a *var*

However, it is not necessary to first make such a declaration. For example, to create an array `fruits`, you can use the assignment

```
fruits=("red apple" "golden banana")
```

or equivalently

```
fruits[0]="red apple"
fruits[1]="golden banana"
```

Thus, Bash indexed arrays are variables with zero-based indexing; that is, the first element of an array has index 0 (`${fruits[0]}` for example), the second element has index 1, and so on. However, the indices do not have to be consecutive. The following examples illustrate array usage (**Ex:** ex05/`arrusage.sh`).

```
#!/bin/bash

br=()                                 # empty array
fruits=("red apple" "golden banana")
fruits+=("navel orange")              # array concatenation    (1)
echo ${fruits[1]}                     # value golden banana
echo ${#fruits[*]} or ${#fruits[@]}   # length of array        (2)
fruits[2]="green pear"                # element assignment
fruits[6]="seedless watermelon"       # gap in index allowed
br+=( "${fruits[@]}" )                # br now same as fruits (3)
```

Note # (line 2) for the length of an array and the += operator (line 1 and 3) for array concatenation.

To go through elements in an array with a loop, you may use

```
for el in "${br[@]}"
do
      ## use $el for some task
done
```

or, if indexing is consecutive,

```
for (( i=0; i < ${#br[@]}; i++ ))
do
    ## do something with ${br[$i]}
done
```

The **read** built-in can also receive words input by the user into an array.

```
echo -n "Please input an array:"
read -a arr
```

If the user enters **gg ff kk bb**, then $*arr* gets four elements.

5.16   Variable Modifiers

Bash provides notations to make variable usage even more flexible for advanced scripting. The value obtained from a variable can be modified before it is introduced into a command or expression. You can

1. Specify the value returned in case a variable is unset (does not exist) or null (Table 5.3).

2. Return a substring of the variable value (Table 5.4).

TABLE 5.3: Variable Testing Modifiers

Modifier	If *var* is unset or null
${*var*:-*word*}	Returns *word*
${*var*:=*word*}	Sets *var* to *word* and returns *word*
${*var*:?*word*}	Exits with standard error message or *word*
${*var*:+*word*}	Returns nothing; otherwise returns *word*

For example, a script requiring one command-line argument may use

```
file=${1:?"Usage: $0 filename"}
```

Note that the : in Table 5.3 can be omitted from the notations in Table 5.3, and it means the test is only for the existence of the variable and not for it being null.

Bash also makes it easy to obtain a substring from the value of a variable (Table 5.4).

Let's look at some examples of substring modifiers (**Ex: ex05/strModifier.sh**).

```
file=/tmp/logo.jpg
${file:3}                  ## is p/logo.jpg
${file:3:5}                ## is p/log
${file#*/}                 ## is tmp/logo.jpg
```

TABLE 5.4: Variable Substring Modifiers

Modifier	Value
${*var*:*offset*:*len*}	Substring of *var*, from *offset* to the end or of length *len* if :*len* is given
${*var*#*pattern*} ${*var*##*pattern*}	*var* with the shortest (#) or longest (##) prefix matching *pattern* deleted
${*var*%*pattern*} ${*var*%%*pattern*}	*var* with the shortest (%) or longest (%%) suffix matching *pattern* deleted
${*var*/*pattern*/*str*}	*var* with the longest substring matching *pattern* replaced by *str*

```
${file##*/}                    ## is logo.jpg (tail)
${file%/*}                     ## is /tmp (dirname or head)
${file%.jpg} or ${file%\.*}    ## is /tmp/logo (root)
${file##*\.}                   ## is jpg (extension)
```

When applied to the positional parameters ($* and $@) or to arrays (${array[*]} and ${array[@]}), the first modifier in Table 5.4 produces a list of words from a subarray. Whereas, the other modifiers in the table each produces a list of words by acting on each value in the given array. Here is how it works (**Ex: ex05/arraymod.sh**).

```
pictures=(a.jpg b.jpg c.jpg d.jpg)
echo ${pictures[*]:2}          ## c.jpg d.jpg
echo ${pictures[*]%.jpg}       ## a b c d
names=( ${pictures[*]%.jpg} )  ## is array (a b c d)
```

As another example of variable modifiers, consider the function

```
function latex ()
{
    /usr/bin/pdflatex ${1%.tex}.tex && \
    /usr/bin/acroread ${1%.tex}.pdf
}
```

The modifier ${1%.tex} makes it possible to use either of the following two ways to invoke the **latex** function

latex *pathname*.tex
latex *pathname*

to create and view the pdf file created from the given LATEX file.

5.17   The Here Document

It is possible to include in a script input that is normally entered interactively. In Shell script terminology, this type of input is known as a *here document.* For example, you may create a script (**Ex: ex05/newyear.sh**) that contains the following:

```
mutt -s 'Happy New Year' <<ABC

Today is `date` and how time flies.
May I wish you a very happy and prosperous NEW YEAR.

                        signed ...
ABC
```

The purpose of this file is to invoke the **mutt** command (for email) and send a message to each name on the alias list called **friends**. The here document consists of all text between the first **ABC** and the second **ABC** on a line without other characters or white space. Having set up this file, you then can issue

at 0010a Jan 1 happynewyear

to schedule the greeting to be sent out at 12:10 A.M. on New Year's Day. The here document is actually a form of input redirection. After the **<<** is an arbitrary word (in this case, **EOF**) followed by a NEWLINE that delimits the beginning and end of the here document. The general form of a here document is

command **<<** *word*
zero or more
lines of input text
included here
word

The delimiter *word* is not variable, filename, or command substituted. The last line must contain only the same *word* and no other characters. The intervening lines are variable and command substituted, but SPACE, TAB, and NEWLINE characters are preserved. The resulting text, up to but not including the line with the end delimiter, is supplied as standard input to the command.

An example is the **timestamp** script (Figure 5.3).

The here document contains a variable substitution and two command substitutions. The **hostname** command displays the name of the host computer. The **date** command displays the date and time (**Ex: ex05/timestamp.sh**).

Substitutions can be suppressed within a here document by quoting all or part of the starting delimiter word with \, ", `, or ', for example,

```
\EOF
'here'
a"b"
`a`b
```

FIGURE 5.3: The **timestamp** Script

```
#!/bin/bash
## script name: timestamp
## usage: timestamp file
## this script stamps date and time on a document


cat >> $1 << here
****************************

RECEIVED by $USER on `hostname`
`date`
here
```

Note that a corresponding end delimiter does not need any quotes.

If <<- is used instead of << for the here document, then any leading TABs in front of the input lines and the delimiter line will be stripped away, making indenting the script source code for easier reading possible.

Also, if the here document is a single string with no white space, you may use instead (**Ex:** ex05/herestr.sh)

```
<<< any_string
```

5.18   More on Functions

We have already seen Bash functions in Chapter 2, Section 2.15. Each function gives rise to a new Shell-level command that can be used just like any other command—interactively on the command line or as part of a Shell script. In a Shell script, you may call functions defined earlier in that script as well as functions made available from the invoking Shell. If the invoking Shell defines a function xyz, then it is made available for Shell scripts with **export -f** xyz. It is recommended that you avoid this feature and make each Shell script self-sufficient by including definitions of all the functions it needs.

Unlike functions in general programming languages such as C or C++, Bash functions have their own way of passing arguments and returning values, as we will explain.

Function Arguments

A Bash function is defined without any named parameter. Thus, the following
is impossible:

```
function compare(str1, str2)     ## wrong, no parameters allowed
{  ...  }
```

Instead, any arguments passed in a function call are accessed from within that
function using the *positional parameters* $1, $2, and so on. Thus, compare (**Ex:
ex05/strcompare.sh**) can be coded as follows:

```
function compare()
{   local str1="$1";           ## 1st argument
    local str2="$2";           ## 2nd argument
    if [[ $str1 == $str2 ]]
       then echo 0;
    elif [[ $str1 > $str2 ]]
       then echo 1;
    else echo -1;
    fi
}
```

The keyword `local` declares variables local to the function (not accessible
from outside the function). Here is a sample call:

compare "apple" "orange";

Arrays can also be passed in function calls. The following function displays
any array that is passed to it (**Ex: ex05/arrusage.sh**).

```
function displayArray()
{   echo -n "(";
    for el    ## iterates over positional parameters (a)
       do echo -n " \"$el\" "
    done
    echo ")";
}
```

Say that we have an array prime=(2 3 5 7 11 13 17), then we can pass all
the array elements in a call

displayArray "${prime[@]}"

resulting in the display

("2" "3" "5" "7" "11" "13" "17")

The function **displArray** works by iterating over the positional parameters passed (line **a**).

Normally, arguments are passed *by value* when a copy of the value of each argument is passed to the called function. However, it is also possible to pass arguments *by reference* when the variable itself (a reference to its value) is passed instead of its value. To illustrate *pass by reference*, consider the function

```
function addOne()
{ let  $1+=1;  }
```

Here is a call to `addOne` with a reference argument n (instead of $n).

```
n=12;
addOne n;        ## function call with reference argument
echo $n          ## 13
```

When we use n, instead of $n, in the call to `addOne`, the $1 inside the function evaluates to the symbol n. Thus, the code `let $1+=1` is the same as `let n+=1` which explains how n becomes 13 after the function call. If we wish to access the value of n inside `addOne`, we can use the *indirect reference evaluation* notation

```
${!1}        ## means eval \$$1 or $n
```

Hence, we might improve the function as follows (**Ex: ex05/addOne.sh**):

```
function addOne()
{    echo ${!1};       ## displays $n
     let  $1+=1;       ## let n+=1
     echo ${!1};       ## displays $n again
}
```

In general, we have

```
x=y;   y="abc"
echo $x              ## displays y
echo ${!x}           ## displays abc
```

Passing by reference can be useful in practice. For example, we can define a function **setenv** to make setting of environmental variables (Chapter 2, Section 2.10) easier (**Ex: ex05/setenv.sh**).

```
function setenv()
{    eval $1=\$2;
     export $1;
}
```

With this function, you can set the command search path (Chapter 2, Section 2.4) with one call:

setenv PATH   *your desired path string*

The indirect reference evaluation also allows you to pass an array by reference, as we will see in the next subsection.

Return Value of a Function

Let's write a function sum that adds together all numbers in any given array and returns the total (**Ex:** ex05/sum.sh).

```
function sum()
{  local total=0;           ## local variable
   for i
      do let total+=$i       ## or (( total+=$i ))
   done
   echo $total               ## return value (I)
}

s=$( sum ${prime[@]} )       ## calling sum and get value (II)
echo $s                      ## 58
```

Note here that we return a value by echoing it (line I) and capture the returned value with command substitution (line II).

Alternatively, we can pass the total back in a reference parameter myTotal. To do that, we revise the function sum to newSum. While we are at it, we also pass the prime array into the function by reference (**Ex:** ex05/newSum.sh).

```
function newSum()
{   local p="$1[@]";         ## $p is "prime[@]"
    for i in "${!p}"         ## evaluates ${prime[@]}
       do let $2+=$i         ## $2 is the symbol myTotal
    done
}

myTotal=0
newSum prime myTotal         ## passing two ref parameters
echo $myTotal
```

The three lines in newSum with comments deserve close study.

A *predicate* is a function that tests for a condition and returns true or false. Here is a predicate that tests if a file is more recently modified than another file (**Ex:** ex05/newer.sh).

```
function newer()
{   if [[ $1 -nt $2 ]]       ## if file $1 is newer than file $2
       then return 0         ## exit status 0 means true
    else
       return 1              ## false
```

```
    fi
}
```

The `return` statement in a function returns an *exit status* (a small integer less than 256). The value of the *exit status* is available in the special variable $? right after the function call. If a function does not call `return`, then its *exit status* is that of the last statement executed before the function ended. A predicate function, such as `newer`, can be used directly in conditional expressions. Here is a call to `newer`.

```
if newer file1 file2
    then ...
fi
```

However, as you may have realized, the predicate function can be simplified to

```
function newer()
{   [[ $1 -nt $2 ]]      ## available also is -ot for older than
}
```

Finally, it is possible for a function to return a value by assigning it to some global variable. Because there is always the danger of some other code using/setting the same global variable for some other purpose, we do not recommend this approach.

5.19   Redefining Bash Built-in Functions

If you define a function whose name coincides with a Bash built-in command or a regular command, then that name invokes the function instead. However, the commands are still accessible:

builtin *commandName args*   (invokes the built-in *commandName*)
command *commandName args*   (invokes the regular *commandName*)

Here is a simple example that redefines **cd** to do a directory listing each time it is called.

```
function cd ()
{    builtin cd "$1"
     /bin/ls -l
}
```

Often, Shell scripts can be written as Shell functions with little change and no impact on how they are called. By implementing a script as a function, you can place it in your Shell initialization file (`.bash_profile` for example) and make it part of your Shell.

5.20 Example Bash Scripts

Now let's consider some more substantial Bash scripts. You can find these scripts in the example code package. Placed the scripts in a folder, $HOME/bin, for example, and open up their execution permissions. Also, make sure that the folder is on the command search PATH.

Example: Removing Unwanted Files

Modern operating systems such as Linux make it easy to create, download, copy, and otherwise manipulate files. However, most users are hesitant about removing files, and the clutter of obsolete files can be a nuisance let alone wasting disk storage. One reason is the sheer tedium of looking through files and discarding those that are no longer needed. Thus, although disk storage is decreasing in cost, new supplies of additional disk space never seem to quite match the demand. The clean.sh script provides some help (**Ex: ex05/clean.sh**). The command

clean.sh *directory*

displays file names in the given *directory*, one at a time, and allows the user to decide interactively whether or not to keep or delete the file. This script is longer and will be explained in sections. The clean.sh script begins with argument checking:

```
#!/bin/bash
## bash script clean.sh
## usage:  clean.sh dir
## helps to rm unwanted files from a directory

if (( $# != 1 ))            ## number of args is not 1
then echo usage: $0 directory
     exit 1
fi
dir="$1"
if ! [[ -d "$dir" && -w "$dir" ]]   ## not a dir or not writable
then echo $dir not a writable directory
     echo usage: $0 directory; exit 1
fi

cd "$dir";
```

After checking for correct input, the script changes the current working directory to the given directory.

A **for** loop is used to treat each file (*) in the current directory (line 1). On any given iteration, if $file is not an ordinary file or not readable, then it is skipped via **continue** (line 2). For a regular file, an infinite loop (line 3) is used to handle its processing. We must break from this inner **while** loop to get to the next file.

For each file, the file name is clearly listed with **ls**, and the user is prompted with

```
***** Delete filename or not?? [y, n, e, m, t, ! or q] :
```

indicating seven possible (single-character) responses (terminated by RE-TURN). User input is received via **read** (line 4) and treated with a **case** construct (line 5).

```
for file in *                                        ## (1)
do
    if ! [[ -f "$file" && -r "$file" ]]
       then continue                                 ## (2)
    fi
    echo " " ## a blank line
    /bin/ls -l "$file"
    while :                                           ## (3)
    do
        echo -n "*** Delete $file or not?? "
        echo -n "[y, n, e, m, t, ! or q]:"
        read c                                        ## (4)
        case $c in                                    ## (5)
            y) if [[ ! -w "$file" ]]
               then echo $file write-protected
               else /bin/rm "$file"
                    if [[ -e "$file" ]]
                    then echo cannot delete $file
                    else echo "+++++ $file deleted"
                    fi
               fi
               break ;;  ## to handle next file
            n) echo "----- $file not deleted"
               break ;;
            e) ${EDITOR:-/bin/vi} "$file"; continue ;;
```

The cases for **y, n** are clear. Note the use of **break** to leave the while loop and process the next file under **for**. The **e** case invokes the user's favorite text editor (set by the environment variable EDITOR) or **vi**.

The choices **m** and **t** offer the user a chance to examine the file before deciding on its disposal. Note the use of **continue** to go back to the **while** loop.

```
            m) /bin/more "$file"; continue ;;
            t) /bin/tail "$file"; continue ;;
            !) echo -n "command: "
```

```
            read cmd
            eval $cmd ;;                              ## (6)
        q) break 2;;   ## break 2 levels
        *)  ## help for user
            echo clean commands: followed by RETURN
            echo "y     yes delete file"
            echo "n     don't delete file, skip to next file"
            echo "e     edit/view file with ${EDITOR:-/bin/vi}"
            echo "m     display file with more"
            echo "t     display tail of file"
            echo "!     Shell escape"
            echo "q     quit, exit from clean"
            ;;
    esac
  done
done
```

In addition to calling on **more** and **tail**, the user may execute any command (with !) to help make the decision. In this case, the script reads a command string from the user and executes it as a Shell command using **eval** (line 6), which executes the string as a line of command. Note that the variable $file can be used in this command string and that there is no restriction as to what command can be used. Some command strings the user may enter are

```
head $file
cp $file ...
mv $file ...
```

The q case quits from the script. For all other cases, we display a menu of single-letter commands for `clean.sh` and proceed to another iteration of **while** for the same $file. If the user mistypes and enters a character other than those expected by the script, the **while** loop is restarted. Also, note that the `clean.sh` script provides feedback, telling the user each action it has taken.

Example: Conditional Copy

The `ccp.sh` (conditional copy) script creates a command that copies files from a *source* directory to a *destination* directory using the following conditions on any ordinary *file* to be copied:

1. If *file* is not in *destination*, copy.

2. If *file* is in *destination* but not as recent as that in *source*, copy.

3. Otherwise, do not copy.

The script (**Ex:** `ex05/cpp.sh`) begins with argument checking:

```
#!/bin/bash
## bash script ccp.sh
## usage: ccp.sh fromDir toDir [ file ... ]

(( $# >= 2 )) && [[ -d "$1" && -d "$2" ]] \
|| { echo usage: $0 fromDir toDir [ file ... ]; exit 1; }   ## (A)

from=$1; to=$2
if (( $# > 2 )) ## files supplied
then filenames=${@:3}                                       ## (B)
else ## all files in fromDir
     pushd $from
     filenames=( * )                                        ## (C)
     popd
fi
```

Unless we have at least two arguments, the first two being directories, we will error out (line A). This works because the next operand of || (logical or) will be evaluated only if the previous operand is false.

Given the correct arguments, the script proceeds to record the from and to directories and to store the files to be processed in the filenames array. If the files to be copied are given on the command line, they are picked up (line B) with a variable modifier (Section 5.16). Otherwise, all files in the from directory are included (line C).

Now the stage is set to process each file to be copied conditionally. A **for** loop is used to go through each element in the array filenames (line D).

```
for file in "${filenames[@]}"                               ## (D)
do
    echo $file;
    if [[ ! -f "$from/$file" ]]    ## not a regular file
        then continue              ## skip
    fi
    if [[ -f "$to/$file" ]]        ## $file in folder $to
        then if [[ "$from/$file" -nt "$to/$file" ]]         ## (E)
                then
                echo /bin/cp \"$from/$file\" \"$to\"
                /bin/cp "$from/$file" "$to"
             fi
        else   ## $file not in folder $to
             echo /bin/cp \"$from/$file\" \"$to\"
             /bin/cp "$from/$file" "$to"
    fi
done
```

If $file is present in $to, then we check to see if the version in $from is newer (line E). Any file copying action is displayed to inform the user. Note the use of double quotes (") throughout the script to guard against multiword file names.

Example: Total File Sizes

In this example (**Ex:** ex05/total.sh) we use a *recursive* Shell function to compute the total number of bytes contained in all files in a certain file hierarchy. The **du** command only provides a rough accounting in kilobytes. The script **total.sh** recursively descends through a directory hierarchy and sums the file sizes by extracting information provided by the **ls** command on each file in the hierarchy.

```
#!/bin/bash
## bash script : total.sh
## compute total bytes in files
##     under any given  directory hierarchy

[[ $# == 1 && -d "$1" ]] \
|| { echo usage: $0 directory; exit 1; }
```

After the checking command-line argument, we proceed to define a function `total` which sums up the file sizes for all files in the current directory and recursively descends the directory hierarchy.

```
function total()
{  local count=0       ## bytes used inside working dir
   for file in .* *    ## all files including hidden ones
   do
     if [[ -f "$file" ]]
     then
       fl=( $(/bin/ls -ld "$file" ) )               ## (a)
       let count+=${fl[4]}                          ## (b)
       continue
     fi
     if [[ "$file" == *\. || "$file" == *\.\. ]]    ## (c)
     then
       continue
     fi
```

For a regular file, the **ls** -l output is captured in the array `fl` (line **a**), and the byte size is added to the total byte count (line **b**). The special files . and .." are excluded (line~\verbc+).

For a subdirectory, we temporarily change to that directory (line **d**), include the sum obtained by a recursive call to `total` (line **e**), and then change the directory back (line **f**).

```
     if [[ -d "$file" ]]
```

```
    then
      pushd "$file"  >/dev/null                              ## (d)
      y=$( total )                                           ## (e)
      let count+=$y
      popd >/dev/null                                        ## (f)
    else
      echo \"$file\" not included in the total >&2
    fi
  done
  echo $count                                                ## (g)
}
```

Note that we redirected **echo** output to `stderr` and output by **pushd** and **pupd** to the data sink `/dev/null`. The only output to `stdout` allowed is the total count (line g). This is the way the function `total` returns a value that is picked up in a call with command substitution (lines e and h).

```
dir="$1"
cd $dir
echo "Total for $dir = " $( total ) Bytes                   ## (h)
```

Example: Secure File Transfer

The need often arises to transfer files between computers. The **sftp** command is commonly used for this purpose. We will a Bash script that helps file upload and download with **sftp**. The script will work smoother if you have already set up password-less SSH and SFTP between your local and remote hosts (Chapter 7, Section 7.6).

The idea now is to set up a special directory for upload and download on a remote computer (say, at work or school) and use the **mput** or **mget** command to invoke the script (**Ex: ex05/mput**) to transfer files to and from it. Here is the script.

```
#!/bin/bash
## upload and download files using sftp
## Usage: mput "*.jpg"  or mget "*.pdf"

#### begin customizable:  user, host, rdir
user=pwang                                                  ## (1)
host=monkey.cs.kent.edu                                     ## (2)
rdir=tmp                                                    ## (3)
#### end customizable

if [[ $0 == *mget* ]]                                       ## (4)
then action=mget
else action=mput
```

```
fi

/usr/bin/sftp $user@$host <<HERE
cd $rdir
$action "$@"
HERE
```

Customizable parameters are **user** (user ID), **host** (remote host), and **rdir** (remote folder) (lines 1-3).

The script is named **mput** with a hard link (Section 6.2) **mget** to it.

ln mput mget

So the script can be invoked as either **mput** or **mget**. The **sftp** action is set according to the value of $0 (line 4).

These values being set, the **sftp** can be invoked with a *here document]* to perform the desired uploading/downloading. For example,

mget memo.pdf	(downloads memo.pdf)
mget*.pdf	(downloads all pdf files)
mput*.jpg	(uploads all jpg files)

Example: Resizing Pictures

Connect your digital camera to your Linux computer and download a set of pictures to your **Pictures** folder. Often, you need to scale down the pictures for emailing or posting on the Web. Here is a script that makes the task easy (**Ex: ex05/resize**).

resize '75

will reduce each .jpg file by 75% under the new names **trip001.jpg** etc.

The **resize** script first processes the command-line arguments.

```
#!/bin/bash
## resize a set of pictures
## Usage:  $0 size-factor newName pic1.jpg pic2.jpg ...
##     scales all pics by size-factor into
##     newname1.jpg, newname2.jpg ...

(( $# < 3 )) \
|| { echo usage:"$0 \"50%\" newName pic.jpg ..."; exit 1; }

sz=$1; name="$2"; declare -i k=1
```

Then, we resize each picture (line i) and save it under sequential numbering after the given new name using three-digit numbers (lines ii and iii). The notation "${@:3}" produces a list of all names on the command line starting from the fourth word (Section 5.16).

```
for pic in "${@:3}"                        ## (i)
do
    if (( $k < 10 ))                       ## (ii)
       then n="00$k"
       elif (( $k < 100 ))                 ## (iii)
       then n="0$k"
    fi
       echo "convert -resize $sz \"$pic\" \"$name$n.jpg\""
       convert -resize $sz "$pic" "$name$n.jpg"
       let k++
done
```

The **convert** command is part of the *ImageMagick* tool that is commonly found on Linux systems. See **man convert** for more details on its usage.

5.21   Debugging Shell Scripts

When a Shell script fails because of syntax problems, the syntax error will cause a display of some unexpected token. You usually will get an error message stating the token and the line number containing the token. This means your syntax problem is on or before that line. Take a close look at your code, and you usually can find the problem or typo and fix it.

You can also place **echo** commands at appropriate places to show values of variables to help you catch the bug. Such **echo** commands can be removed after the debugging is done. Or you may use a conditional echo,

```
function dbecho()
{   [[ ${DEBUG:-off} == off ]] || echo "$*" >&2
}
```

We see that the function `dbecho` produces output to the `stderr` unless the variable `DEBUG` is null, not set, or set to `off`. Thus, you would place `DEBUG=on` at the beginning of your script to enable **dbecho** output and comment the `DEBUG=on` out to disable it.

If you still cannot find the problem, then placing the command

```
set -x           (turns on tracing)
set +x           (turns off tracing)
```

in your script will turn *tracing* on/off from selected places. Tracing will display each command before it is executed.

More tracing information can be display with

bash -x *script.sh*

to run the script with trace turned on within Bash. This will show all commands executed, including any init files.

5.22   Error and Interrupt Handling

An error may occur during the processing of a Shell script at several different stages. A syntax or substitution error will result in the premature termination of the script. However, if a regular command invoked by the script runs into an error, the interpretation of the script continues with the next command in the script. Error messages are produced and sent to the standard error output. To help debugging, the `stderr` can be redirected (Chapter 2, Section 2.5) to a file.

In Linux, when a program terminates (because of either completion or error), an *exit status* is set to a small integer value to provide an indication of the circumstances under which execution was terminated. By convention, the exit status is 0 if termination is normal and greater than 0 if termination is abnormal. A Shell built-in command gives an exit status of 0 when successful and an exit status of 1 when unsuccessful. The special Shell variable $? is set to the exit status after the execution of each command. The value of $? is 0 if the last command was successful and greater than zero (usually 1) if it failed.

To test whether a *command* has failed, the following construct often is used:

if *command*
>   **then**
>>   *commands to execute if command succeeds*
>   **else**
>>   *commands to execute if command fails*

fi

Interrupt Handling

An *interrupt* is an asynchronous *signal* sent to a running program by another process or through the keyboard by the user. The user can send an interrupt signal to a Shell running a script by typing in the *interrupt character*, normally `^C` or DELETE. There are various system-defined signals that can be sent to an executing program using the **kill** command. Signals will be discussed in Chapter 10, Section 10.16. For now, it is sufficient to state that

kill -2 *pid*

sends the interrupt signal 2 to the process *pid*, which causes it to terminate. The process *pid* can be given either as a jobid or as a process number. If this does not terminate the process, use

kill -9 *pid*

which sends signal 9, unconditionally terminating *pid*.

The default response of a Shell executing a script is to terminate if it receives an interrupt signal, but this can be modified. The Bash built-in command **trap** controls the action of the Shell when specific interrupt signals are received or when specific events take place.

trap *command sig*

The given *command* (given as a string in quotes) will be executed when the Shell receives the indicated signal or event. The *sig* is a signal number or signal name (see **man 7 signal**). If *sig* is DEBUG, then the command is executed after each command in the script. If *sig* is EXIT, the command is executed after the Shell script is done. Without any arguments, **trap** displays a list of trapped signals.

Often, a Shell script will create a temporary file that will be removed at the end of the script. For example,

```
...
spell $file >| /tmp/badwords$$
...
...
## at end of script
/bin/rm -f /tmp/badwords$$
```

The value of the special variable $$ is the process number of the running script, and its use here makes the temporary file name unique to the process. However, that file can be left unremoved if the script terminates due to a signal instead of completing all commands. To fix that problem simply add (**Ex: ex05/trap.sh**)

```
trap "/bin/rm -f /tmp/badwords$$" EXIT
```

before creating the temporary file. The action places the given **rm** command as something to execute upon normal or error exit of the script. As a consequence, the **rm** command at the end of the script is no longer necessary.

5.23   The Perl Alternative

Shell scripting is not the only way to write scripts to automate tasks. For more complicated tasks or for problems involving structured data files, many prefer to use *Perl*, the *Practical Extraction and Report Language*, over simple Shell scripts. The Perl language is outside of the scope of this text, and there are many books dedicated to Perl. We will give only a brief introduction here.

Perl is a portable, command-line–driven, interpreted programming/ scripting language. Written properly, the same Perl code will run identically on Linux/UNIX, Windows, and Mac operating systems. Most likely, you'll find Perl pre-installed on your Linux system.

The Perl scripting language is usually used in the following application areas:

- DOS, Linux/UNIX command scripts

- Web CGI programming (Chapter 7, Section 7.17)

- Text input parsing

- Report generation

- Text file transformations and conversions

Perl 1.0 was released December 18, 1987, by Larry Hall with the following description:

> Perl is an interpreted language optimized for scanning arbitrary text files, extracting information from those text files, and printing reports based on that information. It's also a good language for many system management tasks. The language is intended to be practical (easy to use, efficient, complete) rather than beautiful (tiny, elegant, minimal). It combines (in the author's opinion, anyway) some of the best features of C, sed, awk, and sh, so people familiar with those languages should have little difficulty with it. (Language historians will also note some vestiges of csh, Pascal, and even BASIC—PLUS.) Expression syntax corresponds quite closely to C expression syntax. If you have a problem that would ordinarily use sed or awk or sh, but it exceeds their capabilities or must run a little faster, and you don't want to write the silly thing in C, then Perl may be for you....

Perl 5.0, a complete rewrite of Perl adding objects and a modular organization, was released in 1994. The modular structure makes it easy for everyone to develop *Perl modules* to extend the functionalities of Perl.

The Comprehensive Perl Archive Network (CPAN; `www.cpan.org`) was established to store and distribute Perl and Perl-related software.

5.24   For More Information

At the book's companion website (`http://ml.sofpower.com`), you'll find a complete *example code package* containing ready-to-run code files for the examples in this book. The Shell script examples in this chapter are, of course, part of this package.

You can get a quick reference for Bash by

man bash

and you'll see many details including a list of built-in functions.

On the GNU Bash home page (`www.gnu.org/software/bash/`) you can find the Bash Manual which is a complete reference for Bash. You'll also be able to download the latest release of Bash.

POSIX defines standards for utilities, the Shell programming language, the Shell command interface, and access to environment variables. Scripts following the POSIX standard can be much more portable. For additional information, see *Portable Operating System Interface (POSIX) – Part 2: Shell and Utilities*, published by IEEE (IEEE Std 1003.2-1992).

5.25   Summary

Bash provides many features for writing scripts to automate tasks for yourself and others. Proficiency in script writing can make you more efficient and effective on Linux.

A Bash script is an executable text file whose first line must follow a special convention. Such a file can be invoked via explicit or implicit interpretation and is executed by a subshell of the invoking Shell. Arguments are passed into the script and are available in the script as *positional parameters*. Other values can be transmitted to the script by *environment variables*. Upon termination, a Shell script returns an *exit status* to the invoking Shell which can access this value via the special variable $?. A zero exit status indicates successful completion of the script.

Bash provides a good number of constructs for script writing.

- Looping constructs: `for`, `while`, and `until`

- Decision making constructs: `case ... esca`, `if ... then ... else ... fi`

- Test expressions: `[[ ... ]]`, `(( ... ))`, and any command exit status

- Logical operators: `&&`, `||`, `!`

- Arithmetic expressions: `let`, `(( ... ))`

- Glob pattern matching: `==`, `!=`, and `case`

- Regular expression pattern matching: `=~`

- Arrays and functions

- Variable modifications: with `:`, `%`, `#`, `%%`, and `##` (inside `${}`)

Functions are invoked just like commands. A function takes positional parameters and produces an exit status. Arguments can be passed by value or by reference. A value can be returned by echoing it to `stdout`, setting a return-value reference parameter, or setting the exit status.

Many practical scripts have been given as examples. Debugging techniques as well as error trapping for Shell scripts have been discussed.

5.26 Exercises

1. What is the difference between these two ways of invoking a script abc.sh:

   bash   abc.sh
   abc.sh

2. Bash allows the use of $0, $1, $2, and so on to refer to positional parameters. Is it possible to use $10, $15, and so on? Explain.

3. The character * is a special character in Bash.

   (a) Explain how it is used for filename expansion.

   (b) List at least two situations in Bash syntax where the character * is not quoted, but does not serve the function of filename expansion or globbing.

4. Using the **cmdsearch** example in Section 5.6 as a guide, write a Bash script **cmdfind**.

   cmdfind *pattern*

   The script takes a regular expression *pattern* argument and finds all commands on PATH that match the given pattern.

5. The character @ is a special character in Bash.

   (a) Explain the meaning of $*, $@, "$*", and "$@".

   (b) How about $arr[*], $arr[@], "$arr[*]", and "$arr[@]"?

6. Explain how the character # is used in Bash scripts: as a comment character, as the number of positional parameters, and as the number of array elements.

7. Refer to the section on *variable modifiers* (Figure 5.16) and see if it gives a way to change the case of characters in a variable. If not, find out what Bash parameter expansion notations does that.

8. Bash also supports conditional expressions using [...]. Explain the difference between that and the [[...]] conditionals. What about the ((...)) conditionals?

9. Can you suggest ways to improve **clean**? What about cleaning out only old files? Is an undo or undelete feature desirable? What about recursively cleaning out subdirectories as an option? How would you implement the improvements?

10. Write a Shell script to change the names of all files of the form `*.JPG` in a directory (supplied as argument 1) so that they have the same root as before but now end in `.jpg`. Generalize this script so that any two extensions could be used.

11. Write a Shell script **delete** that mimics the way **rm** operates, *but* rather than erasing any files, it would put them in a user's `.Trash` folder. Write an additional Shell script **undelete** to make these files reappear where they were deleted.

12. Reimplement the **delete** script of the previous exercise as a Bash function. Discuss the pros and cons of Shell scripts vs. functions.

13. Write a Shell predicate function **evenp** that takes a integer argument and tests if it is an even number or not.

14. Write a Shell function **findfile** so that

    findfile *name dir1 dir2 ...*

    searches the named file in the directories specified. If the file is found in one of the directories, the current directory is changed to it. Why do we need to implement it as a function in the interactive Shell rather than a regular Shell script?

15. Improve the `mget`/`mput` script so that it can also be invoked as `rv` and will allow you to view a remote PDF (`.pdf`) or MS Word (`.doc`) file locally. No copy of the remote file will be left on the local file system.

16. Recent versions of Bash also supports *associative arrays*. Find out how it works and experiment with (**Ex: ex05/asso.sh**).

Chapter 6

The File System

Storing data as files that can be accessed immediately by programs is essential for modern operating systems. Files are identified by their filenames and may contain many kinds of data. For example, a file may contain a letter, a report, a program written in a high-level language, a compiled program, an organized database, a library of mathematical routines, a picture, or an audio/video clip.

The operating system provides a consistent set of facilities allowing the user to create, store, retrieve, modify, delete, and otherwise manipulate files. The *physical* storage media (usually high-speed magnetic disks) are divided into many *blocks* of *logical* storage areas. A file uses one or more of these blocks, depending on the amount of data in the file. Blocks are used and freed as files are created and deleted. The program that creates, stores, retrieves, protects, and manages files is the *file storage system* (or simply file system) which is part of the kernel of any modern operating system.

Historically, the UNIX operating system evolved from a project to design a new computer data storage system at the then Bell Laboratories. This hierarchical file storage system is a hallmark of UNIX. As UNIX evolved, so did the implementation of its file storage system. Linux basically adopted the same UNIX file storage system implementation. The file system usually consists of one or more self-contained file management units, each is known as a *filesystem*. Also, the Linux file hierarchy usually follows the *File System Standard* (FSSTND), allowing users to find important system files at the same file locations on any compliant Linux system.

The file system affects almost every aspect of the operating system. In this chapter, the file system is discussed in detail, including such topics as type and status of files, access protection, filesystem structure, implementation, quotas, special files, and networked filesystems. A clear understanding of how Linux treats files will be helpful for any Linux user.

6.1   A File Location Road Map

The file system in Linux is much more than a place to store user files. It contains the operating system itself, application programs, compilers, network servers, shared libraries, documentation, system configuration and administration data files, media mount points, log files, temporary scratch areas, and so on. In other words, almost every bit of data and programming that is needed to boot the computer and keep it working must be saved in the file system.

Linux systems generally follow the FSSTND in organizing the file system hierarchy. This makes it easy for Linux users to find their way on different Linux systems. Table 6.1 shows a typical organization of the root folder (/) of

TABLE 6.1: The Root Directory: /

Where	What
bin/	Essential commands—**cat**, **cp**, **rm**, **sh**, **bash**, **vi**, **mount**, etc.
sbin/	Commands for system maintenance
boot/	Everything required at system boot time
dev/	All special files (devices)
etc/	System configuration, data, and maintenance files such as the password file (**passwd**), the filesystem tables (**fstab**, **mtab**), email, printer, X-windows, and network services configuration, and the run-level initialization script folder (**rc.d/**)
home/	Home directories for users
lib/	Kernel modules, shared libraries for essential commands
tmp/	Folder for temporary files by system and users
media/	Mount points for removable media—CD, DVD, USB devices
mnt/	Generic mount points for filesystems and devices
opt/	Optional additions to the Linux distribution
proc/	Kernel run-time data files, off limits for users
usr/	All application programs and their files
var/	Variable data files—mail and printer spool folders, logs, locks

the file tree. From your desktop, clicking on the `Computer` icon then selecting the `File System` link brings you to the root directory. On the command line, **cd /** will do. We already know that files and folders form a tree hierarchy rooted at /. Each file on this file tree is uniquely identified by its *full pathname*, as we already mentioned in Chapter 1, Section 1.4.

Inside each user's home directory, you'll often find these standard folders: `Documents`, `Download`, `Music`, `Pictures`, `Videos`, `Desktop`, and the hidden `.Trash`.

When files and folders accumulate, it can become harder to locate a file that you need. See Section 6.10 and Section 6.11 for helpful commands.

6.2 File Types

The file tree contains different types of files.

1. An *ordinary file* that contains text, programs, or other data

2. A *directory* that contains names and addresses of other files

3. A *special file* that represents an I/O device or a filesystem partition

4. A *symbolic link* that is a pointer to another file

5. A *socket* (or domain socket) that is used for inter-process communication

6. A *named pipe* that is a way for inter-process communication without the socket semantics

The first character in an **ls -l** listing of a file is a *file type symbol*. Table 6.2 lists the different file type symbols.

TABLE 6.2: File Type Symbols

Symbol	Meaning	Symbol	Meaning
-	Regular file	d	Directory
l	Symbolic Link	c	Character special file
b	Block special file	s	Socket
p	Named pipe		

Now, let's describe five of the file types in turn. The socket and named pipe will be discussed later in Chapter 11, Section 11.6.

Ordinary Files

An ordinary file stores data of various *content types*. The entire file storage system is designed to store, retrieve, and manage ordinary files. Your home directory is normally where you store your own files.

Filenames are character strings (it is best not to use any white space). Although Linux filenames do not require them, files of different content types often use different *extensions*. For example, a picture might use the .jpg extension.

The *Multipurpose Internet Mail Extensions* (MIME) provides a standard classification and designation for file *content types*. Files of different content types often use well-known filename extensions for easy recognition and processing. There are hundreds of content types in use today. Many popular types are associated with standard file extensions. Table 6.3 gives some examples.

For a more complete list of content types and file suffixes, see the /etc/mime.types file on your Linux system.

Directories

Files are stored in directories, and that is why they are also known as file folders. A directory is a file whose content consists of *directory entries* for the files placed in the directory. There is one directory entry for each file. Each directory entry contains the filename and the location of its *file information node* (i-node).

TABLE 6.3: Content Types and File Suffixes

Content Type	File Suffix	Content Type	File Suffix
text/plain	txt sh c ...	text/html	html htm
application/pdf	pdf	application/msword	doc, docx
image/jpeg	jpeg jpg jpe	audio/basic	au snd
audio/mpeg	mpga mp2 mp3	application/x-gzip	gz tgz
application/zip	zip	audio/x-realaudio	ra
video/mpeg	mpeg mpg mpe	video/quicktime	qt mov

A filename is a sequence of characters not containing /. The maximum sequence length is dependent on the version of the Linux system. It can be up to 255 characters on most systems, but can be no more than 14 characters on some older versions. The i-node location is an integer index, called the *i-number*, to a table known as the *i-list*. Each entry in the i-list is an *i-node*, which contains status and address information about a file or points to free blocks yet to be used. The entire file system may involve several independent and self-contained parts, each known as a *filesystem*. Each individual filesystem has its own i-list.

Special Files

By representing physical and logical I/O devices such as graphical displays, terminal emulators, printers, CD/DVD drives, and hard drives as special files in the file system, Linux achieves compatible file I/O and device I/O. This means that an application program can treat file and device I/O in the same way, providing great simplicity and flexibility. Under FSSTND, all Linux special files are under the directory /dev. There are two kinds of special files: a *character special file* and a *block special file*. A character special file represents a byte-oriented I/O device such as a display or a printer. A block special file represents a high-speed I/O device that transfers data in blocks (many bytes), such as a hard drive. Typical block sizes are 1024 bytes and 2048 bytes.

Special files usually are owned by the super user (root). The ownership of a terminal emulator special file (under /dev/pts/) is set to the user of the terminal for the duration of the terminal session.

Links

Linux allows a directory entry to be a pointer to another file. Such a file pointer is called a link. There are two kinds of links: a *hard link* and a *symbolic link*. A regular file is an entry in a directory with a name and an i-number. A hard link, or simply a *link*, is an entry in a directory with a name and some other file's i-number. Thus, a hard link is not distinguishable from the original file. In other words, after a hard link is made to a file, you cannot tell the file from the link. The net result is that you have two different directory entries referring to the same i-node. A file may have several links to it. A hard link cannot be made to a directory or to a file on another filesystem.

Thus, hard links allow you to give different names to the same file within the same filesystem. For example, you may have a file called `report` and you enter

ln `report report.txt`

then the report is also under the filename `report.txt`.

The regular command **ln** is used to make links. The general forms of the **ln** command are as follows:

ln *file*	makes a link to *file* in the current folder
ln *file linkname*	establishes *linkname* as a link to existing *file*
ln *file1* ... *dir*	makes links in *dir* to the given file(s)

By default **ln** forms hard links. It is permitted to establish a link to a file even if you are not the owner of the file. When deleting a file (with the **rm** command), the directory entry of the file is deleted. For **rm** *file* to succeed, you need write permission to the parent directory of *file*, not the file itself. A file is only physically deleted from the filesystem when the last link of it is **rm**ed. The total number of hard links to a file is kept as part of the file *status* (Section 6.4).

Symbolic Links

A symbolic link is a directory entry that contains the pathname of another file. Thus, a symbolic link is a file that serves as an indirect pointer to another file. For most commands, if a symbolic link is given as an argument, the file pointed to is to be used. For example, if the file `abc` is a symbolic link to the file `xyz`, then

cat `abc`

displays the contents of `xyz`. There are some exceptions:

rm `abc`

removes the directory entry `abc` (even if it is a symbolic link). As well,

ls `-l` `abc`

displays status information for `abc` (not `xyz`). If you give the command

rm `xyz`

then the symbolic link `abc` points to a non-existent file. If `abc` were a hard link, this situation could not occur.

A symbolic link is distinguishable from the file itself, may point to a directory, and can span filesystems. The `-s` option causes **ln** to create symbolic links:

ln -s *filename linkname*

Unlike a hard link, here *filename* does not even have to be an existing file.

The command **ls -F** displays a symbolic link with a trailing @. The **ls -l** command displays a symbolic link in the form

```
lrwxrwxrwx  1 user            7 Apr 16 17:40 abc@ -> xyz
```

Let's look at an application of symbolic links. Suppose you have the **clean.sh** Shell script in your own home directory, and you wish to make it available to all others on your Linux system. One way to achieve this is to make a link in a system directory to your program. For example, you can issue the following command:

```
ln -s $HOME/cmd/clean.sh /usr/local/bin/clean
```

This establishes the command **clean** as a symbolic link in the system directory `/usr/local/bin` to your `clean.sh`. Assuming the directory `/usr/local/bin` is on users' command search path, then once this link is in place, a new command **clean** is made available to all users. Note that because of file protection, system directories such as `/usr/local/bin` are usually writable only by a super user.

6.3   More on File Access Control

From Chapter 1, we know that files have access control, and the file type and access permissions can be displayed either by the `File Browser` tool or, by using the **ls -l** command. Also, you can change permissions of your own files and folders using the **chmod** command (Chapter 1, Section 1.5 and Figure 1.8) or the `File Browser` (Chapter 3, Section 3.5).

In the following sample **ls** display

```
-rw-r----- 1 pwang faculty 46433 2009-03-06 15:35 report
```

the four *file mode* parts (- rw- r-- ---) show regular file type, read and write permission to u (the file owner), read permission for g (anyone in the faculty group), and no access for o (all others). There are ten positions in the file mode:

Position 1   file type: see Table 6.2

Positions 2-4   r (read), w (write), and x (execute) permission for the owner (u), a - is no permission; the letter s is used instead of x for an executable file with a *set-userid bit* that is on (Section 6.4)

Positions 5-7   r, w, and x permission for g, a - is no permission; the letter s is used instead of x for an executable file with a *set-groupid bit* that is on (Section 6.4)

Positions 8-10 r, w, and x permission for o, a - is no permission

Meaning of Permissions for a Directory

The meaning of read, write, and execute permissions is obvious for a regular file. For a directory, their meanings are different. To access a directory, the execute permission is essential. No execute permission for a directory means that you cannot even perform **pwd** or **cd** on the directory. It also means that you have no access to any file contained in the file hierarchy rooted at that directory, independent of the permission setting of that file. The reason is that you need execute permission on a directory to access the filenames and addresses stored in the directory. Since a file is located by following directories on the pathname, you need execute permissions on all directories on the pathname to locate a file. After locating a file, then the file's own access mode governs whether a specific access is permitted.

To access a directory, you normally need both read and execute permissions. No read permission to a directory simply means that you cannot read the content of the directory file. Consequently, **ls**, for example, will fail, and you cannot examine the filenames contained in the directory. Any filename expansion attempt also will fail for the same reason. However, files in such a directory still can be accessed using explicit names.

The write permission to a directory is needed for creating or deleting files in the directory. This permission is required because a file is created or removed by entering or erasing a directory entry. Thus, write permission on the file itself is not sufficient for deleting a file. In fact, you don't need write permission on a file to delete it from the directory! On the other hand, if you have write permission on a file, but no write permission for its directory, then you can modify the file or even make it into an empty file, but you cannot delete the file.

Default File Protection Settings: umask

When you create a new file, the system gives the file a default protection mode. For most systems, this default setting denies write permission to g and o and grants all other permissions. The default file protection setting is kept in a system quantity known as *umask*. The Shell built-in command **umask** displays the umask value as an octal number. The umask bit pattern specifies which access permissions to deny. The positions of the 1 bits indicate the denied permissions. For example, the umask value 022 (octal 22) has a bit pattern 000010010, and it specifies denial of write permissions for g and o. The command **umask** also sets the umask value. For example,

umask 077

sets the umask to deny all permissions for g and o. If you find yourself using **chmod** go-rwx a lot, you might want to consider putting **umask** 077 into your .bash_profile file.

6.4 File Status

For each file in the Linux file system, a set of *file status* items is kept in the i-node of the file and is maintained by the operating system. The i-node of a file is a data structure that records file meta information (information about the file) that is used by Linux to access and manipulate the file. File status items include

mode	16-bit integer quantity used to represent the file mode
number of links	total number of hard links to this file
owner	user identification of the owner of this file
group	group identification of this file
size	total size in bytes of the data contained in this file
last access	time when this file was last read or written
last content change	time when the contents of this file were last modified; this time is displayed by the **ls -l** command
last status change	time when any status item of this file was changed
i-number	the index number of this i-node
device	hardware device where the file is stored
block size	optimal block size to use for file I/O operations
block count	total number of file blocks allocated to this file

The command

ls -l *file*

displays many status items of a given file. The *system call* **stat** (Chapter 10) can be used in a C program to access file status information.

Many Linux systems also implement the *Second Extended Filesystem* (ext2) or an extension of it (ext3, for example). In an ext2 filesystem, operating system administrators can set additional *ext2 file attributes* kept in the i-node. For example, if the *immutable* (**i**) attribute is set, then the file cannot be altered or deleted, even by a super user.

File Mode

The file mode consists of 16 bits. The four high bits (C-F in Figure 6.1) of the file mode specify the file type. The next three bits define the manner in which an executable file is run. The lowest nine bits of the file mode specify the read, write, and execution permissions for the owner, group, and other. The file type is fixed when a file is created. The *run* and *access* bits are settable

FIGURE 6.1: File Mode Bits

				← Run →											
F	E	D	C	B	A	9	8	7	6	5	4	3	2	1	0
←	Type	→					←				Access				→

by the file owner. You already know how to set the nine *access* bits with the **chmod** command. The run bits can be set together with the *access* bits by the **chmod** command using a numerical mode setting, as in

chmod *mode file*

The numerical *mode* is an octal number that is the logical-or of any number of the settable file modes (Table 6.4). For set-id-on-execution, the symbolic u+s and g-s modes are also available. Only the owner of a file or a super user may change the mode of a file. On most Linux systems, the -R option causes **chmod** to perform the requested mode setting on all files under the given file directories.

TABLE 6.4: Settable File Modes

Mode	Meaning	Mode	Meaning	Mode	Meaning
0001	x for o	0002	w for o	0004	r for o
0010	x for g	0020	w for g	0040	r for g
0100	x for u	0200	w for u	0400	r for u
1000	sticky bit	2000	set g (run)	4000	set u (run)

File Userid and Groupid

In Linux, each file has a *userid* and a *groupid*. The file userid is the userid of the owner who created the file. Each user may belong to one or more (up to a reasonable limit, say, eight) *groups* of users. Each group has a name. The password file (`/etc/passwd`) entry of each user contains a group affiliation. By default, a new user belongs to a group with a groupid the same as the userid. If a user belongs to more than one group, then the additional group affiliations are specified in the file `/etc/group`. The groupid of a file can be set to any group to which the file owner belongs. The group permissions control access to the file by members of the specified group. When a file is first created, it is given by default the groupid of the directory that contains it. The command

chgrp *groupid filename* ...

is used to assign a specified *groupid* to the named files. For example, if `research` is a group name, then

chgrp research *

will change the groupid of each file in the current directory to `research`. The userid of a file can be changed only by the super user. The command

chown *ownerid filename* ...

is used to change the ownership of the named files. For example, the command

chown -R pwang .

changes the ownership of all files in the hierarchy (rooted at .) to **pwang**. Both **chgrp** and **chown** take the **-R** option to process files and folders recursively.

Bash provides a set of queries to determine the file type, access permissions, and so on of a file (Chapter 5, Section 5.13). In addition, the regular Linux command **test** can be used to obtain information about the type and mode of a file. The **test** command is a general conditional command often used in Shell scripts (especially in Sh scripts).

Access Control Enforcement

A file always is accessed through a process, for instance, **ls**, **cat**, **rm**, **vim**, or your Shell (to **cd**, for example). To enforce access control, Linux uses the userid and groupid of a process to grant or deny access to a file according to the file's access mode. The userid and groupid of a process are usually that of the user who invoked the process. A user may belong to more than one group; thus, a process also keeps a *supplementary groupid* list.

Specifically, if the userid of the process is the same as the userid of the file, then the access permissions for u apply. Otherwise, if the groupid of the file matches a groupid of the process, then the g permissions apply. Otherwise, the o settings apply.

Set-userid Mode

To understand the function of the *set-userid* mode, first consider an interesting problem created by controlled access to files. To illustrate, suppose you want to send a piece of electronic mail to another user on the system. To do so, you can use the **mail** command. Your message will be put in a mailbox file that belongs to another user in the mail spool directory. However, the other person's mailbox file is protected against your read or write access. The question is how can you be permitted access to the mailbox through **mail**, but not through **vi**. The answer is in the set-userid bit.

If the set-userid bit is turned on for an executable file, then it effectively assumes the userid of the file owner when it executes. This means that when a user executes a set-userid program, the process is granted the access privileges of the owner of the executable file while running the particular program. The **mail** program is owned by the super user **root** and has its set-userid bit turned on. When a user sends mail by invoking the mail program, the user's process assumes the effective userid **root** for the duration of the **mail** program's execution. This configuration allows you access to another user's mailbox file through **mail**. The *set-groupid* bit works in exactly the same way on the groupid of a process.

The sticky bit, used on older systems to make certain programs load faster, is largely obsolete. In some Linux systems, this bit becomes the *restricted deletion flag* for directories. When set, it prevents a unprivileged user from removing or renaming a file in the directory unless the user is the owner of the directory or the file. In an **ls** listing, a t (T) in the 10th permission position means the sticky bit is on and x for o is on (off).

Establishing a Group

As an example application of the file access control facilities, let's consider establishing a group whose members can collaborate on a project by accessing selected files of one another. To establish the group, you first decide on a name. In this example, the groupid is `projectx`. Next, you must decide who will be members of the group. In this example, the group members are `pwang`, `rsmith`, `jdoe`, `sldog`, and yourself. Now ask your system administrator to create group `projectx`. A system administrator can either edit `/etc/group` directly or use a command such as **groupadd** or **system-config-users** to set up a new group. As soon as this is done, `projectx` exists on your system as a valid group. Once `projectx` is established, members can assign desired access permissions to selected files to allow sharing within the group. One simple way for you to do this is as follows:

1. Establish a directory, `alpha`, say, under your home directory. All files in `alpha` are to be shared with others in `projectx`.

2. Change the groupid of `alpha` to `projectx` by

   chgrp `projectx alpha`

3. Now set the group access permissions for the `alpha` directory. Depending on the access you wish to give, use one of the following:

   chmod `g=rwx alpha`
   chmod `g=rx alpha`
   chmod `g=x alpha`

   The difference between these permissions is described in Section 6.4.

4. Optionally, use

   chmod `+t alpha`

   to set the restricted deletion flag for the `alpha` folder.

5. You must make sure that each file in `alpha` carries the groupid `projectx`, especially files established through **cp** or **mv**. As mentioned, the groupid of a file is displayed with **ls** `-gl`. Depending on the nature of a file, you should assign appropriate group permissions. Give the group write permission only if you allow others in `projectx` to modify a file.

6.5   File System Implementation

As stated earlier in this chapter, a file system is a logical organization imposed on physical data storage media (usually hard disks) by the operating system. This organization, together with the routines supplied by the operating system, allows for systematic storage, retrieval, and modification of files.

Typically for Linux, the entire file system consists of one or more *filesystems*. Each *filesystem* is a self-contained unit consisting of a group of data blocks on a particular hard disk. A file can be viewed as a one-dimensional array of bytes. These bytes are stored in a number of data blocks from a given filesystem.

Modern disk drives offer sizable storage for data. Typical data block sizes are 1024, 2048, and 4096 bytes. A filesystem can gain speed by employing a larger block size. The block size is determined at filesystem creation time.

For each filesystem, the addresses (locations) of the data blocks, the status, and perhaps also the attribute information of a file are stored in a data structure known as the i-node (index node). All the i-nodes of a filesystem are stored in a linear list called the i-list (or i-table), which is stored at a known address on the physical storage medium. I-node and i-list were mentioned in Section 6.4.

The i-node (Figure 6.2) stores meta information for a file including file length (in bytes), device, owner, and group IDs, file mode, and timestamps. The i-node also contains pointers (addresses) to the file's data blocks. For example, an ext2 filesystem allows 12 direct pointers, a single-indirect pointer, a double-indirect pointer, and a triple-indirect pointer.

FIGURE 6.2: The i-Node

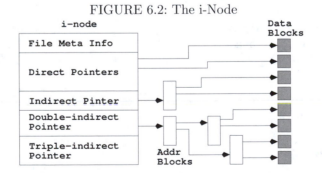

A direct pointer is the address of a block storing the content data of the file. An indirect pointer points to a block of direct pointers. A double indirect pointer points to a block of indirect pointers. A triple indirect pointer points to a block of double indirect pointers. With this arrangement, very large files can be accommodated.

The i-node contains all the vital meta information of a file. Therefore, the implementation of a filesystem centers around access to the i-node. The i-number in a directory entry is used to index the i-list and access the i-node of the file. Thus, a file pathname leads, through a sequence of i-nodes, to the i-node of the file. Figure 6.3 shows how the pathname /bin/ls leads from the root directory / to the file ls through a sequence of i-nodes and directory entries.

FIGURE 6.3: File Address Mapping

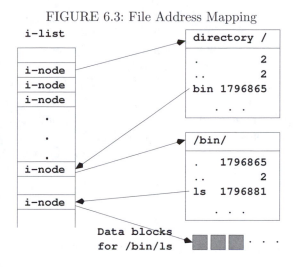

A hard link to a file can be seen as simply another directory entry containing the same i-number. Once the i-node of a file is located, it is read into primary memory and kept on the *active i-node table* until access to the file is closed. The i-list also contains *free* i-nodes that are used to create new files.

Mounted Filesystems

In Linux, a *filesystem* refers to the logical storage device represented by a single i-list. The complete Linux file system may contain one or more filesystems. One of these is the *root filesystem*; the others are *mounted filesystems*. The location of the i-list of the root filesystem is always known to the operating system. A mounted filesystem is attached (mounted) to the root filesystem at any directory in the root filesystem. A mounted filesystem can be removed by unmounting it with the **umount** command.

A super user may use the command

mount [-r] *devfile directory*

to mount the filesystem stored on the block special file *devfile* at the the given *directory*, which is usually an empty directory created for this purpose. This directory is called the *root directory* of the mounted filesystem. If the option -r is given, the filesystem is mounted as read-only. The **mount** command without any arguments displays the names of all mounted filesystems and the points on the file tree where they are mounted. The command **df** displays file system space usage and the free disk spaces on all the filesystems. Here is a typical **df** display.

```
df -h
```

```
Filesystem           Size  Used Avail Use% Mounted on
/dev/sda6            140G   18G  115G  14% /
/dev/sda3            99M   20M   75M  21% /boot
tmpfs               376M   68K  376M   1% /dev/shm
/dev/sda2           146G   32G  115G  22% /media/ACER
```

showing a Linux/Windows® dual-boot computer with four filesystems.

The `/media/ACER` is the mount point of an NTFS (NT Filesystem) for the Windows® side. Most Linux systems have built-in support for NTFS so files and folders in an NTFS partition are usable from either Linux or Windows®. This can be very convenient. Do a **man -k ntfs** to see Linux support for NTFS on your system.

Filesystem Super Block and Block Groups

A Linux ext2 filesystem consists of a number of *block groups*. Each block group also contains a duplicate copy of crucial filesystem control information (super block and group descriptors) in addition to the block group's own block bitmap, i-node bitmap, i-list, and, of course, data blocks.

The *super block* defines a filesystem. It records vital information about the configuration, organization, and operations of a filesystem:

- The filesystem type and a block device reference

- The overall size and block size of the filesystem

- The length of the i-node list

- Free blocks and free i-nodes

- Read, write, and other methods for i-nodes

The group descriptor stores the location of the block bitmap, i-node bitmap and the start of the i-node table for every block group; and these, in turn, are stored in a group descriptor table. The super block and the group descriptor table are critical for a filesystem, and they are stored at the beginning of each block group to provide redundancy.

6.6 The Filesystem Table

Each different filesystem on Linux has its own block-type special file. The names of the these special files, together with other information for control and management of the entire file system, are kept in the *filesystem table* (typically, `/etc/fstab`). This file contains one line for each filesystem specifying the block special filename, the directory name where mounted, the filesytem type (local, NFS,[1] or for memory swapping), mount/swap options, and other information.

Of all the filesystems contained in the filesystem table, all or a subset may be mounted at any given time. The *mount table* (`/etc/mtab`) contains a list of currently mounted filesystems. The mount table is modified by the commands **mount** and **umount**.

6.7 File Storage Quotas

The file quota mechanism is designed to allow restrictions on disk space usage for individual users and/or groups. A separate quota can be set for each user/group on each filesystem. Quotas can be enforced on some filesystems and not on others. For example, in a computer science department, one filesystem for students may have quota enforced; at the same time, another filesystem for professors may have no quota enforced. The quota specifies limits on the number of files and disk blocks a user may occupy. There are two kinds of limits: *soft* limits and *hard* limits. If a user-initiated operation causes the soft limit to be exceeded, a warning appears on the user's terminal. The offending operation is allowed to continue if the hard limit is not exceeded. The idea is to encourage users to stay within their soft limits between login sessions. In other words, exceeding the soft limit temporarily is all right, as long as the user releases file space and returns within the soft limit before logout. At login time, a warning is provided if any soft limits still are violated. After a few such warnings, the user's soft limits can be enforced as hard limits.

The filesystem table indicates which filesystems need to support quotas. The quotas for users and groups are kept in files (`aquota.user` and `aquota.group`, for example) located in the root directory of the filesystem. For a mounted filesystem, its root directory is its mount point on the file tree. The command **edquota** is used to set and change quotas. Only a super user can invoke **edquota**. The command **quota** displays your disk usage and your quota limits. A super user can give this command an optional *userid* to display the information of a specific user. A super user also can turn on and off quota enforcing for an entire filesystem using the commands

quotaon *filesys* ...
quotaoff *filesys* ...

---

[1] See Section 6.9.

6.8   Creating Special Files

As previously mentioned, the Linux system uses special files to represent physical and logical I/O devices, and achieves uniform file I/O and device I/O. Special files normally are created exclusively under the system directory /dev. The command

mknod *filename* [b or c] *major minor*

is used to establish a special file by the given filename. The character b is used if the device is a block I/O device (hard disk). The character c is used for a character I/O device such as a terminal emulator or a printer. Each physical I/O device on Linux is assigned a major device number according to the type of device it is and a minor device number indicating the unit number within the same type of devices. These numbers are integers. For example, the two printers lp0 and lp1

```
crw-rw----  1 root lp      6,   0 2009-03-06 11:48 lp0
crw-rw----  1 root lp      6,   1 2009-03-06 11:48 lp1
```

have major device number 6 and minor device numbers 0 and 1, respectively. Only a super user can create special files.

6.9   Network Filesystem

Many Linux systems allow file operations not only on local filesystems stored on the host computer, but also on *remote filesystems* stored on other computers connected by a network. The *Network Filesystem* (NFS) allows *transparent* access to remote files. In other words, there is no difference between user requests for operations on remote and local files. NFS brings many advantages to file organization for businesses and organizations. For example, duplicate storage of the same files on different hosts can be avoided by centralizing them on *file server* machines accessible via NFS.

To make things even more convenient, NFS can work with different hardware and operating systems. A filesystem on a local host is made remotely accessible by *exporting* it. The file /etc/exports specifies local filesystems that can be exported and any restrictions on each filesystem. The command **exportfs** must be run after modifying /etc/exports.

The file /var/lib/nfs/etab (or xtab) lists the filesytems currently being exported. A filesystem can be exported to a list of allowed clients or to all and can allow read-only or read-write access.

A client host makes a remote filesystem accessible by the **mount** command

mount *remote-filesystem local-directory*

which mounts a remote filesystem, specified by *host*:*directory*, onto a local directory of choice.

On most Linux systems, even the mounting and unmounting of remote filesystems are automated through the *autofs* mechanism assisted directly by the Linux kernel. The kernel calls the *automount program* to mount a remote filesystem when an actual file access to its mount point occurs. Automounted filesystems are dismounted after a time period with no access.

6.10   Searching the File Tree: `find`

We know the Linux file system is organized into a tree structure. It is sometimes necessary to search a part of this tree and visit all nodes in a subtree. This means visiting all files in a given directory and, recursively, all files contained in subdirectories of the given directory. The **find** command provides just such a tree searching mechanism.

The **find** command visits all files in a subtree and selects files based on given *Boolean expressions*. The selection feature allows us to find the desired files and apply operations to them. Any file in the subtree for which the given Boolean expressions evaluate to `true` will be selected.

The **find** command can be used to locate (display the pathname of) files whose names match a given pattern in the subtree. For example,

find . -name *.c -print

In this example, the **find** command is given two Boolean expressions, -name *.c and -print. The command searches the subtree rooted at the current directory visiting each file. The file that currently is being visited is referred to as the *current file*. If the name of the current file matches the pattern *.c (the filename ends in .c), then the next expression (-print) is evaluated. The -print expression simply displays the pathname of the current file on the standard output. Thus, the effect of the preceding example is to find all C source files under the current directory and display their pathnames.

The general form of the **find** command is

find *filename* ... *expression* ...

The command name is followed by one or more filenames, each either an ordinary file or a directory, and then by one or more expressions. The tree search is conducted on each file and directory given. Each expression is a predicate on the current file and always produces a true/false value, although the expression also may have other effects. An expression is evaluated only if all preceding expressions are true. In other words, expression evaluation for the current file terminates on the first false expression, and the search process then goes on to the next file in the subtree.

The expressions used in **find** are *primary expressions* or a Boolean combination of primary expressions. Some important primary expressions are explained here. The effect and the Boolean value of each also is described. (Since some expressions may involve concepts and features we have not covered yet, you may skip those expressions for now if you wish.) In the descriptions, the argument n is used as a decimal integer that can be specified in one of three ways: an integer, an integer preceded by +, or an integer preceded by −. Specifying $+n$ means more than n, $-n$ means less than n, and n means exactly n.

-atime n	True if the file has last been accessed in n days
-name *pattern*	True if the name of the current file matches the given Glob *pattern* (for example, -name 'chapter.*')
-newer *file*	True if the current file has been modified more recently than the given *file*
-print	Always true; causes the pathname of the current file to be displayed
-exec *cmd args* ;	Executes the given Shell command *cmd* and returns value true if *cmd* returns a zero exit status, (for *cmd*, an argument {} stands for the current file and the last argument must be ;)
-type t	True if the type of the file is t, where t can be b (block special), c (character special), d (director), f (regular file), l (symbolic link), p (named pipe), and s (socket)
-user *userid*	True if the file belongs to the user *userid* (login name or numeric userid)

The following Boolean operations (in order of decreasing precedence) can be used to combine any valid expressions *e1* and *e2*.

(*e1*)	True if *e1* is true (the parentheses enforce precedence)
! *e1*	True if *e1* is not true (the *not* operation)
e1 e2	True if *e1* and *e2* are both true (the *and* operation)
e1 -o *e2*	True if at least one of *e1* and *e2* is true (the Boolean *or* operation)

Here are some additional examples. To remove all files, under your home directory, named either **a.out** or *.o that have not been accessed for at least four weeks, type in (**Ex: ex06/findrm**)

find ~ \(-name a.out -o -name '*.o' \)   -atime +28 \
        -exec **rm** '{}' \;

You can avoid the line continuation by entering everything on one command line.

Consider another example. To display the names of all files not owned by smith under the current directory, type in

find . \! -user smith -print

Note that many characters used in these examples have been quoted to avoid Shell interpretation.

Now, for a third example (**Ex: ex06/findstr**), suppose you have several HTML files under your personal Web space (`$HOME/public_html`) that contain the word `Linux`, but you are not sure exactly which files. You can use **find** to apply **fgrep** to each HTML file.

```
find public_html -name '*.html' -exec fgrep Linux \{\} \; -print
```

6.11   The `locate` Command

While **find** is nice and powerful, the **locate** command can be easier to use and faster. You give **locate** a Glob pattern or a regular expression and it can display all pathnames, in the file tree, that contain a node whose name matches. For example,

locate gnome	(pathname containing gnome)
locate -b \gnome	(base filename exactly gnome)
locate --regex \.html$	(filename ending in .html)

The **locate** command runs faster because it searches a database of files and folders on your system called an *udpatedb* which is regularly updated automatically daily.

6.12   Saving, Compressing, and Distributing Files

Sometimes the need arises to pack a number of files into a neat package and send them to another computer. The **tar** command is used to collect a set of files onto a single file, called a *tar file* (the name came from *tape archive*). The **tar** command copies entire *directory hierarchies*. A directory hierarchy refers to all files and directories contained in a subtree of the file tree. It works by packing multiple files into a single file in the *tar format* which can later be unpacked by **tar** preserving the original file and folder structure. The tar file can be saved as a backup or transferred easily by email or ftp (Chapter 7, Section 7.6). The **tar** command is often used together with common file compression schemes such as **gzip** (GNU Zip) and **bzip2**. The latter generally provides better compression.

Let's first look at the simplest uses of **tar**.

tar cvf *tarfile.*tar *name1 name2 ...*	(A)
tar zcvf *tarfile.*tgz *name1 name2 ...*	(B)
tar jcvf *tarfile.*tbz *name1 name2 ...*	(C)

saves the named file hierarchies to the given *tarfile* with no compression (A), gzip compression (B), or bzip2 compression (C). The options are c (create tarfile), v (verbose), f (tarfile name follows), z (use gzip), and j (use bzip2).

The corresponding commands

tar xvf *tarfile*.tar
tar zxvf *tarfile*.tgz
tar jxvf *tarfile*.tbz

extract the files contained in *tarfile*. If you wish to preserve the file permissions and other attributes, use the p option when packing and unpacking with **tar**. Many software packages in tar format are available for download to your Linux system.

The ZIP utility commonly used on Windows platforms is also available on Linux. The **zip** and **unzip** commands make it convenient to exchange archive files with other platforms.

zip -r archive.zip *name1 name2 ...*

packs files and folders into the given archive, while **unzip** unpacks.

When providing an archive file for downloading, it is good practice to also provide a finger print file to check the integrity of the download. Creating an MD5 (*Message-Digest algorithm 5*) finger print for your archive file is simple. The command

md5sum *archivefile* > *archivefile*.md5

places the name of the archive file and its MD5 finger print in the finger print file *archivefile*.md5.

Packing Files with shar

The **tar** is the regular command for saving and retrieving files because it restores all file attributes such as ownership and access protection modes. The **shar** command is another way to pack multiple files into one which does not worry about retaining file attribute information, and it can be easier to use.

Basically, **shar** packs the files into a single file of *sh* commands. The packed file is unpacked by letting **sh** process the file.

The command

shar *file1 file2 ...* > *outfile*.sh

packs the named files (including directories) into one file and sends that to standard output. The resulting *outfile*.sh sent by email or uploaded to another Linux/UNIX computer.

To unpack simply do

sh < *outfile*.sh

6.13   More File-Related Commands

Some additional commands that are useful in dealing with files and managing the filesystem are listed here. The function of each command is indicated, but no full explanations are given. For more detailed information and options on these commands, refer to the respective manual pages.

- **basename** removes prefixes and suffixes from a filename.

- **cmp** compares two files to see if they are identical.

- **comm** selects or rejects lines common to two sorted files.

- **df** displays disk space free on all filesystems.

- **diff** compares two files or directories and outputs the differences.

- **du** displays all file sizes in kilobytes in a directory hierarchy.

- **size** displays the size of an object file.

- **split** splits a file into pieces.

- **touch** updates the last modified time of a file; if a file does not exist, it creates an empty one.

- **uniq** reports repeated lines in a file.

- **wc** counts the number of words, lines in given files.

6.14   For More Information

For the File System Standard (FSSTD), see the *Linux Journal* article by Daniel Quinlan available on the Web from ACM:

`portal.acm.org/citation.cfm?id=324517`

For complete information on the Linux file hierarchy, see the Linux Documentation Project online article:

`tldp.org/LDP/Linux-Filesystem-Hierarchy/html`

For more details on filesystem internals and implementations, refer to *Design and Implementation of the Second Extended Filesystem* and to *Linux NFS-HOWTO* at Source Forge `SourceForge.net`.

For NFS, see the following Redhat document:

`www.redhat.com/docs/manuals/linux/RHL-9-Manual/ref-guide/ch-nfs.html`

6.15 Summary

The file system is central to any operating system and is part of the Linux kernel. The Linux file system hierarchy contains files and directories arranged in a tree structure that grows down from the root directory /. The Linux file hierarchy largely follows the FSSTND.

Different file types are directories, special files, links, regular files, sockets, and named pipes. There are two kinds of links: hard links and symbolic links. A symbolic link can link to a directory and can span filesystems. Access to files and directories is governed by `rwx` permissions for the file owner (`u`), for users in the file group (`g`), and for others (`o`).

The set-userid bit for executable files is an important concept. When a process executes a set-userid file, its effective userid becomes that of the file owner.

The entire file system consists of a root filesystem and possibly additional mountable filesystems. Linux supports different filesystem implementations, including `ext2` and its extensions. Each filesystem is organized by an i-list, which is a list of i-nodes that contains status and address information for each file and all free space in the filesystem. File status information includes userid, access groupid, mode, timestamps, and disk addresses. Part of the file mode specifies file access permissions.

The NFS allows transparent access to remote (NFS) and local filesystems, making it easy to share files across a network.

To do a systematic search through a file hierarchy, use the **find** command. To quickly locate files/folders based on their names, use the **locate** command. Use the simple **shar** command or the more efficient **tar** command (with **gzip** or **bzip2** file compression) to pack and compress multiple files into an archive for easy transport. Use **zip** to manage archive files across different computer systems.

6.16 Exercises

1. Try the **umask** command. What does it tell you about the files you create? Try setting the `umask` value and then creating some files. Look at their protection bits.

2. If you have not done it yet, download the most recent HTML version of the Linux man pages from `www.tldp.org/manpages/man-html/` to your computer. Unpack it so that you can use it with your Web browser.

3. How many mountable filesystems are there in the system you use? How many are mounted at this time? How much free space is there in the file structure where most of your files are stored? How does the filesystem table file correspond to the actual files mounted on the system?

4. The term *filesystem* is different from the phrase "file system." Can you clearly specify their meaning?

5. Why is a hard link indistinguishable from the original file itself? What happens if you **rm** a hard link? Why is it not possible to have a hard link to a file in a different filesystem?

6. Clearly state the meaning of the **rwx** permissions for a directory. What would happen if you perform **ls** *dir* with read permission to *dir* but no execute permission? Why?

7. Write a Shell script **forweb** which takes the name of a folder *fname* and makes all files **o+r** and all folders **o+rx** in the file hierarchy rooted at **fname**.

8. How would you go about figuring out the size of the largest file a Linux file system can accommodate with its i-node structure?

9. What command displays the i-number of a file/directory?

10. It is clear how commands **rm** and **ls** work on ordinary files. Describe how they work on symbolic links. Must a symbolic link point to an existing file? What happens if the actual file of a symbolic link is deleted? Is it possible for a symbolic link to point to another symbolic link?

11. Consider the . and .. special files. Is it correct to say that these files are system-created hard links to directories?

12. Consider the Bash script **clean.sh** (Chapter 5, Section 5.20). Does the script still work correctly if there are symbolic links in the directory it is trying to clean? If there is a problem, how would you fix it?

13. Try to **rm** a file to which you have no write permission. What message does **rm** give? How did you respond? Were you able to delete the file? Why?

14. When an executable file is invoked, does the new process always assume the userid of its invoker? Explain.

15. You are looking for a file somewhere under your home directory that contains the string **zipcode** in it. Describe how you can locate the file if you do/don't know which directory contains the file. What if the file may be a hidden file whose name begins with a dot.

16. How exactly does one create a .tgz file? How does one extract from a .tgz file? What about .tbz files?

17. Compare the pros and cons of the three file compression schemes: ZIP, gzip, and bzip2.

Chapter 7

Networking, Internet, and the Web

Early packet-switched computer networking, involving a few research institutions and government agencies, started in the late 1960s and early 1970s. Today, it is hard to tell where the computer ends and the network begins. The view "The Network is the Computer" is more valid than ever. Most people cannot tolerate even a few minutes of Internet connection outage.

A *computer network* is a high-speed communications medium connecting many, possibly dissimilar, computers or *hosts*. A network is a combination of computer and telecommunication hardware and software. The purpose is to provide fast and reliable information exchange among the hosts. Typical services made possible by a network include

- Electronic mail

- On-line chatting and Internet phone calls

- File transfer

- Remote login

- Distributed databases

- Networked file systems

- Audio and video streaming

- Voice and telephone over a network

- World Wide Web, E-business, E-commerce, and social networks

- Remote procedure and object access

In addition to host computers, the network itself may involve dedicated computers that perform network functions: hubs, switches, bridges, routers, and gateways. A network extends greatly the powers of the connected hosts.

A good understanding of basic networking concepts, commands, information security, and how the Web works will be important for any Linux user/programmer.

7.1   Networking Protocols

For programs and computers from different vendors, under different operating systems, to communicate on a network, a detailed set of rules and conventions must be established for all parties to follow. Such rules are known as *networking protocols*. We use different networking services for different purposes; therefore, each network service follows its own specific protocols. Protocols govern such details as

- Address format of hosts and processes

- Data format

- Manner of data transmission

- Sequencing and addressing of messages

- Initiating and terminating connections

- Establishing services

- Accessing services

- Data integrity, privacy, and security

Thus, for a process on one host to communicate with another process on a different host, both processes must follow the same protocol. The *Open System Interconnect* (OSI) *Reference Model* (Figure 7.1) provides a standard layered view of networking protocols and their interdependence. The corresponding layers on different hosts, and inside the network infrastructure, perform complementary tasks to make the connection between the communicating processes (P1 and P2 in Figure 7.1).

FIGURE 7.1: Networking Layers

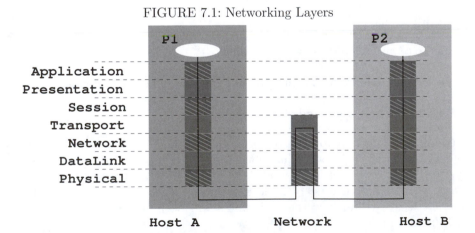

Among common networking protocols, the Internet Protocol Suite is the most widely used. The basic IP (*Internet Protocol*) is a *network layer* protocol. The TCP (*Transport Control Protocol*) and UDP (*User Datagram Protocol*) are at the *transport layer*. The Web is a service that uses an *application layer* protocol known as HTTP (the *Hypertext Transfer Protocol*).

Networking protocols are no mystery. Think about the protocol for making a telephone call. You (a client process) must pick up the phone, listen for the dial tone, dial a valid telephone number, and wait for the other side (the server process) to pick up the phone. Then you must say "hello," identify yourself, and so on. This is a protocol from which you cannot deviate if you want the call to be made successfully through the telephone network, and it is clear why such a protocol is needed. The same is true of a computer program attempting to talk to another computer program through a computer network. The design of efficient and effective networking protocols for different network services is an important area in computer science.

Chances are your Linux system is on a *Local Area Network* (LAN) which is connected to the Internet. This means you have the ability to reach, almost instantaneously, across great distances to obtain information, exchange messages, upload/download files, interact with others, do literature searches, and much more without leaving the seat in front of your workstation. If your computer is not directly connected to a network but has a telephone or cable modem, then you can reach the Internet through an Internet service provider (ISP).

7.2   The Internet

The Internet is a global network that connects computer networks using the *Internet Protocol* (IP). The linking of computer networks is called *internetworking*, hence the name Internet. The Internet links all kinds of organizations around the world: universities, government offices, corporations, libraries, supercomputer centers, research labs, and individual homes. The number of connections on the Internet is large and growing rapidly.

The Internet evolved from the ARPANET,[1] a U.S. Department of Defense Advanced Research Projects Agency (DARPA) sponsored network that developed the IP as well as the higher level *Transmission Control Protocol* (TCP) and *User Datagram Protocol* (UDP) networking protocols. The architecture and protocol were designed to support a reliable and flexible network that could endure wartime attacks.

The transition of ARPANET to the Internet took place in the late 1980s as NSFnet, the U.S. National Science Foundation's network of universities and supercomputing centers, helped create an explosive number of IP-based local

---

[1]The ARPANET was started in the late 1960s as an experimental facility for reliable military networking.

and regional networks and connections. The Internet is so dominant now that it has virtually eliminated all historical rivals such as BITNET and DECnet.

The *Internet Corporation for Assigned Names and Numbers* (ICANN; www.icann.org) is a nonprofit organization responsible for IP address space allocation, protocol parameter assignment, domain name system management, and maintaining root server system functions.

Network Addresses

An address to a host computer is like a phone number to a telephone. Every host on the Internet has its own network address that identifies the host for communication purposes. The addressing technique is an important part of a network and its protocol. An Internet address (IP address) is represented by 4 bytes in a 32-bit quantity. For example, monkey, a host at Kent State, has the IP address 131.123.41.83 (Figure 7.2). This *dot notation* (or *quad*

FIGURE 7.2: IP Address

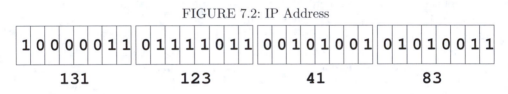

notation) gives the decimal value (0 to 255) of each byte.[2] The IP address is similar to a telephone number in another way: the leading digits are like area codes, and the trailing digits are like local numbers.

Because of their numerical nature, the dot notation is easy on machines but hard on users. Therefore, each host may also have a *domain name* composed of words, rather like a postal address. For example, the domain name for monkey is monkey.cs.kent.edu (at the Department of Computer Science, Kent State University). The Linux command **host** displays the IP and domain name of any given host. For example,

host monkey.cs.kent.edu

displays

```
monkey.cs.kent.edu is an alias for monkey.zodiac.cs.kent.edu.
monkey.zodiac.cs.kent.edu has address 131.123.41.83
```

With domain names, the entire Internet name space for hosts is recursively divided into disjoint domains in a hierarchical tree (Figure 7.3). The address for monkey puts it in the cs local domain, within the kent subdomain, which is under the edu *top-level domain* (TLD) for U.S. educational institutions.

---

[2]To accommodate the explosive growth, the Internet is moving to IPv6, which supports 128-bit addresses.

FIGURE 7.3: The Domain Name Hierarchy

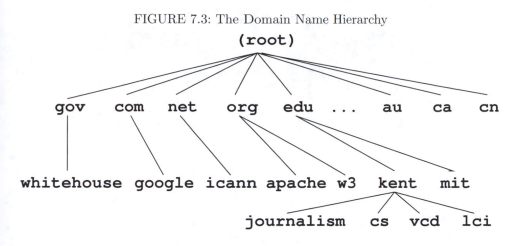

Other TLDs include `org` (nonprofit organizations), `gov` (U.S. government offices), `mil` (U.S. military installations), `com` (commercial outfits), `net` (network service providers), `uk` (United Kingdom), `cn` (China), and so forth. Within a local domain (for example, `cs.kent.edu`), you can refer to machines by their hostname alone (for example, `monkey`, `dragon`, `tiger`), but the full address must be used for machines outside. Further information on Internet domain names can be found in Section 7.16.

The ICANN accredits *domain name registrars*, which register domain names for clients so they stay distinct. All network applications accept a host address given either as a domain name or as an IP address. In fact, a domain name is first translated to a numerical IP address before being used.

Packet Switching

Data on the Internet are sent and received in *packets*. A packet envelops transmitted data with address information so the data can be routed through intermediate computers on the network. Because there are multiple routes from the source to the destination host, the Internet is very reliable and can operate even if parts of the network are down.

Client and Server

Most commonly, a network application involves a server and a client (Figure 7.4).

- A *server* process provides a specific service on a host machine that offers such a service. Example services are email (`SMTP`), secure remote host access (`SSH`), secure file transfer (`SFTP`), and the World Wide Web (`HTTP`). Each *Internet standard service* has its own unique *port number* that is identical on all hosts. The port number together with the Internet

FIGURE 7.4: Client and Server

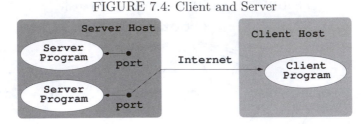

address of a host identifies a particular server program (Figure 7.4) anywhere on the network. For example, SFTP has port number 115, SSH has 22, and HTTP has 80. On your Linux system, the file /etc/services lists the standard and additional network services, indicating their protocols and port numbers.

- A *client* process on a host connects with a server on another host to obtain its service. Thus, a client program is the agent through which a particular network service can be obtained. Different agents are usually required for different services.

A Web browser such as Firefox is an HTTP client. It runs on your computer to access Web servers on any Internet hosts. The Linux **wget** command is another useful client that can download files from the Internet using the HTTP or the FTP protocol.

7.3   The Domain Name System

As stated in Section 7.2, every host on the Internet has a unique IP address and a domain name. The *network name space*, the set of all domain names with their associated IP addresses, changes dynamically with time due to the addition and deletion of hosts, regrouping of local work groups, reconfiguration of subparts of the network, maintenance of systems and networks, and so on. Thus, new domain names, new IP addresses, and new domain-to-IP associations can be introduced in the name space at any time without central control. The *domain name system* (DNS) is a network service that supports dynamic update and retrieval of information contained in the distributed name space (Figure 7.5). A network client program (for example, the Firefox browser) will normally use the DNS to obtain IP address information

FIGURE 7.5: Domain to IP

for a target host before making contact with a server. The dynamic DNS also supplies a general mechanism for retrieving many kinds of information about hosts and individual users.

Here are points to note about the DNS name space:

- The DNS organizes the entire Internet name space into a big tree structure. Each node of the tree represents a *domain* and has a label and a list of resources.

- Labels are character strings (currently not case sensitive), and sibling labels must be distinct. The root is labeled by the empty string. Immediately below the root are the TLDs: `edu`, `com`, `gov`, `net`, `org`, `info`, and so on. TLDs also include country names such as `at` (Austria), `ca` (Canada), and `cn` (China). Under `edu`, for example, there are subdomains `berkeley`, `kent`, `mit`, `uiuc`, and so on (Figure 7.3).

- A full domain name of a node is a dot-separated list of labels leading from the node to the root (for example, `cs.kent.edu.`).

- A relative domain name is a prefix of a full domain name, indicating a node relative to a domain of origin. Thus, `cs.kent.edu` is actually a name relative to the root.

- A label is the formal or canonical name of a domain. Alternative names, called *aliases*, are also allowed. For example, the main Web server host `info` has the alias `www`, so it is also known as `www.cs.kent.edu`. To move the Web server to a different host, a local system manager reassigns the alias to another host.

See Section 7.16 for more information on the DNS and name servers.

7.4   Networking in Nautilus

We first introduced the GNOME Nautilus file manager in Chapter 3, Section 3.5. By the command

nautilus network:///

or by simply clicking on the `Networking` icon (or the `Go->Networking` menu option), you can bring up a list of all systems on your network and access files on them. Linux systems are listed individually. Systems running other operating systems are grouped under different icons such as the `Windows Network` icon. Of course, you can browse only machines with permission. Normally, login will be required unless you have arranged a no-password login (Section 7.6).

For example, these `Locations` work:

- `sftp://pwang@monkey.cs.kent.edu`—Secure FTP, home directory of `pwang` on `monkey.cs.kent.edu`)

- `ssh://pwang@monkey.cs.kent.edu`—Secure FTP, same as above

- `sftp://pwang@monkey.cs.kent.edu/Pictures`—Secure FTP, `Pictures` folder of `pwang` (Figure 7.6)

- `ftp://pwang@monkey.cs.kent.edu`—Regular FTP

FIGURE 7.6: SFTP via Nautilus

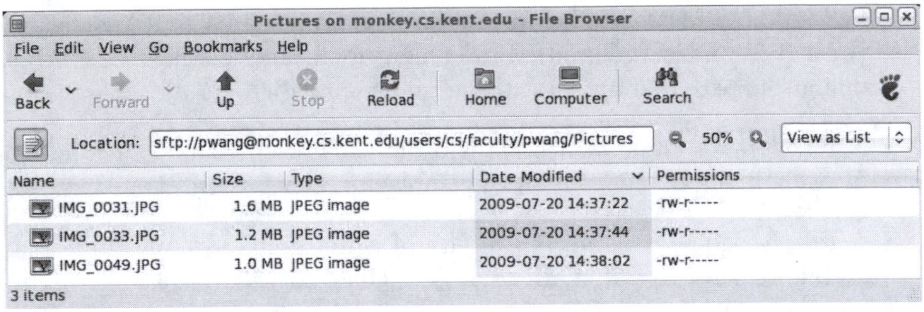

Accessing Samba Shared Files

Usually, you'll find Linux and MS Windows® systems on the same in-house network. Nautilus makes it easy to access shared files from MS Windows®. Just enter the `Location`

smb://*host*/*share_folder*

to reach the target shared folder via the *Common Internet File System* protocol, the successor of *Server Message Block* (SMB). Linux systems use *SaMBa*, a free, open-source implementation of the CIFS file sharing protocol, to act as server and client to MS Windows® systems. Use an IP for the *host* to be sure. Here ae some `Location` examples on a home network.

```
smb://192.168.2.102/SharedDocs
smb://192.168.2.107/Public
```

7.5   Networking Commands

Linux offers many networking commands. Some common ones are described here to get you started. In earlier chapters, we mentioned briefly several networking commands. For example, we know that

hostiname

displays the domain name of the computer you are using. If given an argument, this command can also set the domain name (when run as root), but the domain name is usually only set at system boot time. To get the IP address and other key information from the DNS about your computer or another host, you can use

host $(hostname)   (for your computer)
host targetHost   (for target host)

For example, **host** google.com produces

```
google.com has address 74.125.45.100
google.com has address 74.125.67.100
google.com has address 209.85.171.100
google.com mail is handled by 10 smtp4.google.com.
google.com mail is handled by 10 smtp1.google.com.
google.com mail is handled by 10 smtp2.google.com.
google.com mail is handled by 10 smtp3.google.com.
```

For any given host, its DNS data provide IP address, canonical domain name, alias domain names, DNS server hosts, and email handling hosts. Other commands that help you access the DNS data from the command line include **nslookup** and **dig** (*DNS Information Groper*). For example,

dig monkey.cs.kent.edu

gives

```
; <<>> DiG 9.5.0-P2 <<>> monkey.cs.kent.edu

;; QUESTION SECTION:
;monkey.cs.kent.edu.      IN      A

;; ANSWER SECTION:
monkey.cs.kent.edu. 1800 IN  CNAME  monkey.zodiac.cs.kent.edu.
monkey.zodiac.cs.kent.edu. 43200 IN  A  131.123.41.83

;; AUTHORITY SECTION:
zodiac.cs.kent.edu.  300  IN   NS   ns.cs.kent.edu.
zodiac.cs.kent.edu.  300  IN   NS   ns.math.kent.edu.

;; Query time: 152 msec
;; SERVER: 192.168.2.1#53(192.168.2.1)
```

The desired information (**ANSWER** section) together with the identity of the name server (**SERVER**) that provided the data is displayed.

The command **dig** is very handy for verifying the existence of hosts and finding the IP address or domain name aliases of hosts. Once the name of a host is known, you can also test if the host is up and running, as far as networking is concerned, with the **ping** command.

ping *host*

This sends a message to the given remote host requesting it to respond with an echo if it is alive and well.

To see if any remote host is up and running, you can use **ping**, which sends an echo *Internet control message* to the remote host. If the echo comes back, you'll know that the host is up and connected to the Internet. You'll also get round-trip times and packet loss statistics. When successful, the **ping** commands continues to send echo packets. Type CTRL+C to quit.

7.6   SSH with X11 Forwarding

Networking allows you to conveniently access Linux systems remotely. Most Linux distributions come with OpenSSH installed. As mentioned in Chapter 1, Section 1.2, you can **ssh** to a remote Linux and use it from the command line. Furthermore, you can

ssh -X   *userid* @*remoteHostname*

to log in to the given remote host with *X11 forwarding/tunneling*, which allows you to start any X applications, such as **gedit** or **gnome-terminal**, on the remote host and have the graphical display appear on your local desktop.

This works if your local host is a Linux/UNIX/MacOS system. It can also work from MS Windows®. Follow these steps:

1. Obtain and install an X11 server on Windows, such as the *Xming* or the heavier duty *Cygwin*.

2. Assuming you have downloaded and installed Xming, click the Xming icon to launch the X11 server. The X11 server displays an icon on your start panel so you know it is running.

3. Set up SSH or Putty on your Windows® system:

   - Putty Settings—Go to `Connection->SSH->X11` and check the `Enable X11 forwarding` box. Also set `X display location` to `127.0.0.1:0.0`.

   - SSH Settings—Check the `Tunneling->Tunnel X11 Connections` box. Also check the `Authentication->Enable SSH2 connections` box.

4. Use either Putty or SSH to connect to a remote Linux/Unix computer. Make sure your remote account login script, such as `.bash_profile`, does not set the `DISPLAY` environment variable. It will be set for you to something like `localhost:10.0` automatically when you connect via SSH.

5. Make sure your X11 server (Xming for example) is running. Now, if you start an X application on the remote Linux system, that graphical application will then SSH tunnel to your PC and use the X11 server on your PC to display a graphical user interface (GUI). For example, you can start **gedit**, **nautilus --no-desktop**, or even **firefox**.

Note, using an application with a remote GUI can be slow due to much heavier networking load as compared a remote CLI.

No Password `ssh`, `sftp`, and `scp`

The commands **ssh**, **sftp**, and **scp** are for remote login, secure ftp, and secure remote cp, respectively. When using any of these you usually need to enter the password for the remote system interactively. When you need to perform such tasks frequently, this can be a bother. Fortunately, you can easily avoid having to enter the password. Just follow these steps.

Most Linux systems come with OpenSSH installed. This means you already have the SSH suite of commands. These enable you to securely communicate from one computer (as $user_1$ on $host_1$) to another (as $user_2$ on $host_2$). We will assume you are logged in as $user_1$ on $host_1$ (this is your local host), and you wish to arrange secure communication with your account $user_2$ on $host_2$, which we will refer to as the remote host.

SSH can use public-key encryption for data security and user authentication (Section 7.7). If you have not done it yet, the first step in arranging for password-less login is to generate your own SSH keys. Issue the command

ssh-keygen

You'll be asked for a folder to save the keys and a passphrase to access them. In this case, don't provide any input in response to these questions from **ssh-keygen**. Simply press the ENTER key in response to each question.

Key generation takes a little time. Then you'll see a message telling you that your identity (private key) is `id_rsa` and your public key is `id_rsa.pub` saved under the standard folder `.ssh.` in your home directory.

The second step is to copy your `id_rsa.pub` to your account on the desired *remote-host*. Issue the command

ssh-copy-id `-i` `~/.ssh/id_rsa.pub` *your_userid* `@` *remote-host*

to append your public SSH key to the file `~userid/.ssh/authorized_key` on the *remote-host*.

Now you are all set. You can log in to *remote-host* without entering a password.

ssh *userid@remote-host*

The same setup avoids a password when you use **sftp** or **scp**.

Remote File Synchronization

The **rsync** command makes it easy to keep files in sync between two hosts. It is very efficient because it uses a remote-update protocol to transfer just the differences between two sets of files across the network connection. No updating is performed for files with no difference. With the commands

rsync **-az** *userid@host*:*source destDir*   (remote to local sync)
rsync **-az** *source userid@host*:*destDir*   (local to remote sync)

the given *source* file/folder is used to update the same under the destination folder *destDir*. When *source* is a folder, the entire hierarchy rooted at the folder will be updated.

The **-az** option indicates the commonly used *archive mode* to preserve file types and modes and **gzip** (Chapter 6, Section 6.12) data compression to save networking bandwidth. The **rsync** tool normally uses **ssh** (Section 7.6) for secure data transfer and does not require a password if you have set up password-less SSH between the two hosts (Section 7.6). For example,

```
rsync -az pwang@monkey.cs.kent.edu:~/linux_book ~/projects/
```

updates the local folder `~/projects/linux_book` based on the remote folder `~/linux_book` by logging in as `pwang` on the remote host `monkey.cs.kent.edu`. See the **rsync** man page for complete documentation.

7.7   Public-Key Cryptography and Digital Signature

Security is a big concern when it comes to networking. From the user's viewpoint, it is important to keep data and network transport secure and private. *Public-key cryptography* is an essential part of the modern network security infrastructure to provide privacy and security for many networking applications. Before the invention of public-key cryptography, the same secret key had to be used for both encryption and decryption of a message (*symmetric-key cryptography*). Symmetric-key is fine and efficient, and remains in widespread use today. However, a secret key is hard to arrange among strangers never in communication before; for example, parties on the Internet. The public-key cryptography breakthrough solves this *key distribution* problem elegantly.

GnuPG (GNU Privacy Guard), part of OpenGP, supports *public-key cryptography*. The Linux command for *GnuPG* is **gpg** or the largely equivalent **gpg2**. With **gpg**, you can generate a public key that you share with others and a private key you keep secret. You and others can use the public key to encrypt files and messages which only you can decrypt using the private key (Figure 7.7).

FIGURE 7.7: Public-Key Cryptography

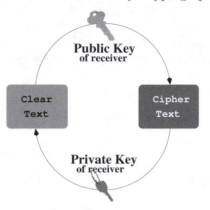

Using your private key, you can also attach a digital signature to any message/file. A receiver can verify the integrity (not altered) and authenticity (really from the sender) of the the signed message. To do all that, make sure you first set up GnuPG and your personal keys.

If your Linux distribution does not already provide **gpg**, you can easily install the **gnupg** package (Section 8.24) with either of the following commands:

sudo apt-get install gnupg   (Ubuntu/Debian)
sudo yum install gnupg   (CentOS/Fedora)

If you like to use a GUI for **gpg**, install also the **gpa** package. However, the command-line interface is entirely adequate.

Setting Up GnuPG Keys

To use **gpg**, you first need to generate your public-private key pair

```
gpg --gen-key
```

You'll be prompted to enter your choices for *keytype* (pick the default), *keysize* (pick 2048), and a *passphrase* (pick something you won't forget, but will be very hard for anyone to guess). The passphrase is required each time you access your private key, thus preventing others from using your private key.

You'll get a keyid displayed when your key pair is generated. Your keys and other info are stored by default in the folder $HOME/.gnupg. Use

```
gpg --list-public-keys
```

to display your public keys. For example,

```
pub    1024D/FCF2F84D 2009-07-25
uid    Paul Wang (monkeykia) <pwang@cs.kent.edu>
sub    1024g/B02C4B40 2009-07-25
```

The pub line says that Paul's public master key (for signature) is a 1024-bit DSA key with id FCF2F84D and that his public subkey (for data encryption) is a 1024-bit ElGama key.

To enable others to encrypt information to be delivered for your eyes only, you should send your public keys to a public key server. The command

gpg --send-keys *your_keyid*

sends your public key to a default gpg key server, such as

```
hkp://subkeys.pgp.net
```

Also, you can send your public keys to anyone by sending them an *ASCII armored* file generated by

gpg --armor --export *your_keyid* > mykey.asc

The .asc suffix simply indicates that a file is an ASCII text file. The mykey.asc contains your key encoded using *base64*, a way to use 64 ASCII characters (**AZ**, **az**, **09** and **+/**) to encode non-ASCII files for easy communication over networks, especially via email. The Linux **base64** command performs this encoding/decoding on any file. See **man base64** for more information.

Such ASCII armored key files can be emailed to others or sent to another computer and imported to another GnuPG key ring with a command such as

gpg --import mykey.asc

Also, edit your $HOME/gnupg/gpg.conf file and append the line

```
default-key   your_keyid
```

Encryption/Decryption with GnuPG

To encrypt a file using a public key of *uid*,

gpg --encrypt -r *uid*

resulting in an encrypted file *filename.*gpg that can be sent to the target user who is the only one that can decrypt it.

Even if you are not going to send a file to anyone, you can still keep secrets in that file of yours protected in case someone gains unauthorized access to your computer account. You can

gpg --encrypt -r "*your_uid*" *filename*
rm *filename*

generating the encrypted *filename*.`gpg` and removing the original *filename*.
You can easily view the encrypted version with

nano < (**gpg** –decrypt *filename*.`gpg`)

Note that the Bash process expansion (Chapter 2, Section 2.7) is handy here.

To make maintaining an encrypted file even easier, you may configure **vi/vim** to work transparently with **gpg**, allowing you to use **vim** to view and edit clear as well as **gpg** encrypted files. The VIM extension *tGpg* (yet another plug-in for encrypting files with gpg) is a good choice for this purpose.

7.8   Secure Email with Mutt and GnuPG

The Linux email client **mutt** works well with GnuPG (Section 7.7) to support *s/mime* (Secure/Multipurpose Internet Mail Extensions), allowing you to send and receive encrypted/signed email.

Assuming that you have arranged for your keys and sent your public keys to a key server as described in Section 7.7 and that your email correspondents are also set up with GnuPG or some other public-key system for their s/mime, you can easily use `mutt` to exchange emails securely with them.

Follow these steps to set up **mutt**.

1. Locate the file `gpg.rc` for **mutt** on your Linux. Usually, you'll find it at

   `/usr/share/doc/mutt-version/gpg.rc`

2. Edit your **mutt** configuration file `$HOME/.muttrc` and add at the end a line to include the `gpg.rc`

   `source /usr/share/doc/mutt-version/gpg.rc`

3. Import your secure email correspondents' keys into your GnuPG key ring. Get your email correspondent to send you an ASCII armored key file or search for the key on the key server the `--search-keys` option:

   ```
   gpg  --import someKey.asc
   gpg  --search-keys targetEmailAddress
   ```

Now, you can encrypt/sign email after composing the email message by using the P key within **mutt** to select from the following options:

```
* encrypt
* sign
* both
* sign as
```

Receiving encrypted/signed email with **mutt** is just a matter of following on-screen instructions.

The popular email client *Thunderbird* also works with GnuPG if you install the *Enigmail* extension (via the `tools->add-ons`).

7.9   Message Digests

A *message digest* is a *digital fingerprint* of a message or file. Various algorithms have been devised to take a message (file) of any length and reduce it
to a short fixed-length hash known as the *digest* of the original message or file
(Figure 7.8).

FIGURE 7.8: MD5 Message Digest

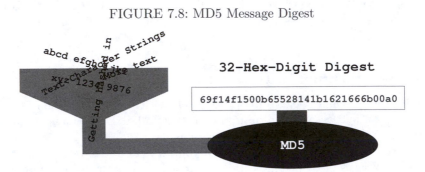

These algorithms are designed to produce a different digest if any part of
the message is altered. It is almost impossible to deduce the original message
from knowledge of the digest. However, because there are an infinite number of
possible messages but only a finite number of different digests, vastly different
messages may produce the same digest.

Message digests are therefore useful in verifying the *integrity* (unaltered-
ness) of files. When software is distributed online, a good practice is to display
a fingerprint for the file, allowing you to check the integrity of the download
and to avoid any *Trojan horse* code.

MD5 is a popular algorithm producing 128-bit message digests. An MD5
hash is usually displayed as a sequence of 32 hexadecimal digits. On Linux,
you can produce an MD5 digest with the **md5sum** command

md5sum *filename* > digestFile

You'll get a digestFile file containing only the hash and the name *filename*.
After downloading both *filename* and digestFile, a user can check file integrity with

md5sum digestFile

Other digest algorithms in wide use include SHA-1 and others. The Linux
command **sha1sum** is an alternative to **md5sum**.

Message Signing with GnuPG

To digitally sign a particular message, a message digest is created first. The
message digest is then encrypted using your private key to produce a digital

FIGURE 7.9: Digital Signature

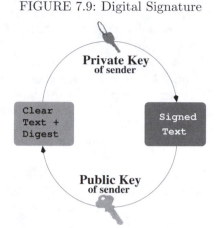

signature which is attached to the message. Any receiver of a signed message can generate a message digest from the received message and check it against the digest obtained by decrypting the digital signature with the signer's public key. A match verifies the integrity and the authenticity of the received message.

Here is how to use **gpg** for digital signature (Figure 7.9).

gpg --sign *file*   (produces signed binary *file*.gpg)
gpg --clearsign *file*   (produces signed ASCII *file*.asc)

The --decrypt option automatically verifies any attached signature.

7.10   The Web

Out of all the networking applications, the Web is perhaps one of the most important and deserves our special attention.

There is no central control or administration of the Web. Anyone can potentially put material on the Web and retrieve information from it. The Web consists of a vast collection of documents that are located on computers throughout the world. These documents are created by academic, professional, government, and commercial organizations, as well as by individuals. The documents are prepared in special formats and delivered through *Web servers*, programs that return documents in response to incoming requests. Linux systems are often used to run Web servers. An introduction to the Web is provided in this chapter. Chapter 8 discusses serving the Web.

Primarily, Web docuemnts are written in Hypertext Markup Language (HTML, Section 7.10). Each HTML document can contain (potentially many) links to other documents served by different servers in other locations and therefore become part of a *web* that spans the entire globe. New materials are

put on the Web continuously, and instant access to this collection of information can be enormously advantageous. As the Web grew, MIT (Massachusetts
Institute of Technology, Cambridge, MA) and INRIA (the French National
Institute for Research in Computer Science and Control) agreed to become
joint hosts of the *W3 Consortium*, a standards body for the Web community.

A *Web browser* is a program that helps users obtain and display information from the Web. Given the location of a target document, a browser
connects to the correct Web server and retrieves and displays the desired document. You can click *links* in a document to obtain other documents. Using a
browser, you can retrieve information provided by *Web servers* anywhere on
the Internet.

Typically, a Web browser, such as Firefox, supports the display of HTML
files and images in standard formats. Helper applications or plug-ins can aug-
ment a browser to treat pages with multimedia content such as audio, video,
animation, and mathematical formulas.

Hypertext Markup Language

A Web browser communicates with a Web server through an efficient HTTP
designed to work with hypertext and hypermedia documents that may contain
regular text, images, audio, and video. Native Web pages are written in the
HTML (Section 7.13) and usually saved in files with the `.html` (or `.htm`)
suffix.

HTML organizes Web page content (text, graphics, and other media data)
and allows *hyperlinks* to other pages anywhere on the Web. Clicking such a
link causes your Web browser to follow it and retrieve another page. The
Web employs an open addressing scheme that allows links to objects and
services provided by Web, email, file transfer, audio/video, and newsgroup
servers. Thus, the Web space is a superset of many popular Internet services.
Consequently, a Web browser provides the ability to access a wide variety of
information and services on the Internet.

URLs

The Web uses *Uniform Resource Locators* (URLs) to identify (locate) resources (files and services) available on the Internet. A URL may identify
a host, a server port, and the target file stored on that host. URLs are used,
for example, by browsers to retrieve information and by HTML to link to
other resources.

A full URL usually has the form

scheme://*server*:*port*/*pathname*

The *scheme* part indicates the information service type and therefore the protocol to use. Common schemes include `http` (Web service), `ftp` (file transfer

service), `mailto` (email service), `file` (local file system), `https` (secure Web service), and `sftp` (secure file transfer service). For example,

`sftp://pwang@monkey.cs.kent.edu/users/cs/faculty/pwang`

gets you the directory list of `/users/cs/faculty/pwang`. This works on Firefox and on the Linux file browser **nautilus**, assuming that you have set up your SSH/SFTP (Section 7.6). Many other schemes can be found at `www.w3.org/addressing/schemes`.

For URLs in general, the *server* identifies a host and a server program. The optional port number is needed only if the server does not use the default port (for example, 21 for `FTP` and 80 for `HTTP`). The remainder of the URL, when given, is a *file pathname*. If this pathname has a trailing / character, it represents a directory rather than a data file. The suffix (`.html`, `.txt`, `.jpg`, etc.) of a data file indicates the file type. The pathname can also lead to an executable program that dynamically produces an HTML or other valid file to return.

Within an HTML document, you can link to another document served by the same Web server by giving only the *pathname* part of the URL. Such URLs are *partially specified*. A partial URL with a / prefix (for example, `/file_xyz.html`) refers to a file under the *server root*, the top-level directory controlled by the Web server. A partial URL without a leading / points to a file relative to the location of the document that contains the URL in question. Thus, a simple `file_abc.html` refers to that file in the same directory as the current document. When building a website, it is advisable to use a URL relative to the current page as much as possible, making it easy to move the entire website folder to another location on the local file system or to a different server host.

Accessing Information on the Web

You can directly access any Web document, directory, or service by giving its URL in the `Location` box of a browser. When given a URL that specifies a directory, a Web server usually returns an *index file* (typically, `index.html`) for that directory. Otherwise, it may return a list of the filenames in that directory.

You can use a search engine such as *Google* to quickly look for information on the Web.

7.11   Handling Different Content Types

On the Web, files of different *media types* can be placed and retrieved. The Web server and Web browser use standard *content type* designations to indicate the media type of files in order to process them correctly.

The Web borrowed the content type designations from the Internet email system and uses the same MIME (Multipurpose Internet Mail Extensions) defined content types. There are hundreds of content types in use today. Many popular types are associated with standard file extensions. Chapter 6, Table 6.3 gives some examples.

When a Web server returns a document to a browser, the content type is indicated. The content type information allows browsers to decide how to process the incoming content. Normally, HTML, text, and images are handled by the browser directly. Others types such as audio and video are usually handled by plug-ins or external helper programs.

7.12 Putting Information on the Web

Now let's turn our attention to how information is supplied on the Web. The understanding sheds more light on how the Web works and what it takes to serve up information.

The Web puts the power of publishing in the hands of anyone with a computer connected to the Internet. All you need is to run a Web server on this machine and establish files for it to service.

Major computer vendors offer commercial Web servers with their computer systems. **Apache** is a widely used open-source Web server that is freely available from the *Apache Software Foundation* (www.apache.org).

Linux systems are especially popular as Web hosting computers because Linux is free, robust, and secure. Also, there are many useful Web-related applications such as Apache, PHP (active Web page), MySQL (database server), and more available free of charge.

FIGURE 7.10: Web Server Function

Once a Web server is up and running on your machine, all types of files can be served (Figure 7.10). On a typical Linux system, follow these simple steps to make your personal Web page.

1. Make a file directory in your home directory (~*userid*/public_html) to contain your files for the Web. This is your *personal Web directory*. Make this directory publicly accessible:

   chmod a+x ~*userid*/public_html

When in doubt, ask your system managers about the exact name to use for your personal Web directory.

2. In your Web directory, establish a home page, usually `index.html`, in HTML. The home page usually functions as an annotated table of contents. Make this file publicly readable:

chmod `a+r` ~*userid*/`public_html/index.html`

3. Place files and directories containing desired information in your personal Web directory. Make each directory and each file accessible as before. Refer to these files with links in the home page and other pages.

4. Let people know the URL of your home page, which is typically

`http://`*your-sever*/~*your-userid*/

In a Web page, you can refer to another file of yours with a simple link containing a relative URL (`<a href="`*filename*`">`), where *filename* can be either a simple name or a pathname relative to the current document.

Among the Web file formats, hypertext is critical because it provides a means for a document to link to other documents.

7.13   What Is HTML?

HTML (the Hypertext Markup Language) is used to markup the content of a Web page to provide page structure for easy handling by Web clients on the receiving end. Since HTML 4.0, the language has become standardized. XHTML (XML compatible HTML) is the current stable version. However, a new standard HTML5 is fast approaching.

A document written in HTML contains ordinary text interspersed with *markup tags* and uses the `.html` filename extension. The tags mark portions of the text as title, section header, paragraph, reference to other documents, and so on. Thus, an HTML file consists of two kinds of information: contents and HTML tags. A browser follows the HTML tags to layout the page content for display. Because of this, line breaks and extra white space between words in the content are mostly ignored. In addition to structuring and formatting contents, HTML tags can also reference graphics images, link to other documents, mark reference points, generate forms or questionnaires, and invoke certain programs. Various visual editors or *page makers* are available that provide a GUI for creating and designing HTML documents. For substantial website creation projects, it will be helpful to use *integrated development environments* such as Macromedia Dreamweaver (Chapter 11). If you don't have ready access to such tools, a regular text editor can create or edit Web pages.

TABLE 7.1: Some HTML Tags

Marked As	HTML Tags
Entire document	`<html>...</html>`
Header part of document	`<head>...</head>`
Document title	`<title>...</title>`
Document content	`<body>...</body>`
Level n heading	`<hn>...</hn>`
Paragraph	`<p>...</p>`
Unnumbered list	`<ul>...</ul>`
Numbered list	`<ol>...</ol>`
List item	`<li>...</li>`
Comment	`<!--...-->`

An HTML tag takes the form <*tag*>. A *begin tag* such as `<h1>` (level-one section header) is paired with an *end tag*, `</h1>` in this case, to mark content in between. Table 7.1 lists some frequently used tags.

The following is a sample HTML page (**Ex: ex07/Fruits**):

```
<html>
<head> <title>A Basic Web Page</title> </head>
<body>
    <h1>Big on Fruits</h1>
    <p>Fruits are good tasting and good for you ...</p>
    <p> There are many varieties, ...
    and here is a short list: </p>
    <ol>
        <li> Apples </li>
        <li> Bananas </li>
        <li> Cherries </li>
    </ol>
</body></html>
```

Figure 7.11 shows the Big on Fruits page displayed by Firefox.

7.14   Web Hosting

Web hosting is a service to store and serve ready-made files and programs so that they are accessible on the Web. Hence, publishing on the Web involves

1. Designing and constructing the pages and writing the programs for a website

2. Placing the completed site with a hosting service

FIGURE 7.11: A Sample Web Page

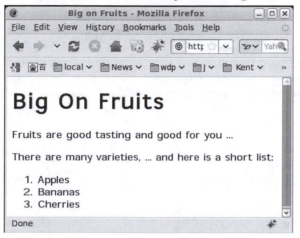

Colleges and universities host personal and educational sites for students and faculty without charge. Web hosting companies provide the service for a fee.

Commercial Web hosting can provide secure data centers (buildings), fast and reliable Internet connections, specially tuned Web hosting computers (mostly Linux boxes), server programs and utilities, network and system security, daily backup, and technical support. Each hosting account provides an amount of disk space, a monthly network traffic allowance, email accounts, Web-based site management and maintenance tools, and other access such as FTP and SSH/SFTP.

To host a site under a given domain name, a hosting service associates that domain name to an IP number assigned to the hosted site. The domain-to-IP association is made through DNS servers and Web server configurations managed by the hosting service.

7.15 Domain Registration

To obtain a domain name, you need the service of a *domain name registrar*. Most will be happy to register your new domain name for a very modest yearly fee. Once registered, the domain name is property that belongs to the *registrant*. No one else can register for that particular domain name as long as the current registrant keeps the registration in good order.

ICANN accredits commercial registrars for common TLDs, including .com, .net, .org, and .info. Additional TLDs include .biz, .pro, .aero, .name, and .museum. Restricted domains (for example, .edu, .gov, and .us) are handled by special registries (for example, net.educause.edu, nic.gov and nic.us). Country-code TLDs are normally handled by registries in their respective countries.

Accessing Domain Registration Data

The registration record of a domain name is often publicly available. The standard Internet *whois* service allows easy access to this information. On Linux systems, easy access to whois is provided by the **whois** command

whois *domain_name*

which lists the domain registration record kept at a registrar. For example,

whois kent.edu

produces the following information

```
Registrant:
   Kent State University
   500 E. Main St.
   Kent, OH 44242
   UNITED STATES

Technical Contact:
Administrative Contact:
   Bob Hart
   Mgr., Network & Telecomm
   Kent State University
   120 Library Bldg
   Kent, OH 44242
   UNITED STATES
   (330) 672-0385
   pki-admin@kent.edu

Name Servers:
   NS.NET.KENT.EDU          131.123.1.1
   DHCP.NET.KENT.EDU        131.123.252.2

Domain record activated:    19-Feb-1987
Domain record last updated: 17-Mar-2009
Domain expires:             31-Jul-2009
```

On Linux systems, the **whois** command is sometimes called **jwhois**.

7.16   The DNS

DNS provides the ever-changing domain-to-IP mapping information on the Internet. We mentioned that DNS provides a distributed database service that supports dynamic retrieval of information contained in the name space. Web browsers and other Internet client applications will normally use the DNS to obtain the IP of a target host before making contact with a server over the Internet.

There are three elements to the DNS: the DNS name space (Section 7.2), the DNS servers, and the DNS resolvers.

DNS Servers

Information in the distributed DNS is divided into *zones*, and each zone is supported by one or more name servers running on different hosts. A zone is associated with a node on the domain tree and covers all or part of the subtree at that node. A name server that has complete information for a particular zone is said to be an *authority* for that zone. Authoritative information is automatically distributed to other name servers that provide redundant service for the same zone. A server relies on lower level servers for other information within its subdomain and on external servers for other zones in the domain tree. A server associated with the root node of the domain tree is a *root server* and can lead to information anywhere in the DNS. An authoritative server uses local files to store information, to locate key servers within and without its domain, and to cache query results from other servers. A boot file, usually `/etc/named.boot`, configures a name server and its data files.

The management of each zone is also free to designate the hosts that run the name servers and to make changes in its authoritative database. For example, the host `ns.cs.kent.edu` may run a name server for the domain `cs.kent.edu`.

A name server answers queries from resolvers and provides either definitive answers or referrals to other name servers. The DNS database is set up to handle network address, mail exchange, host configuration, and other types of queries, with some to be implemented in the future.

The ICANN and others maintain *root name servers* associated with the root node of the DNS tree. In fact, the VeriSign host `a.root-servers.net` runs a root name server. Actually, the letter `a` ranges up to `m` for a total of 13 root servers currently.

Domain name registrars, corporations, organizations, Web hosting companies, and other Internet service providers (ISPs) run name servers to associate IPs to domain names in their particular zones. All name servers on the Internet cooperate to perform domain-to-IP mappings on the fly.

DNS Resolvers

A DNS resolver is a program that sends queries to name servers and obtains replies from them. On Linux systems, a resolver usually takes the form of a C library function. A resolver can access at least one name server and use that name server's information to answer a query directly or pursue the query using referrals to other name servers.

Resolvers, in the form of networking library routines, are used to translate domain names into actual IP addresses. These library routines, in turn, ask prescribed name servers to resolve the domain names. The name servers to use for any particular host are normally specified in the file `/etc/resolv.conf` or `/usr/etc/resolv.conf`.

The DNS service provides not just the IP address and domain name information for hosts on the Internet. It can provide other useful information as well. Table 7.2 shows common DNS record and request types.

TABLE 7.2: DNS Record/Request Types

Type	Description
A	Host's IP address
NS	Name servers of host or domain
CNAME	Host's canonical name, and an alias
PTR	Host's domain name, IP
HINFO	Host information
MX	Mail exchanger of host or domain
AXFR	Request for zone transfer
ANY	Request for all records

7.17   Dynamic Generation of Web Pages

Documents available on the Web are usually prepared and set in advance to supply some fixed content, either in HTML or in some other format such as plain text, PDF, or JPEG. These fixed documents are *static*. A Web server can also generate documents on the fly that bring these and other advantages:

- Customizing a document depending on when, where, who, and what program is retrieving it

- Collecting user input (with HTML forms) and providing responses to the incoming information

- Enforcing certain policies for outgoing documents

- Supplying contents such as game scores and stock quotes, which are changing by nature

Dynamic Web pages are not magic. Instead of retrieving a fixed file, a Web server calls another program to compute the document to be returned. As you may have guessed, not every program can be used by a Web server in this manner. There are two ways to add server-side programming:

- Load programs directly into the Web server to be used whenever the need arises.

- Call an external program from the server, passing arguments to it (via the program's **stdin** and environment variables) and receiving the results (via the program's **stdout**) thus generated. Such a program must conform to the Common Gateway Interface (CGI) specifications governing how the Web server and the external program interact (Figure 7.12).

FIGURE 7.12: Common Gateway Interface

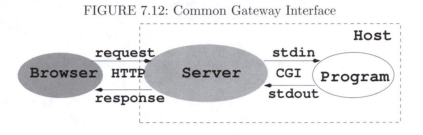

Dynamic Server Pages

The dynamic generation of pages is made simpler and more integrated with Web page design and construction by allowing a Web page to contain active parts that are treated by the Web server and transformed into desired content on the fly as the page is retrieved and returned to a client browser.

The active parts in a page are written in some kind of notation to distinguish them from the static parts of a page. The ASP (Active Server Pages), JSP (Java Server Pages), and the popular PHP (Hypertext Preprocessor; Section 8.16) are examples.

Because active pages are treated by modules loaded into the Web server, the processing is faster and more efficient compared to CGI programs. Active page technologies such as PHP also provide form processing, HTTP sessions, and easy access to databases. Therefore, they offer complete server-side support for dynamic Web pages.

Both CGI and server pages can be used to support HTML forms, the familiar fill-out forms you often see on the Web.

7.18   HTTP Briefly

On the Web, browser-server communication follows HTTP. A basic understanding of HTTP is important for Linux programmers because Linux systems are very popular Web server hosts.

The start of HTTP traces back to the beginning of the Web in the early 1990s. HTTP/1.0 was standardized early in 1996. Improvements and new features have been introduced and HTTP/1.1 is now the stable version.

Here is an overview of an HTTP transaction:

1. *Connection*—A browser (client) opens a connection to a server.

2. *Query*—The client requests a resource controlled by the server.

3. *Processing*—The server receives and processes the request.

4. *Response*—The server sends the requested resource back to the client.

5. *Termination*—The transaction is finished, and the connection is closed unless another transaction takes place immediately between the client and server.

HTTP governs the format of the query and response messages (Figure 7.13).

FIGURE 7.13: HTTP Query and Response Formats

initial line	(different for query and response)
HeaderKey1: value1	(zero or more header fields)
HeaderKey2: value2	
	(an empty line with no characters)
Optional message body contains query or response data.	
Its data type and size are given in the headers.	

The header part is textual, and each line in the header should end in RETURN and NEWLINE, but it may end in just NEWLINE.

The initial line identifies the message as a query or a response.

- A query line has three parts separated by spaces: a *query method* name, a local path of the requested resource, and an HTTP version number. For example,

```
GET    /path/to/file/index.html    HTTP/1.1
```

or

```
POST   /path/script.cgi   HTTP/1.1
```

The GET method requests the specified resource and does not allow a message body. A GET method can invoke a server-side program by specifying the CGI or active-page path, a question mark, and then a *query string*:

```
GET /cgi-bin/newaddr.cgi?name=value1&email=value2    HTTP/1.1
Host: monkey.cs.kent.edu
```

Unlike GET, the POST method allows a message body and is designed to work with HTML forms for collecting input from Web users.

- A response (or status) line also has three parts separated by spaces: an HTTP version number, a status code, and a textual description of the status. Typical status lines are

```
HTTP/1.1   200   OK
```

for a successful query or

```
HTTP/1.1   404   Not Found
```

when the requested resource cannot be found.

- The HTTP response sends the requested file together with its content type (Section 7.11) and length (optional) so the client will know how to process it.

7.19   A Real HTTP Experience

Let's manually send an HTTP request and get an HTTP response. To do that we will use the **nc** commnad. The command **nc** provides command-line (and scripting) access to the basic TCP and UDP (Section 7.2) and therefore allows you to make any TCP connections or send any UDP packets. Such abilities are usually reserved to programs at the C-language level that set up *sockets* (Chapter 11, Section 11.6) for networking.

For example, the simple Bash pipeline (**Ex: ex07/poorbr.sh**)

echo $'GET /WEB/test.html HTTP/1.0\n' |
nc monkey.cs.kent.edu 80

retrieves the Web page monkey.cs.kent.edu/WEB/test.html. In this example, we applied the Bash *string expansion* (Chapter 2, Section 2.7).

Note the HTTP get request asks for the file /WEB/test.html under the document root folder managed by the Web server on monkey. The request is terminated by an empty line, as required by the HTTP protocol.

Try this and you'll see the result display.

```
HTTP/1.1 200 OK
Date: Tue, 07 Apr 2009 19:45:03 GMT
Server: Apache/2.0.54 (Fedora)
X-Powered-By: PHP/5.0.4
```

```
Cache-Control: max-age=86400
Expires: Wed, 08 Apr 2009 19:45:03 GMT
Vary: Accept-Encoding
Content-Length: 360
Connection: close
Content-Type: text/html; charset=UTF-8

<!DOCTYPE HTML PUBLIC "-//W3C//DTD HTML 4.0 //EN">
<html>  AND THE REST OF THE HTML PAGE
</html>
```

As you can see from the HTTP response, the Web server on `monkey` is `Apache` version 2 running under Fedora, a Linux system.

For downloading from the Web, you don't need to rely on our little pipeline. The **wget** command takes care of that need nicely. Wget supports HTTP, HTTPS, and FTP protocols and can download single files or follow links in HTML files and recursively download entire websites for offline viewing. The **wget** command can continue to work after you log out so you can download large amounts of data without waiting.

7.20   For More Information

- IPv6 is the next-generation Internet protocol. See `www.ipv6.org/` for an overview.

- The official website for Gnu Privacy Guard is `www.gnupg.org`, and for OpenSSH, is `www.openssh.com`.

- Public-Key Cryptography Standards (PKCS) can be found at RSA Laboratories (`www.rsa.com/rsalabs`).

- HTML5 is the new and coming standard for HTML. See the specification at W3C.

- The DNS is basic to keeping services on the Internet and Web running. Find our more about DNS at `www.dns.net/dnsrd/docs/`.

- HTTP is basic to the Web. See RFC 1945 for HTTP 1.0 and RFC 2068 for HTTP 1.1.

7.21   Summary

In the modern computing environment, computers and networks are inseparable. Networking is an important aspect of any operating system, especially

Linux because the Internet has its origins in UNIX/Linux, and Linux systems are excellent server hosts.

On the Internet, each host computer is identified by its IP address as well as by its domain name. The TCP/IP and UDP/IP protocols are basic to the Internet. Network-based services often follow the client-and-server model, where client programs (such as Web browsers) communicate with server programs (such as Web servers) using well-defined protocols (such as HTTP). A particular server program running on a specific host is identified by the host's IP or domain name together with the server program's port number (such as 80 for Web servers).

The ICANN manages the IP address space and the DNS. The distributed Domain Name Service is a fundamental networking service because it dynamically maps domain names to IP addresses and also provides important information for sending/receiving email. The commands **host**, **nslookup**, and **dig** can be used to obtain DNS data for target hosts.

With networking you can upload/download files with **ftp** and **sftp**; log in to remote computers with **telnet** and **ssh**; copy and synch files with **rcp**, **scp**, and **rsync**; check if a remote system is alive/connected with **ping**; test protocols with **nc**; access the Web; send and receive emails, and perform many other operations.

When it comes to networking, security and privacy are important concerns. Increasingly, computer systems require SSH, SFTP, and SCP for better protection. Automatic file sync can also use SSH for data transfer. The Gnu Privacy Guard (GnuPC) supports secure email, and digital signature, as well as data/file encryption, with public-key cryptography. Message digest algorithms such as MD5 can produce *digital fingerprints* for data/programs to guard their integrity.

Linux systems are often used to run Web servers and to provide Web hosting for individuals and organizations. Basic Web documents are coded in HTML. Hyper references use URLs to link to other documents. MIME content types indicate the media type served on the Web.

The stateless HTTP is a request-response protocol whose messages may have a number of headers and an optional message body.

7.22 Exercises

1. What is a computer network? Name the major components in a computer network.

2. What is a networking client? What is a networking server? What is a networking protocol?

3. What addressing scheme does the Internet use? What is the format of an IP address? What is the quad notation?

4. Consider the IP address

   123.234.345.456

   Is there anything wrong with it? Please explain.

5. Refer to Section 7.6 and set up your own password-less SSH and SFTP.

6. You can schedule commands to be executed automatically by Linux at regular intervals. Find out about the *crontab* and the **crontab** command. Then set up your *crontab* to **rsync** some important folder from one system to another. Show your *crontab* code in full and explain.

7. Refer to Section 7.7 and set up you your GnuPG keys.

8. Write a script that will encrypt/decrypt with **gpg** a file and leave it in the same place as before (with the same filename).

9. What is DNS? Why do we need it?

10. What do name servers do? Why do we need them?

11. What is the relation between the Web and the Internet? What is the relation between HTTP and TCP/IP?

12. What are the major components of the Web? Why is HTML central to the Web?

13. What is the difference between a Web server and a Web browser? Is the Web server a piece of hardware or software? Explain.

14. How does a Web page get from where it is to the computer screen of a user?

15. What is a URL? What is the general form of a URL? Explain the different URL schemes.

16. What are content types? How are they useful?

17. What is the difference between a static Web page and a generated Web page?

18. What is an HTTP transaction? What is an HTTP query? What is an HTTP response?

19. Take the domain name sofpower.com and write the full URL that will access its Web server. Add /linux to the end of that URL. Where does that lead?

20. Take the domain name sofpower.com and find its IP address. Use this IP address instead of the domain name to visit the site. Write the bit pattern for this IP address.

21. Search on the Web for ICANN. Visit the site and discover its mission and services.

22. Find the domain record for `sofpower.com`. Who is the owner of this domain name? Who are the administrative and technical contacts?

23. Find the DNS record for `sofpower.com`.

24. Find out and describe in your own words what the special domain `in-addr.arpa` is.

25. Refer to Section 7.19. Explain the notation

    `$'GET /WEB/test.html HTTP/1.0\n'`

26. Refer to Section 7.19. Use the **nc** command to write a *poor man's Web browser* script `poorman.sh`.

    poorman.sh *path host*

    retrieves the page `http://host/path`.

Chapter 8

Web Hosting: Apache, PHP, and MySQL

Started in the early 1990s as a file sharing system among physicists, the World Wide Web (WWW or simply Web) has grown rapidly to a globe-spanning information system that modern societies won't do without even for a short while. In a real sense, the Web has leveled the playing field and empowered individuals all over the world.

A key factor for this great success is the low cost of putting information on the Web. You simply find a Web hosting service to position your files and programming for your website on the Web. Any Internet host can provide Web hosting if it has a good Internet connection and runs a *Web server* and other related programs.

According to `netcraft.com`'s March 2009 survey, among all Web servers, a full 66.65% are Apache, and a majority of Apache servers run on Linux systems. A Linux-Apache Web hosting environment usually also supports PHP for *active pages* and MySQL for database-driven websites. The Linux, Apache, MySQL, and PHP combination (known as LAMP) works well to support Web hosting. An introduction to these programs, together with their configuration, and operation is presented.

In addition to understanding the big picture and the underlying principles, a practical hands-on approach guides you through the installation, configuration, testing, and administration of Apache, PHP, and MySQL so you can learn Linux Web hosting through doing. Root access on your Linux is convenient, but not necessary.

8.1   What Is a Web Server?

A *Web server* is a piece of software that runs on a particular host to supply documents to the Web. The host computer is called a *server host* and often provides many network-based services including the Web. Linux systems are widely used to run Web servers, and it is important for Linux programmers to become familiar with operations related to the Web server.

A Web server listens to a specific networking *port* on the host and follows the Hypertext Transfer Protocol to receive HTTP requests and send HTTP responses. The standard port is 80, but can be some other designated port such as 8080.

In response to an incoming request, a server may return a static document from files stored on the server host, or it may return a document dynamically generated by a program indicated by the request (Figure 8.1).

FIGURE 8.1: Web Server Overview

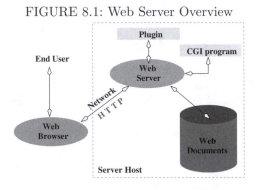

A single-thread server handles one HTTP request at a time, while a multi-thread server can handle multiple concurrent requests. A server host may have multiple copies of a Web server running to improve the handling of requests.

Many different brands of Web servers are available from companies and from open-source organizations. *GlassFish* is a free Web server that comes with the Java EE distribution from `java.sun.com`. The *Apache* Web server, available free from the *Apache Software Foundation* (`apache.org`), is widely used on Linux systems. The popular Apache usually comes pre-installed on Linux distributions.

8.2   URL and URI

An important cornerstone of the Web is the *Universal Resource Locator* (URL, Chapter 7, Section 7.10) that allows Web clients to access diverse resources located anywhere on the Web. For example, the HTTP URL

`http://ml.sofpower.com`

leads to the companion website for this textbook. An HTTP URL (Figure 8.2) identifies a Web server running on a particular host computer and provides the following information:

- A *Universal Resource Identifier* (URI) that corresponds to a local pathname leading to a target resource (a file or program) stored on the server host

- An optional *pathinfo* indicating a target file/folder location as input data to the target resource

- An optional *query string* providing *key=value* pairs as input data to the target resource

FIGURE 8.2: HTTP URL Structure

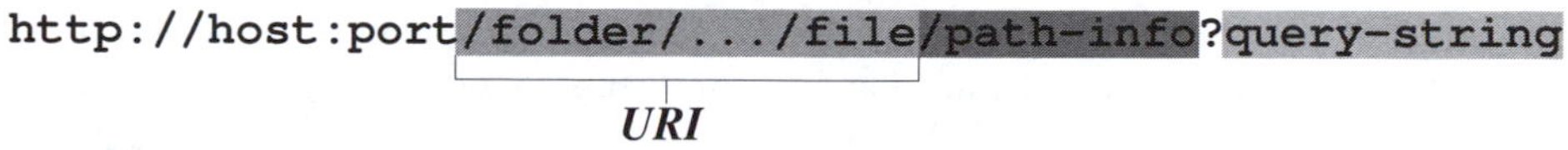

`http://host:port`/folder/.../file/path-info`?query-string`

URI

The part of the URL immediately after the host:port segment (Figure 8.2) is referred to as the URI. The Web server uses the URI to locate the target resource, which can be a static page, an active page, or an executable program. A static page is returned directly in an HTTP response. Any pathinfo and query string is made available, as input, to an active page or an executable program. The resulting output is then returned in an HTTP response.

The set of files and directories made available on the Web through a Web server is known as its *document space*. The *document root* is the root directory for the document space, and it corresponds to the URI /. In addition to the document root hierarchy, there can be other files and directories in the document space, for example, the `/cgi-bin` and the ˜*userid* usually map to directories outside the document root hierarchy.

A Web server also works with other special directories (outside of its document space) for server configuration, passwords, tools, and logs. An URI is interpreted relative to the document root, cgi-bin, or another directory, as appropriate. The Web server can enforce access restrictions, specified in the Web server configuration files, on any file/folder in the document space.

8.3 Request Processing

For each incoming HTTP request, a Web server executes the following request processing cycle:

1. Accepts client connection (via TCP/IP Section 7.2)

2. Processes request (fetches and processes page or invokes program)

3. Sends response

4. Closes connection (or keeps it alive under HTTP1.1)

While processing a request, a busy website often will receive many new requests. It is normal to use multiple servers (multiprocessing) and/or multiple threads within the same server (multithreading) to handle concurrent requests.

8.4   Response and Content Types

For each incoming HTTP request, the Web server sends back an HTTP response containing the requested resource or an indication of error or some other condition.

An HTTP response has two parts: the headers and the body. The server specifies the `Content-Type` header to indicate the media type of the response body. Standard MIME (Multipurpose Internet Mail Extensions) content types (Chapter 6, Table 6.3) are used. The most common content type is `text/html`, but there are many other types. For a static file, the Web server uses the filename extension to infer its media type using a list often found in the file `/etc/mime.types`. The location of this content type list is configurable.

In case of dynamic content, those generated by server-side programs, the Web server relies on that program to set content type.

8.5   The Apache Web Server

Apache is the most popular Web server, especially on Linux systems. You can download and install the Apache HTTP server (Apache) from the Apache Software Foundation (`apache.org`) free of charge (Section 8.14).

However, your Linux will most likely have Apache already installed. Apache is derived from the NCSA[1] httpd project and evolved through a series of code *patches* (thus, *a patchy* server). Apache, written in the C language, is open source and runs on almost all platforms. Apache is fast, reliable, multi-threaded, full-featured, and HTTP/1.1 compliant. Although Apache 1.3 is still available, the most recent stable Apache 2 version is the one to use.

Apache has many components, including

- *Server executable*—The runnable program **httpd**

- *Utilities*—For server control, passwords, and administration

- *Files*—Including server configuration files, log files, password files, and source code files

- *Dynamic loadable modules*—Pre-compiled library modules that can be loaded into the **httpd** at run-time

- Documentation

---

[1] National Center for Supercomputing Applications at the University of Illinois, Urbana-Champaign.

8.6   Apache on Linux

Because of its importance, most popular Linux distributions come with Apache already installed. Networking servers, the Web server included, are automatically started as Linux boots and stopped as Linux shuts down. The boot-time *init scripts* for these services are normally kept in the folder /etc/init.d/. For Apache, the script is /etc/init.d/httpd. The **system-**

FIGURE 8.3: Service Control/Configuration Tool

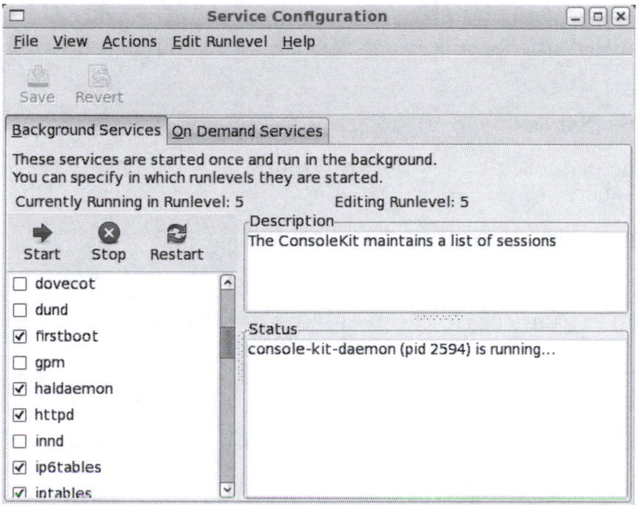

config-services is a tool (Figure 8.3) to turn network services on/off as well as to start/stop/restart any particular service. You'll find **httpd** among the service entries listed. The same operations can also be done with the **service** command from a Command Line Interface (CLI). For example,

service httpd start
service httpd graceful   (restarts without service interruption)
service httpd stop

Both **system-config-services** and **service** invoke the proper /etc/init.d/ scripts to do the job.

On Ubuntu/Debian, you may have to first install this package (Section 8.24) to use the **service** command.

sudo apt-get install sysvconfig

The command **apt-get** is the automatic software package installer/updater on Ubuntu/Debian. Normally, it can only be run by **root**. The **sudo** command is a way of accessing commands normally off limits (see Section 8.14 for more information).

On CentOS/Fedora, you can add/delete any program to this services list with

chkconfig --add *serviceName*
chkconfig --del *serviceName*

On Debian/Ubuntu, use **sysv-rc-conf** instead.

You'll usually find the document root at **/var/www/html/** and the Apache main configuration file at

/etc/httpd/conf/httpd.conf   (CentOS/Fedora)
/etc/apache2/apache2.conf   (Ubuntu/Debian)

Often, the main configuration file will include other component configuration files such as php.conf and ssl.conf.

To check if **httpd**, or any other process, is running, you can use

pidof httpd
pidof *process_Name*

and see if one or more process ids are found.

Controlling the Apache Server

The command **apachectl** (CentOS/Fedora) or **apache2ctl** (Ubuntu/Debian), usually found in /usr/sbin, can be used to control the **httpd**

apachectl *action*
apache2ctl *action*

Possible *actions* are listed in Table 8.1. The init script **/etc/init.d/httpd**

TABLE 8.1: Actions of **apachectl**

Action	Meaning
start	Starts **httpd** if not already running
stop	Stops **httpd** if running
restart	Starts/restarts **httpd**
graceful	Restarts **httpd**, respecting ongoing HTTP requests
configtest or -t	Checks the syntax of configuration files

checks a few things before it actually calls **apachectl** to take care of business.

8.7   Apache Run-Time Configuration

Features and behaviors of the Apache **httpd** can be controlled by *directives* kept in configuration files. The main configuration file is usually httpd.conf (or apache2.conf). When **httpd** starts, it reads the configuration files first. After making changes to the configuration, the **httpd** needs to be restarted before the new configuration takes effect. Unless you have installed your own Apache as an ordinary user (Section 8.15), you'll need root privilege to modify the Apache configuration or to restart it.

Apache Configuration File Basics

An Apache configuration file (`httpd.conf`, for example) is a text file that contains *configuration directives*. Each directive is given on a separate line which can be continued to the next line by a \ character at end of the line.

Lines that begin with the char # are comments and are ignored. A comment must occupy the entire line. No end-of-line comments are allowed. There are many different directives. Directive names are not case sensitive, but their arguments often are. A directive applies globally unless it is placed in a *container* which limits its scope. When in conflict, a local directive overrides a global directive.

The main configuration file is `httpd.conf`, and other *component configuration files* may exist and are included by the main file with the `Include` directive. For example, on many Linux systems the configuration directory `/etc/httpd/conf.d/` stores component configuration files such as `ssl.conf` for SSL (secure socket layer to support HTTPS, Chapter 7, Section 7.10) and `php.conf` for PHP (Section 8.16). The directive

```
Include conf.d/*.conf
```

is used to include all such component configuration files.

To test your Apache configuration for syntax errors, use either one of the following commands:

```
service httpd configtest
apachectl configtest
httpd -t
```

In addition to the central (main and component) configuration files, there are also *in-directory configuration files* known as *access files*. An access file, often named `.htaccess`, is placed in any Web-bound folder (your `public_html`, for example) to provide configuration settings applicable for the file hierarchy rooted at that particular folder. Directives in an access file *override* settings in the central configuration files. The possibility of an access file and what directives it may contain are controlled by the `AllowOverride` directive in the main configuration file. The `.htaccess` files are especially useful for individual users to configure their own Web spaces, usually the `public_html` under their home directories.

About Configuration Directives

Configuration directives control many aspects of the Apache Web server. The `httpd.conf` file has three main parts: *Global Environment, main server configurations,* and *virtual hosts configurations.* Comments are provided for each configuration directive to guide its usage. Apache has reasonable and practical default settings for all the directives, making it easy to configure a *typical server.* Additional directives specify how loaded components work. Commonly used directives include

- Server properties: host identification (`ServerName` *name*), file locations (`ServerRoot`, `DocumentRoot`, `ScriptAlias`), network parameters (`Listen` [*IP:*]*port*), and resource management (`StartServers`, `KeepAlive`)

- Enabling optional server features (`Options`) and in-directory configuration overrides (`AllowOverride`)

- Access restrictions and user authentication (`Allow`, `Deny`, `Require`, `Satisfy`, `AuthName`, `AuthType`, `AuthFile`)

- Content handling (`AddHandler`, `AddType`, `AddOutputFilter`)

- HTTP caching and content deflation (`ExpiresActive`, `ExpiresByType`, `DeflateCompressionLevel`, `AddOutputFilterByType DEFLATE`)

- Virtual hosts (`NameVirtualHost`)

For example, the directive

```
DirectoryIndex index.html  index.php
```

says `index.html` (or `index.php`) is the *directory index file* which is displayed if the folder containing it is the target resource of an incoming URI. Without an index file, a listing of filenames in that folder is generated (*index generation*) for display only if the `Indexes` option has been enabled. Otherwise, an error is returned.

Loading Modules

Apache is a modular server. Only the most basic functionalities are included in the core **httpd**. Many extended features are implemented as dynamically loadable modules (`.so`) that can be selectively loaded into the core server when it starts. This organization is very efficient and flexible.

The loadable modules are placed in the **modules** folder under the *server root* directory, which is defined in the main configuration file with the `ServerRoot` directive. To load a certain module, use the directive

```
LoadModule  name_module modules/moduleFileName.so
```

For example,

```
LoadModule dir_module modules/mod_dir.so       (loads module dir)
LoadModule php5_module modules/libphp5.so      (loads module php5)
```

The `dir` module enables Apache to generate a directory listing. The `php5` module supports dynamic Web pages using the PHP scripting language (Section 8.16).

Configuration directives may be included conditionally, depending on the presence of a particular module, by enclosing them in an `<IfModule>` container. For example,

```
<IfModule mod_userdir.c>
UserDir public_html
</IfModule>
```

says if we are using the `userdir` module, then the Web folder for each Linux user is `public_html`.

Global Directives

Table 8.2 shows some more directives relating to how the Apache server works globally (**Ex:** ex08/apacheGlobal.conf). The `Alias` and `ScriptAlias` directives map an incoming URI to a designated local folder.

TABLE 8.2: Apache Global Directives

Directive	Effect
ServerRoot "/etc/httpd"	
KeepAlive On	Keeps connection for next request
MaxKeepAliveRequests 100	
KeepAliveTimeout 15	
User apache	Server userid is `apache`
Group apache	Server groupid is `apache`
ServerName monkey.cs.kent.edu	Domain name of server host
ServerAdmin pwang@cs.kent.edu	Email of administrator
DocumentRoot "/var/www/html"	Server document space root
UserDir public_html	Folder name of per-user Web space
AccessFileName .htaccess	In-directory configuration file name
TypesConfig /etc/mime.types	MIME types file
ScriptAlias /cgi-bin/	
"/var/www/cgi-bin/"	CGI program folder
Alias /special/ "/var/www/sp/"	Special URI-to-folder mapping

Container Directives

Configuration directives can be placed inside a container directive to subject them to certain conditions or to limit their scope of applicability to particular directories, files, locations (URLs), or hosts. Without being limited, a directive applies globally.

For example,

```
<IfModule mod_userdir.c>
   UserDir public_html
</IfModule>
```

enables the per-user Web space (**Ex:** ex08/peruser.conf) and designates the user folder to be `public_html` only if the `userdir` module is loaded.

Also, consider these typical settings (**Ex:** ex08/docroot.conf) for the document root /var/www/html:

```
<Directory "/var/www/html">
    Options Indexes FollowSymLinks              (1)
    Order allow,deny                            (2)
    Allow from all                              (3)
    AllowOverride None                          (4)
</Directory>
```

Within the directory /var/www/html, we allow index generation and the following of symbolic links (line 1). The order to apply the access control directives is `allow` followed by `deny` (line 2), and access is allowed for all incoming requests (line3+) unless it is denied later.

The `AllowOverride` (line 4) permits certain directives in .htaccess files. Its arguments can be `None`, `All`, or a combination of the keywords `Options`, `FileInfo`, `AuthConfig`, and `Limit`. We'll return to this topic when we discuss access control in detail (Section 8.8).

You'll also find the following typical setting (**Ex:** ex08/htprotect.conf) in your httpd.conf:

```
<Files ~ "^\.ht">
    Order allow,deny
    Deny from all
</Files>
```

It denies Web access to any file whose name begins with .ht (Chapter 4, Section 4.3). This is good for security because files such as .htaccess are readable by the Apache Web server, but we don't want their contents exposed to visitors from the Web.

As <Directory> and <Files> work on the file pathnames on your computer, the <Location> container works on URIs. We also have <DirectoryMatch>, <FileMatch>, and <LocationMatch> that use regular expressions as defined for **egrep** (Section 4.3).

8.8   Access Control under Apache

What Is Access Control?

Running a Web server on your Linux system means that you can make certain files and folders accessible from the Web. However, you also want to control how such files can be accessed and by whom.

To make a file/folder accessible from the Web, you must place it somewhere in the document space configured for your Web server. This usually means placing a file/folder under the document root or inside your own public_html and also making the file readable (the folder readable and executable) by the

Web server via **chmod a+r** *file* (**chmod a+rx** *folder*). Files on your system not placed under the server document space or not having the right access modes (Chapter 6, Section 6.4) will not be accessible from the Web.

The Web server can be configured to further limit access. Access control specifies who can access which part of a website with what HTTP request methods. Access control can be specified based on IP numbers, domains, and hosts, as well as passwords. Access restrictions can be applied to the entire site, to specific directories, or to individual files.

Apache access control directives include `Allow`, `Deny`, `Order`, `AuthName`, `AuthType`, `AuthUserFile`, `AuthGroupFile`, `Require`, `Satisfy`, `<Limit>`, and `<LimitExcept>`.

Access Control by Host

If a file in the server document space has no access control, access is granted. The `order` directive specifies the order in which `allow` and `deny` controls are applied. For example,

```
order allow,deny
```

only access allowed but not denied are permitted. In the following, if access is first denied then allowed, it is allowed.

```
order deny,allow
deny from all
allow from host1 host2 . . .
```

On `monkey.cs.kent.edu`, we have a set of pages reserved for use inside our departmental local area network (LAN). They are placed under the folder `/var/www/html/internal`. Their access has the following restriction (**Ex:** `ex08/folderprotect.conf`):

```
<Location /internal>
  order deny,allow
  deny from all
  allow from .cs.kent.edu
</Location>
```

Thus, only hosts in the `.cs.kent.edu` domain are allowed to access the location `/internal`. The IP address of a host can be used. For example,

```
allow from 131.123
```

grants access to requests made from any IP with the prefix `131.123`.

To enable users to control access to files and folders under their per-user Web space (`public_html`), you can use something such as (**Ex:** `ex08/htaccess.conf`)

```
<Directory /home/*/public_html>
   AllowOverride All
   Order allow,deny
   Allow from all
</Directory>
```

in `httpd.conf`. This means users can place their own access control and other directives in the file `~/public_html/.htaccess`.

8.9   Requiring Passwords

Allowing access only from certain domains or hosts is fine, but we still need a way to restrict access to registered users either for the whole site or for parts of it. Each part of a site under its own password control is known as a *security realm*. A user needs the correct userid and password to log in to any realm before accessing the contents thereof. Thus, when accessing a resource inside a realm, a user must first be *authenticated* or verified as to who the user is. The Apache Web server supports two distinct HTTP authentication schemes: the *Basic Authentication* and the *Digest Authentication*. Some browsers lack support for Digest Authentication which is only somewhat more secure than Basic Authentication.

Let's look at how to set user login.

Setting Up User Login under Apache

To illustrate how to password protect Web files and folders, let's look at a specific example where the location `/WEB/csnotes/` is a folder we will protect.

We first add the following authentication directives to the `httpd.conf` file (**Ex:** ex08/validuser.conf):

```
<Location "/WEB/csnotes/">
    AuthName  "WDP-1 Notes"
    AuthType  Basic
    AuthUserFile /var/www/etc/wdp1p
    require valid-user
</Location>
```

The `AuthName` gives a name to the realm. The realm name is displayed when requesting the user to log in. Thus, it is important to make the realm name very specific so that users will know where they are logging into. The `AuthType` can be either `Basic` or `Digest`. The `AuthUserFile` specifies the full pathname of a file containing registered users. The optional `AuthGroupFile` specifies the full pathname of a file containing group names and users in those groups. The `Require` directive defines which registered users may access this realm.

```
valid-user                        (all users in the AuthUserFile)
user id1 id2 id3 ...              (the given users)
group grp1 grp2 ...               (all users in the given groups)
```

The `AuthUserFile` lists the userid and password for each registered user, with one user per line. Here is a sample entry in `/var/www/etc/wdp1`.

`PWang:RkYf8U6S6nBqE`

The Apache utility **htpasswd** (**htdigest**) helps create password files and add registered users for the Basic (Digest) authentication scheme. (See the man page for these utilities for usage.) For example,

htpasswd -c /var/www/etc/wdp1 PWang

creates the file and adds an entry for user `PWang`, interactively asking for `PWang`'s password. If you wish to set up a group file, you can follow the format for `/etc/group`, namely, each line looks like

group-name: userid1 userid2 ...

It is also possible to set up login from an `.htaccess` file. For example, put in `.htaccess` under user `pwang`'s `public_html`

```
AuthUserFile /home/pwang/public_html/.htpassword
AuthName "Faculty Club"
AuthType Basic
Require valid-user
```

Then, place in `.htpassword` any registered users.

If more than one `Require` and/or `allow from` conditions is specified for a particular protected resource, then the `satisfy any` (if any condition is met) or `satisfy all` (all conditions must be met) directive is also given. For example (**Ex:** ex08/flexibleprotect.conf),

```
<Location /internal>
    order deny,allow
    deny from all
    allow from .cs.kent.edu
    AuthName  "CS Internal"
    AuthType  Basic
    AuthUserFile /var/www/etc/cs
    require valid-user

    satisfy any
</Location>
```

means resources under the `/internal` can be accessed by any request originating from the `cs.kent.edu` domain (no login required) or a user must log in.

8.10   How HTTP Basic Authentication Works

Upon receiving an unauthorized resource request to a realm protected by *Basic Authentication*, the Web server issues a *challenge*:

```
HTTP/1.0 401 Unauthorized
WWW-Authenticate: Basic realm="CS Internal"
```

Upon receiving the challenge, the browser displays a login dialog box request-ing the userid and password for the given realm. Seeing the login dialog, the user enters the userid and password. The browser then sends the same resource request again with the added authorization HTTP header

```
Authorization: Basic QWxhZGRpbjpvcGVuIHNlc2FtZQ==
```

where the base64 (Chapter 7, Section 7.7) encoded *basic cookie* decodes to *userid:password*. From this point on, the browser automatically includes the basic cookie with every subsequent request to the given realm. This behavior persists until the browser instance is closed.

8.11   How HTTP Digest Authentication Works

Unless conducted over a secure connection, such as SSL (secure socket layer) used by HTTPS, the Basic Authentication is not very secure. The userid and password are subject to easy eavesdropping over HTTP. The *Digest Authentication* is an emerging HTTP standard to provide a somewhat more secure method than Basic Authentication.

With Digest Authentication, the server sends a challenge (on a single line)

```
HTTP/1.1 401 Unauthorized WWW-Authenticate:
 Digest realm="Gold Club" nonce="3493u4987"
```

where the *nonce* is an arbitrary string generated by the server. The recom-mended form of the nonce is an *MD5 hash* (Chapter 7, Section 7.9), which includes the client's IP address, a timestamp, and a private key known only to the server.

Upon receiving the challenge, the browser computes

```
str1 = MD5(userid + password)
str2 = MD5(str1 + nonce + Resource_URI)
```

The browser then sends the authorization HTTP header (on one line)

```
Authorization: Digest realm="Gold Club", nonce="...",
 username="pwang", uri="/www/gold/index.html",
 response="str2"
```

The server verifies the response by computing it using the stored password.

From this point on, the browser includes the Digest Authentication header with every request to the same realm. The server may elect to rechallenge with a different nonce at any time.

Basic vs. Digest Authentication

Basic Authentication is simple and works with all major browsers. Digest Authentication is somewhat more secure, but browser support is less complete. Web servers, including Apache, tend to support both authentication schemes. When security is a concern, the best practice is to move from Basic Authentication over HTTP directly to Basic Authentication over HTTPS (Secure HTTP over SSL).

8.12   Password Encryption

The Apache-supplied **htpasswd** tool uses the same Linux/UNIX password/data encryption method as implemented by the C library function `crypt`. In this encryption scheme, a *key* is formed by taking the lower 7 bits of each character from the password to form a 56-bit quantity. Hence, only the first 8 characters of a password are significant. Also, a randomly selected 2-character *salt* from the 64-character set [a-zA-Z0-9./] is used to perturb the standard Data Encryption Algorithm (DEA) in 4096 possible different ways. The key and salt are used to repeatedly encrypt a constant string, known only to the algorithm, resulting in an 11-character code The salt is prepended to the code to form a 13-character encrypted password which is saved in the password file for registered users. The original password is never stored.

When verifying a password, the salt is extracted from the encrypted password and used in the preceding algorithm to see if the encrypted password is regenerated. If so, the password is correct.

8.13   Automatic File Deflation

Apache takes advantage of many HTTP 1.1 features to make Web pages faster to download. One such feature is automatic compression of a page before network transfer, resulting in significantly reduced file size and delivery time. This is especially true for textual pages whose compression ratio can reach 85% or more. A compressed page is uncompressed by your browser automatically.

The `mod_deflate` module for Apache 2.0 supports automatic (dynamic) file compression via the HTTP 1.1 `Content-Encoding` and `Accept-Encoding` headers. These two configuration directives (**Ex:** `ex08/deflate.conf`)

```
DeflateCompressionLevel 6
AddOutputFilterByType DEFLATE text/html text/plain \
    text/xml text/css application/x-javascript \
    application/xhtml+xml application/xslt+xml \
    application/xml application/xml-dtd image/svg+xml
```

indicate a list of content types for dynamic compression (using zlib) at the indicated compression level. Deflation adds a bit of processing load on the server side and the higher the compression level, the heavier the processing load.

Compression will only take place when the incoming HTTP request indicates an acceptable compression encoding. The detection of browser compression preferences and the sending of compressed or uncompressed data are automatic. Of course, any compressed outgoing page will carry an appropriate `Content-Encoding` response header.

The `AddOutputFilterByType` directive needs `AllowOverride FileInfo` to work in `.htaccess`.

8.14 Installing Apache with Package Management

We have mentioned that most Linux distributions come with Apache installed. With root access, you can use the Linux package management (Section 8.24) commands

CentOS/Fedora:
yum `install httpd`
yum `update httpd`

Ubuntu/Debian:
sudo apt-get `install apache2`
sudo apt-get `update apache2`

to install/update your Apache server.

If you wish to have the very latest Apache release, or if you don't have root access, you can install Apache manually as described in Section 8.15.

Sudo

Linux administration tasks such as setting up new user accounts, installing and updating system-wide software, and managing network services must usually be performed by privileged users such as `root`. This is secure but not very flexible.

Sudo is a method to allow regular users to perform certain tasks temporarily as `root` or as some other more privileged user. The command name **sudo** comes from the command **su** (substitute user) which allows a user to become another user. Putting **sudo** in front of a complete command you wish to execute says: "*allow me enough privilege to execute the following command.*" If the given command is allowed, **sudo** sets the real and effective `uid` and `gid` (Chapter 6, Section 6.4) to those of a specific privileged user for the duration of the execution of the given command. All **sudo** commands are logged for security.

The file `/etc/sudoers` contains data governing who can gain what privileges to execute which commands on exactly what host computers and whether your password is required or not. Thus, the same `sudoers` file can be shared by many hosts within the same organization. The file can only be modified via the privileged command **sudoedit**. You can read about `sudoers` and its syntax rules by

man 5 sudoers

The general form of a user entry in `sudoers` is

r_user hosts=(*s_user*) *commands*

meaning *r_user* can execute the given *commands* as *s_user* on the listed *hosts*. The (*s_user*) part can be omitted if *s_user* is `root`.

Here are some example `sudoers` entries.

```
pwang   localhost=/sbin/shutdown -h now

pwang   localhost=/sbin/service httpd start, \
                  /sbin/service httpd start, \
                  /sbin/service httpd restart, \
                  /sbin/service httpd graceful

pwang   localhost=/sbin/apt-get update apache, \
                  /sbin/apt-get install apache, \
                  /sbin/apt-get remove apache

root    ALL=(ALL)        ALL
```

Each entry must be given on one line. The `root` entry is always there to give `root` the ability to **sudo** all commands on hosts as any user.

Even if you log in (or **su**) as `root`, you may prefer to use **sudo** so as to leave log entries for the tasks performed.

8.15   Manual Installation of Apache

If you prefer not to install Apache with package management, you may install Apache manually. The installation procedure follows the standard Linux configure, make, install sequence.

If you have `root` access, you will be able to install Apache in a system directory such as `/usr/local` and assign port 80 to it. If not, you still can install Apache for yourself (for experimentation) in your own home directory and use a non-privileged port, such as 8080. Let `$DOWNLOAD` be the download folder, for example, either `/usr/local/apache_src` or `$HOME/apache_src`, and let `$APACHE` be the installation folder, for example, `/usr/local/apache` or `$HOME/apache`.

To download and unpack the Apache HTTP server distribution, follow these steps.

1. Download—Go to `httpd.apache.org/download.cgi` and download the

   `httpd-`*version*`.tar.gz`

   or the `.tar.bz2` file, as well as its MD5 fingerprint file, into your `$DOWNLOAD` folder.

2. Integrity check—Use **md5sum** on the fingerprint file to check the downloaded file.

3. Unpack—From the `$DOWNLOAD` folder unpack with one of these commands.

   tar zxvpf `httpd-`*version*`.tar.gz`
   tar jxvpf `httpd-`*version*`.tar.bz2`

   You'll find a new Apache source folder, `httpd-`*version*, containing the unpacked files.

Configure and Compile

Now you are ready to build and install the Apache Web server. Follow the `INSTALL` file and the *Compiling and Installing* section of the Apache documentation `httpd.apache.org/docs/`*version-number*. You'll need an ANSI C compiler (**gcc** preferred) to compile, Perl 5 to make tools work, and DSO (Dynamic Shared Object) support. These should already be in place on newer Linux distributions.

From the Apache source folder, issue the command

`./`**configure** *options*

to automatically generate the compilation and installation details for your computer. The `INSTALL` file has good information about configuration. To see all the possible options, give the command

`./`**configure** `--help`.

For example, the `--prefix=`*serverRoot* option specifies the pathname of the server root folder, and the option `--enable-mods-shared=all` elects to compile all Apache modules into dynamically loadable shared libraries.

The recommended method (**Ex:** `ex08/makeapache.bash`) to configure and compile is

`./`**configure** `--prefix=$APACHE --enable-mods-shared=all \`
               otherOptions
make
make install

Here the Apache server root folder has been set to your installation folder $APACHE as the destination for the results of the installation. The recommended *otherOptions* are

```
--enable-cache --enable-disk-cache \
--enable-mem-cache --enable-proxy \
--enable-proxy-http --enable-proxy-ftp \
--enable-proxy-connect --enable-so \
--enable-cgi --enable-info \
--enable-rewrite --enable-spelling \
--enable-usertrack --enable-ssl \
--enable-deflate --enable-mime-magic
```

Each of the preceding three commands will take a while to run to completion.

After successful installation, it is time to customize the Apache configuration file $APACHE/conf/httpd.conf. Follow these steps:

1. Check the **ServerRoot** and **DocumentRoot** settings. These should be the full pathnames as given by $APACHE and $APACHE/htdocs, respectively.

2. Set the listening port:

```
Listen 80           (requires root privilege)
Listen 8080         (no need for root privilege)
```

3. Make any other configuration adjustments as needed.

Now you can start the Apache server with

$APACHE/bin/apachectl start

If the start is successful, you can then use a Web browser on the same host computer to visit

http://localhost.localdomain:*port*

and see the Apache welcome page, which is the file

$APACHE/htdocs/index.html

It is recommended that you install PHP together with Apache. See Section 8.17 for details.

8.16   What Is PHP?

PHP, a recursive acronym for *PHP: Hypertext Preprocessor*, represents a powerful and widely used program for generating dynamic Web content. It evolved from an earlier project by Rasmus Lerdorf, and PHP 3.0 was released in mid-1998. PHP has matured as an important server-side tool and is marching toward version 6 at the time of this writing. In addition to serving the Web, PHP can also be used as a Linux command for general-purpose text processing.

Although PHP runs on multiple platforms, we will focus on PHP as an Apache server module on Linux. As such, PHP executes as part of Apache and interprets code embedded in Web-bound pages to dynamically generate content for those pages. For example, an HTML document containing

```
<p>It is <?php echo(date("l M. d, Y")); ?>,
<br />do you know where your project is?</p>
```

generates the text

```
It is Thursday June. 18, 2009,
do you know where your project is?
```

The date displayed depends on the exact time of access.

Any PHP code is given within the PHP bracket `<?php ... ?>` and interleaved (embedded) within normal HTML code, or other types of code as the case may be. Pages containing such embedded codes are often called active (or dynamic) pages, because they are not static and contain information generated on the fly by the embedded code. The embedded code is never seen by the receiver of the resulting document; it gets replaced by any information it generates (Figure 8.4).

FIGURE 8.4: PHP Code Interpretation

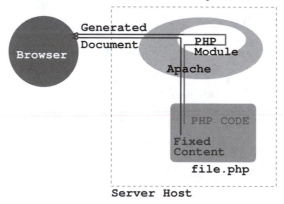

8.17   The PHP Module for Apache

An Apache server is generally expected to support PHP, and it is not hard to add the PHP module for Apache. With the PHP module, the Apache Web server will be able to interpret PHP codes embedded in textual documents of any type as they are being delivered to the Web (Figure 8.4). Most Linux distributions will have Apache installed with PHP already. For example, you may find the PHP module `libphp5.so` already in the Apache modules folder (usually `/etc/httpd/modules`).

You can also use the Linux package management facility to install/update Apache+PHP:

yum `install httpd php`   (CentOS/Fedora)
yum `update httpd php`   (CentOS/Fedora)
sudo apt-get `install apache2 php5 \`
         `libapache2-mod-php5`   (Ubuntu/Debian)
sudo apt-get `update apache2 php5 \`
         `libapache2-mod-php5`   (Ubuntu/Debian)

Installing the PHP Module

This section describes how to install the PHP module manually and add it to your Apache server. If you already have Apache+PHP installed, please skip this section.

First, download the current PHP release (**php-***version*`.tar.gz` or `.tar.bz2`) from `www.php.net/downloads.php`, check the MD5 fingerprint, and unpack into your $DOWNLOAD folder as before (Section 8.15).

Next, go to the the PHP source code folder $DOWNLOAD/php-*version* to configure the PHP module. For example (**Ex: ex08/makephp.bash**),

cd $DOWNLOAD/php-*version*
./configure `--with-apxs2=$APACHE/bin/apxs \`
  `--prefix=$APACHE/php  --enable-shared=all \`
  `--with-gd --with-config-file-path=$APACHE/php \`
  `--enable-force-cgi-redirect --disable-cgi \`
  `--with-zlib --with-gettext --with-gdbm \`
  `> /tmp/conf.output 2>&1`

Then check the conf.output to see if you get these lines:

```
checking if libtool supports shared libraries... yes
checking whether to build shared libraries... yes
checking whether to build static libraries... no
```

If you need to redo the configuration step, please first clean things up with

make `distclean`

After successful configuration, you are ready to create the PHP module. Enter the command

make

It will take a while. After it is done you should check the `.libs/` folder to see if the PHP module `libphp5.so` has been created. If so, then issue the command

make `install`

The install directory is `$APACHE/php` as specified by the `--prefix` option. The install process also moves `libphp5.so` to the folder `$APACHE/modules/` and modifies `$APACHE/conf/httpd.conf` for the **httpd** to load the PHP module when it starts by adding the Apache configuration directive

```
LoadModule php5_module modules/libphp5.so
```

In addition, you also need to add a few other directives to tell Apache what files need PHP processing:

```
AddHandler php5-script .php
AddType text/html .php
DirectoryIndex index.php index.html
```

As stated, any time a change is made to the configuration, you need to restart Apache (Section 8.6) in order to get the new configuration to take effect.

8.18   Testing PHP

To test your Apache+PHP installation, you can create the page `info.php` (**Ex:** `ex08/info.php`)

```
<html><head><title>php info</title></head>
<body> <?php phpinfo(); ?>
</body></html>
```

and place it under the document root folder. Then, visit

```
http://localhost.localdomain/info.php
```

from your Web browser. The `phpinfo()` function generates a page of detailed information about your PHP installation, including version number, modules loaded, configuration settings, and so on.

As Apache starts, it loads the PHP module and also any PHP-specific configuration in a file usually named `php.ini`. The location of this file (usually `/etc/php.ini`) is given as the *Loaded Configuration File* in the `phpinfo()` generated display.

8.19 PHP Configuration

The configuration file (**php.ini**) is read when the PHP module is loaded as the Web server (**httpd**) starts. Any changes made to **php.ini** will only take effect after Apache is restarted (Section 8.6).

PHP has toggle (on/off) and value configuration directives. You edit the **php.ini**, which contains a set of reasonable defaults, to make any adjustments.

For example, if you are running a Web development site where seeing error messages will help debugging PHP scripts, then you would set (**Ex: ex08/php.ini**)

```
;;;; Enables error display output from PHP
display_errors = On
display_startup_errors = On
```

For a production Web server, you would definitely want to change these to

```
display_errors = Off
display_startup_errors = Off
log_errors = On
;;;; Enables all error, warning, and info msg reporting
error_reporting = E_ALL
;;;; Sends msgs to log file
error_log = <pathname of a designated error.txt file>
```

PHP also allows you to open any local or remote URL for generating page content. However, if your site has no need for opening remote URLs from PHP, you may increase security by setting

```
allow_url_fopen = Off
```

PHP also has very good support for HTTP file uploading. If you wish to allow that, then use

```
file_uploads = On
```

```
;;;; Use some reasonable size limit
upload_max_filesize = 2M
```

PHP *extensions* provide optional features for many different purposes. For example, the **gd** extension supports manipulation of fonts and graphics from PHP, and the **mysql** extension provides a PHP interface to MySQL databases. Dynamically loadable extensions are collected in a PHP modules folder (usually **/usr/lib/php/modules**), but are set in the **php.ini** by the **extension_dir** directive. On many Linux systems, the extensions are loaded by default through extension-specific **.ini** files in the folder **/etc/php.d/**. By editing these files you control which extensions are loaded when Apache+PHP starts.

To examine the setting of all PHP configurations directives, you can simply look at the **phpinfo()** display (Section 8.18).

8.20   Database Support for the Web

A computer database is a system for conveniently storing, retrieving, updating, and inquiring information for concurrent access by many users. Modern databases are *relational*; information is stored in multiple *tables* (Figure 8.5) that are interrelated.

FIGURE 8.5: The EMPLOYEES Table

SS	Last	First	Hiredate	Email	
f1	f2	f3	f4	f5	

A database system is SQL-compliant if it supports the *Structured Query Language* standard API (Application Programming Interface). For example, the following SQL SELECT query retrieves all rows from table EMPLOYEES where the field LAST is Wang:

```
SELECT * FROM EMPLOYEES WHERE LAST = "Wang";
```

Programs written in SQL can access and manipulate any SQL-compliant database. Databases can be used for decision support, online transaction processing, personnel records, inventory control, user accounts, multi-user online systems, and many other purposes.

A database can also make websites easier to construct, maintain, and update. On the other hand, the Web can make databases accessible from any computer connected to the Internet.

PHP provides excellent support for using databases for and from the Web. The *SQLite* extension of PHP is a fast SQL interface to a flat file database that comes with PHP (version 5 or later). For many simple Web applications, SQLite is just the right solution.

8.21   MySQL

More complicated websites with larger data loads will need heavier duty database systems than SQLite. For that, the free *MySQL* is often the right choice, especially in combination with Linux and PHP because PHP also has excellent built-in support for connecting and querying MySQL databases.

MySQL is a freely available open-source relational database management system that supports SQL. It runs on Linux, MS Windows®, Mac OS X®, and other systems and can be used from many programming languages, including C/C++, Eiffel, Java, Perl, PHP, Python, and Tcl. The MySQL database server supports both local and network access. It supports a *privilege and password system* to specify who can access/modify what in the database system.

Most Linux distributions come with MySQL installed. If you can locate the command **mysql** (often in /usr/bin) on your system, then you have MySQL already. If not, or if you wish to install the latest version of MySQL, please refer to Section 8.23.

Initializing, Starting, and Stopping MySQL

MySQL uses a **default database** named mysql for its own purposes, such as recording registered users (userid and password), managing databases, and controlling access privileges. The command **mysql_install_db** (in usr/bin/) is run once to initialize the MySQL default database (usually located in /var/lib/mysql/mysql/) and is done automatically when the MySQL server **mysqld** is started for the very first time. The **mysql_install_db** script contains many initialization settings for MySQL, and adjusting these settings allows you to customize various aspects of MySQL.

Starting **mysqld** can be done with the **system-config-services** GUI tool or the command

service mysqld start

The same GUI and command-line tools can be used to stop/restart the **mysqld**.

With **mysqld** started, MySQL client programs can communicate with it to access/manipulate databases served by it (Figure 8.6).

FIGURE 8.6: MySQL Server and Clients

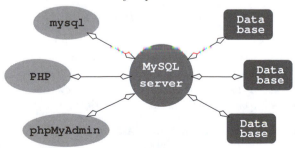

MySQL Run-Time Configuration

As **mysqld** (the database server) starts, it reads configuration values in my.cnf (usually kept in /etc or /etc/mysql). Specified are the data folder, the socket (Chapter 11, Section 11.6) location, the userid of **mysqld**, and possibly many other settings. Edit `my.cnf`, and delete the line `bind-address = 127.0.0.1` if present.

It is also recommended that you consider running a local-access-only MySQL server rather than one that is network enabled. The latter allows MySQL clients to access the server via a network which can mean security problems. The former will limit access to MySQL clients on the same host, making it much more secure. To do this, add the configuration setting

```
skip-networking
```

to both the `[mysqld]` and the `[mysqld_safe]` sections in `my.cnf`. You need to restart **mysqld** after making changes to the configurations. See the MySQL documentation for details about MySQL configuration.

After starting **mysqld**, you can use **netstat**, a command to display networking status and activity on your system, to double check. Run the command

netstat -tap | grep mysqld

If you see a display, it means **mysqld** is allowing network access. If you see no display, then only local clients are allowed access. The **-tap** option tells **netstat** to display all information related to TCP with names of programs involved.

Administering MySQL

MySQL protects databases by requiring a userid and password, and, depending on what privileges the user has, various operations/accesses are allowed or denied.

At the beginning, MySQL has an administrator (`root`) and a blank password. The very first administrative task is to set a password for `root`.[2]

mysqladmin -u root password *new_password*

The option -u specifies the MySQL userid `root` and the admin operation is password setting. Make sure you save the password for future use. Let's assume the `root` password is `foobar`.

The MySQL `root` is the user who can create new databases, add users, and set privileges for them. Let's create a new database `lxux`.

mysqladmin -h localhost -u root -pfoobar create lxux

The commandindexRegular command!mysqladmin@**mysqladmin** takes the hostname, userid, and password information and carries out the creating new database operation. The new database `lxux` is usually placed in `/var/lib/mysql`.

---

[2]Not to be confused with the Linux super user which is also `root`.

Now we can add **pwang** as a user with all privileges to use **lxux**. One way is to use the **mysql** tool which is a command-line interface to the MySQL database server. Give the command

mysql -h localhost -u root -pfoobar

then you are working within **mysql**, and you may enter SQL queries. Do the following (**Ex**: ex08/adduser.sql):

```
mysql> USE mysql;                    (setting database name to mysql)
mysql> SHOW TABLES;                  (listing names of tables)
+-----------------+
| Tables_in_mysql |
+-----------------+
| columns_priv    |
| db              |
| func            |
| host            |
| tables_priv     |
| user            |
+-----------------+
mysql> INSERT INTO user (Host, User, Password, Select_priv)
    -> VALUES ('', 'pwang', password('thePassword'), 'Y');
mysql> FLUSH PRIVILEGES;
mysql> GRANT ALL PRIVILEGES ON lxux.* TO pwang
    -> IDENTIFIED BY 'thePassword';
mysql> FLUSH PRIVILEGES;
mysql> quit
```

Then inform user **pwang** about his userid, password, and database name. See the MySQL documentation for more information on setting user privileges. To reset the password for **pwang** use the SQL

```
mysql> USE mysql;
```

```
mysql> update user set Password=PASSWORD('newOne')
    -> WHERE User='pwang';
```

Because PHP is often available on the same host, the free *phpMyAdmin* tool (**phpmyadmin.net**) is often also installed to enable MySQL administration over the Web. *PhpMyAdmin* (Section 8.22) supports a wide range of operations with MySQL. The most frequently used operations are supported by the Web browser supplied GUI (managing databases, tables, fields, relations, indexes, users, permissions, and so on). Other operations are always doable via direct SQL statements. Both the **root** user and any user for a specific database can do database administration through *phpMyAdmin* from anywhere on the Web.

Resetting the MySQL Root Password

It is important to not forget the MySQL root password. However, if you find yourself in a such a situation, you can reset it. As Linux root, first stop the **mysqld**:

service mysqld stop

Then run **mysqld** in safe mode without security checking:

/usr/bin/mysqld_safe --skip-grant-tables &

Then run **mysql** on the default database mysql:

mysql -u root mysql

Then update the password for `root`:

```
mysql> update user set Password=PASSWORD('anything')
    -> WHERE User='root';
Query OK, 2 rows affected (0.04 sec)
Rows matched: 2  Changed: 2  Warnings: 0

mysql> flush privileges; exit;
```

Now kill the **mysqld_safe** process and restart the **mysqld**.

8.22   Installing phpMyAdmin

First, download the latest version from `phpmyadmin.net` and unpack in your Web document root folder (usually `/var/www/html`). For example (**Ex: ex08/myadmin.install**),

```
cd /var/www/html
tar jxvpf phpMyAdmin-3.3.4-english.bz2
rm phpMyAdmin-3.3.4-english.bz2
mv phpMyAdmin-3.3.4-english phpMyAdmin
```

The resulting **phpMyAdmin** folder is now in place under the Web document root and you can display installation instructions and other documentation with the URL

`http://localhost.localdomain/phpMyAdmin/Documentation.html`

To install **phpMyAdmin**, you only need to do a few things. In the **phpMyAdmin** folder create a configuration file `config.inc.php` by copying and editing the sample file `config.sample.inc.php`.

It is recommended that you pick the `cookie` authentication method and set up a *control user*, as indicated by the sample configuration file, on your

MySQL server so anyone who has a MySQL login can use `phpMyAdmin` to manage databases accessible to that particular user. See the `phpMyAdmin` documentation for configuration details.

After installation, the URL

`http://`*host*`/phpMyAdmin`

reaches the on-Web MySQL admin tool for any valid user to manage the database server. (Figure 8.7).

FIGURE 8.7: phpMyAdmin Tool

8.23 Installing MySQL

MySQL comes with most Linux distributions. In case there is a need, the Linux package management system makes installation/update easy. For CentOS/Fedora, do as root one of

```
yum install mysql-server mysql
yum update mysql-server mysql
```

For Ubuntu/Debian, do one of

```
sudo apt-get install mysql-server
sudo apt-get update mysql-server
```

Now proceed to edit the `my.cnf` file (Section 8.21) and then start/restart the **mysqld** (Section 8.21).

If you wish to install/update Apache+PHP+MySQL to achieve LAMP all at once, use these commands.

CentOS/Fedora:
```
yum install httpd php  mysql-server mysql
yum update httpd  php  mysql-server mysql
```

Ubuntu/Debian:
sudo
apt-get install apache2 php5 libapache2-mod-php5 mysql-server
sudo
apt-get update apache2 php5 libapache2-mod-php5 mysql-server

It may be even easier if you install *XAMPP for Linux*, which is LAMP made easier to install on Linux. Remember these installations are very nice as developmental systems, but not secure enough as production systems. Enterprise editions of Linux will most likely include a production Web server with LAMP and more. What you learn here will apply directly to such production servers.

Refer to `dev.mysql.com/downloads/` at the MySQL site for manual installation.

8.24   Linux Package Management

A *package management system* automates the installation and maintenance of software applications for any given operating system. For Linux we have two major systems: the *Advanced Packaging Tool* (**apt**) for the the Debian family and the *Yellow dog Updater, Modified* (**yum**) for the Red Hat family.

Using the package management tools, you can easily install/remove, configer, and update packages made available by developers in on-line *repositories*. The checking of software dependencies and placement/replacement of files and commands are performed automatically.

We have used YUM and APT commands to install/update packages, especially in this chapter. Let's give a brief summary of package management commands here.

YUM and RPM

On CentOS/Fedora, the **yum** command is used for package management. It is basically a front end for the lower level *rpm* tool.

- **yum** install *package-name* ...—Installs the specified packages along with any required dependencies.

- **yum** groupinstall *group-name* ...—Installs the specified package groups along with any required dependencies.

- **yum** erase *package-name* ...—Removes the specified packages from your system.

- **yum** search *string*—Searches the list of packages for names and descriptions that contain the fixed *string* and displays the matching package names, with architectures and a brief description of the package contents.

- **yum** `deplist` *package-name*—Displays a list of all libraries and modules on which the given package depends.

- **yum** `check-update`–Checks and lists available updates to installed packages.

- **yum** `info` *package-name*—Displays the name, description, version, size, and other useful information of the software.

- **yum** `reinstall` *package-name ...*—Removes and then installs a new copy of each given package.

- **yum** `localinstall` *local-rpm-file*—Installs without having to download.

- **yum** `update` *package-name ...*—Downloads and installs all updates including bug fixes, security releases, and upgrades, as provided by the distributors of your Linux. If no package name is given, all packages will be updated.

- **yum** `groupupdate` *group-name ...*—Downloads and installs all updates for the named group.

- **yum** `upgrade`–Upgrades all packages installed in your system to the latest release.

FIGURE 8.8: GUI for YUM

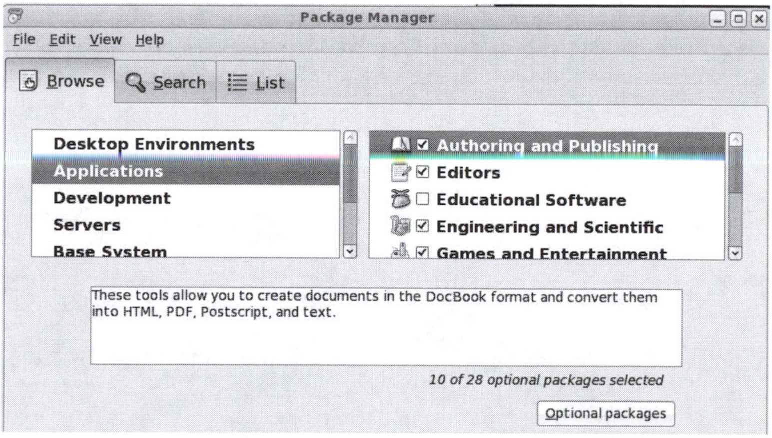

The command **pirut** provides a GUI (Figure 8.8) for **yum** and can be easier to use.

APT

On Ubuntu/Debian, use the **apt-get** command for package management.

- **sudo apt-get** `install package-name` ...—Installs the given packages, along with any dependencies.

- **sudo apt-get** `remove package-name` ...—Removes the packages specified, but does not remove dependencies.

- **sudo apt-get** `autoremove`—Removes any dependencies which remain installed but are not used by any applications.

- **sudo apt-get** `clean`–Removes downloaded package files for software already installed.

- **sudo apt-get** `purge package-name` ...—Combines the functions of `remove` and `clean` for specified packages. Also removes their configuration files.

- **sudo apt-get** `update`—Reads the `/etc/apt/sources.list` file and updates the system's database of packages available for installation. Run this after editing `sources.list`.

- **sudo apt-get** `upgrade`—Upgrades all packages if there are updates available. Run this after the command **apt-get** `update`.

The **aptitude** command is an interactive command-line front end for **apt-get** and can be more convenient to use.

8.25   For More Information

- Complete information for the Apache Web server can be found at `httpd.apache.org/`.

- The latest releases and documentation for PHP are at `php.net/index.php`.

- The site `www.mysql.com` contains all current releases and other information for MySQL.

- There is also a site for building LAMP servers at `www.lamphowto.com`.

- Linux package repositories can be easily found on the Web. For example, see

  `http://rpm.pbone.net`

  for RPM packages and

http://www.debian.org/distrib/packages

for APT packages.

- There are many textbooks on website development and design. *An Introduction to Web Design and Programming*, by Paul S. Wang and Sanda Katila, is particularly good because it combines Web programming with graphical design.

8.26   Summary

A Web server follows HTTP to receive requests and send responses. Its main function is to map incoming URIs to files and programs in the document space designated for the Web.

The Apache **httpd** Web server supports dynamic module loading and runtime configuration, making it very easy to customize and fit the requirements of a wide range of Web hosting operations. Configuration directives can be placed in central files and in *access files* under individual folders within the document space.

In addition to controlling features and behaviors of **httpd**, Apache configurations can specify access limitations to parts of the document space and can require login with HTTP Basic or Digest Authentication.

PHP is a popular active page language that can generate dynamic Web pages. PHP scripts are embedded in textual files within any number of <?php ... ?> brackets. PHP can be installed as an Apache module and will interpret embedded PHP scripts as the Apache **httpd** delivers a response page. PHP can be dynamically configured via the php.ini file.

PHP supplies a wide range of capabilities for the Web, including file inclusion, form processing, local/remote file operations, file uploading, image processing, session control, cookie support, and database access.

PHP has a built-in lightweight database, but also works well with the heavy-duty MySQL database system. MySQL supports multiple databases protected by userid and password. Different database users may have different access privileges and can be managed easily using Linux commands (**mysqladmin**, **mysql**, and so on) or the Web-based phpMyAdmin tool.

The combination Linux, Apache, MySQL, and PHP (LAMP) forms a popular and powerful Web hosting environment. The freely available LAMP makes a great developmental system, but should not be used as part of a production Web server for security reasons.

The YUM/APT package management tools are handy for installing and maintaining Linux software packages.

8.27   Exercises

1. Assuming your Linux is running the Apache Web server, find the version of Apache server, the `httpd.conf` file, and the document root folder.

2. How does one go about finding out if your Linux system supports per-user Web space?

3. Install your own Apache server with PHP support under your home directory (Hint: use a non-privileged port). After installation, start your own **httpd** and test it.

4. How does one find out if your Apache has PHP support? If so, where is the file `php.ini` and for what purpose?

5. Set up your Apache to automatically deflate `.html`, `.css`, and `.js` files.

6. Look at your `php.ini` and figure out how to enable/disable php error output.

7. Configure your Apache to require a password on some Web folder. Create some valid users and test your setting to make sure that it works.

8. Set up some database tables using the PHP built-in SQLite. Test your set up with PHP code in a Web page.

9. Install your own MySQL under your home directory. You'll be the root database user. Create a new test database and some tables using the **mysql** tool.

10. Install the phpMyAdmin tool. Use it to manage your MySQL database.

11. Set up some database tables for the Web in your MySQL using your phpMyAdmin tool. Test your set up with PHP code in a Web page.

12. Find out about the PEAR library for PHP. Install it if it is not already installed.

Chapter 9

C Programming in Linux

With a basic understanding of commands, Shell usage and programming, structure of the file system, networking, and Web hosting, you now are ready to explore Linux system programming itself, which is the subject of Chapters 9, 10, and 11.

Early on, in Chapter 1 (Section 1.10), we briefly mentioned creating, compiling, and running a program written in C. Linux supports C, C++,[1] Java, Fortran, and other languages, but C remains special for Linux.

The Linux system and many of its commands are written in the C language. C is a compact and efficient general-purpose programming language that has evolved together with UNIX and Linux. Thus, C is regarded as the native language for Linux. The portability of Linux is due, in large part, to the portability of C.

Because of its importance, C has been standardized by the American National Standards Institute (ANSI) and later by the International Organization for Standardization (ISO). The latest standard is known as ISO C99. The C99 standard specifies language constructs and a *Standard C Library* API (Application Programming Interface) for common operations, such as I/O (input/output) and string handling. Code examples in this book are compatible with ISO C99.

On most Linux distributions, you'll find

- **gcc** (or **g++**)—The compiler from GNU that compiles C (or C++) programs. These include support for ISO C99 and ISO C++ code.

- glibc—The POSIX[2]-compliant C library from GNU. A library keeps common code in one place to be shared by many programs. The glibc library package contains the most important sets of shared libraries: the standard-compliant C library, the math library, as well as national language (locale) support.

On Linux, it is easy to write a C program, compile it with **gcc**, and run the resulting executable. For creating and editing short programs, such as examples in this book, simple text editors like **gedit** and **nano** are fine. More capable editors such as **vim** and **emacs** have C editing modes for easier coding. *Integrated Development Environments* (IDEs) for C/C++ on Linux,

---

[1]C++ is a super set of C that supports *Object-Oriented Programming* (OOP).
[2]Portable Operating System Interface for UNIX.

such as **kdevelop**, Anjula, and Borland C++, are also available to manage larger programming projects.

In this and the next two chapters, we will look at facilities for programming at the C-language level and write C code to perform important operating system tasks including I/O, file access, piping, process control, inter-process communications, and networking. The material presented will enable you to implement new commands in C, as well as control and utilize the Linux kernel through its C interface.

A collection of basic topics that relates to writing C code under Linux is explored in this chapter:

- Command-line argument conventions

- Actions of the C compiler

- Standard C Libraries

- Use and maintenance of program libraries

- Error handling and recovery

- Using the **gdb** debugger

9.1   Command-Line Arguments

Commands in Linux usually are written either as Shell scripts or as C programs. Arguments given to a command at the Shell level are passed as character strings to the `main` function of a C program. A `main` function expecting arguments is normally declared as follows:

```
int main(int argc, char *argv[])
```

The parameter `argc` is an integer. The notation

```
char *argv[]
```

declares the formal array parameter `argv` as having elements of type `char *` (character pointer). In other words, each of the array elements `argv[0]`, `argv[1]`, ..., `argv[argc-1]` points to a character string. The meanings of the formal arguments `argc` and `argv` are as follows:

`argc`—The number of command-line arguments, including the command name

`argv[n]`—A pointer to the *n*th command-line argument as a character string

If the command name is **cmd**, and it is invoked as

cmd *arg1 arg2*

then

argc	is 3
argv[0]	points to the command name **cmd**
argv[1]	points to the string *arg1*
argv[2]	points to the string *arg2*
argv[3]	is 0 (NULL)

The parameters for the function **main** can be omitted (int main()) if they are not needed.

Now let's write a program that receives command-line arguments (**Ex:** ex09/echo.c). To keep it simple, all the program does is echo the command-line arguments to standard output.

```
/****** the echo command ******/
#include <stdlib.h>
#include <stdio.h>

int main(int argc, char *argv[])
{   int i = 1;                      /* begins with 1          */
    while (i < argc)
    {   printf("%s", argv[i++]);    /* outputs string         */
        printf(" ");                /* outputs SPACE          */
    }
    printf("\n");                   /* terminates output line */
    return EXIT_SUCCESS;            /* returns exit status    */
}
```

The program displays each entry of **argv** except **argv[0]**, which is actually the command name itself. The string format **%s** of **printf** is used. To separate the strings, the program displays a SPACE after each **argv[i]**, and the last argument is followed by a NEWLINE.

Exit Status

Note that **main** is declared to return an **int** and the last statement in the preceding example returns a constant defined in <stdlib.h>

```
return EXIT_SUCCESS;
```

When a program terminates, an integer value, called an *exit status* (Chapter 5, Section 5.7), is returned to the invoking environment (a Shell, for example) of the program. The exit status indicates, to the invoker of the program, whether the program executed successfully and terminated normally. An exit status **EXIT_SUCCESS** (0 on Linux) is normal, while **EXIT_FAILURE** (1 on Linux), or any other small positive integer, indicates abnormal termination. At the Linux Shell level, for example, different actions can be taken depending on the exit status (value of $?) of a command. For a C program, the return value of **main**, or the argument to a call to **exit**, specifies the exit status. Thus, `main` should always return an integer exit status even though a program does not need the quantity for its own purposes. (See Chapter 10, Section 10.14 for more discussion on the exit status.)

Compile and Execute

To compile C programs, use **gcc**. For example,

`gcc echo.c -o myecho`

Here, the executable file produced is named `myecho`, which can be run with

myecho `To be or not to be`

producing the display

`To be or not to be`

The `argv[0]` in this case is `myecho`.

The command **gcc** runs the GNU C Compiler (GCC). See Section 9.3 for more information on GCC.

9.2   Linux Command Argument Conventions

Generally speaking, Linux commands use the following convention for specifying arguments:

command [*options*] [*files*]

Options are given with a single or double hyphen (-) prefix.

-char
--word

where *char* is a single letter and *word* is a full word. For example, the **ls** command has the single-letter `-F` and the full-word `--classify` option. A command may take zero or more options. When giving more than one option, the single-letter options sometimes can be combined by preceding them with a single -. For example,

`ls -l -g -F`

can be given alternatively as

`ls -lgF`

Some commands such as **ps** and **tar** use options, but do not require a leading hyphen. Other options may require additional characters or words to complete the specification. The **-f** (script file) option of the **sed** command is an example.

A file argument can be given in any one of the three valid filename forms: simple name, relative pathname, and full pathname. A program should not expect a restricted filename or make any assumptions about which form will be supplied by a user.

9.3   The GCC Compiler

To program in C, it is important to have a clear idea of what the C compiler does and how to use it. A compiler not only translates programs into machine code to run on a particular computer, it also takes care of arranging suitable *run-time support* for the program by providing I/O, file access, and other interfaces to the operating system. Therefore, a compiler is not only computer hardware specific, but also operating system specific.

On Linux, the C compiler will likely be GCC, which is part of the GNU compiler collection. The C compiler breaks the entire compilation process into five phases (Figure 9.1).

FIGURE 9.1: Linux C Compilation Phases

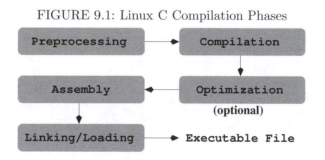

1. *Preprocessing*—The first phase is performed by the **cpp** (C preprocessor) program (or **gcc -E**). It handles constant definition, macro expansion, file inclusion, conditionals, and other preprocessor directives.

2. *Compilation*—Taking the output of the previous phase as input, the compilation phase performs syntax checking, parsing, and assembly code (.s file) generation.

3. *Optimization*—This optional phase specializes the code to the computer's hardware architecture and improves the efficiency of the generated code for speed and compactness.

4. *Assembly*—The assembler program **as** takes .s files and creates object

(.o) files containing binary code and relocation information to be used by the linker/loader.

5. *Linking*—The **collect2/ld** program is the linker/loader which combines all object files and links in necessary library subroutines as well as run-time support routines to produce an executable program (a.out).

The **gcc** command can automatically execute all phases or perform only designated phases.

The gcc Command

Because of the close relationship between C and Linux, the **gcc** command is a key part of any Linux system. The **gcc** supports traditional as well as the standard ISO C99.

Typically, the **gcc** command takes C source files (.c and .h), assembly source files (.s), and object files (.o) and produces an executable file, named a.out by default. The compiling process will normally also produce a corresponding object file (but no assembly file) for each given source file.

Once compiled, a C program can be executed. The command name is simply the name of the executable file (if it is on the command search PATH). For all practical purposes, an executable file *is* a Linux command.

Options for gcc

You can control the behavior of **gcc** by command-line options. A select subset of the available options is described here.

Please note that some options, such as -D and -I, have no space between the option and the value that follows it.

-E	Performs preprocessing only, outputs to stdout.
-S	Produces assembly code files (.s).
-c	Produces object (.o) files. No linking or a.out is done. This option is used for separate compilation of component modules in a program package.
-g or -ggdb	Includes debugging information in object/executable code for **gdb** and other debuggers.
-o *filename*	Names the executable file *filename* instead of a.out.
-O, -O2, -O3	Activates the optimization phase and performs level 1, 2, or 3 optimization. The generated code will have increasingly improved speed and, most likely, also a smaller size. Optimization algorithms slow the compiler down considerably. Apply this option only after your code has been tested and debugged and the code is ready for *production* use.

-l*libname*	Specifies *libname* as a library file to use when linking and loading the executable file. This option is passed by **gcc** to the linker/loader.
-L*dir*	Adds *dir* to the library search path.
-std=*standard*	Uses the given *standard* for C such as **ansi** or **c99**.
-v	Displays the names and arguments of all subprocesses invoked in the different phases of **gcc**. (The verbose mode.)
-D*name*=*str*	Initializes the **cpp** macro *name* to the given string *str*. This command-line option is equivalent to inserting **#define** *name str* at the beginning of a source file. If =*str* is omitted, *name* is initialized to 1.
-I*dir*	Adds the directory *dir* to the directory list that **gcc** searches for **#include** files. The compiler searches first in the directory containing the source file, then in any directories specified by the **-I** option, and then in a list of standard system directories. Multiple **-I** options establish an ordered sequence of additional **#include** file directories.
-pg	Prepares to generate an execution profile to be used with the Linux **gprof** utility.

The C Preprocessor

The C preprocessor (the **cpp** command) performs the first phase of the compilation process. The preprocessor provides important facilities that are especially important for writing system programs. Directives to the C preprocessor begin with the character # in column one. The directive

```
#include
```

is used to include other files into a source file before actual compilation begins. The included file usually contains constant, macro, and data structure definitions that usually are used in more than one source code file. The directive

```
#include "filename"
```

instructs **cpp** to include the entire contents of *filename* (note that the " marks are part of the command). If the filename is not given as a full pathname, then it is first sought in the directory where the source code containing the **#include** statement is located; if it is not found there, then some standard system directories are searched. If you have header files in non-standard places, use the **-I** option to add extra header search directories. The directive

```
#include <filename>
```

has the same effect, except the given filename is found in standard system directories. One such directory is **/usr/include**. For example, the standard header file for I/O is usually included by

```
#include <stdio.h>
```

at the beginning of each source code file. As you will see, an important part of writing a system program is including the correct header files supplied by Linux in the right order.

The **cpp** directive `#define` is used to define constants and macros. For example, after the definitions

```
#define TRUE 1
#define FALSE 0
#define TABLE_SIZE 1024
```

these names can be used in subsequent source code instead of the actual numbers. The general form is

```
#define identifier token ...
```

The preprocessor will replace the identifier with the given tokens in the source code. If no tokens are given, *identifier* is defined to be 1. Macros with parameters also can be defined using the following form:

```
#define identifier(arg1, arg2, ...) token ...
```

For example,

```
#define MIN(x,y)   ((x) > (y) ? (y) : (x))
```

defines the macro MIN, which takes two arguments. The macro call

```
MIN(a + b, c - d)
```

is expanded by the preprocessor into

```
((a+b) > (c-d) ? (c-d) : (a+b))
```

The right-hand side of a macro may involve symbolic constants or another macro. It is possible to remove a defined identifier and make it undefined by

```
#undef identifier
```

The preprocessor also handles *conditional inclusion*, where sections of source code can be included in or excluded from the compiling process, depending on certain conditions that the preprocessor can check. Conditional inclusion is specified in the general form

```
#if-condition
            source code lines A
[#else
            source code lines B ]
#endif
```

TABLE 9.1: Conditional Inclusion

If Condition	Meaning
#if constant-expression	True if the expression is non-zero
#ifdef *identifier*	True if *identifier* is #defined
#ifndef *identifier*	True if *identifier* is not #defined

If the condition is met, source code A is included; otherwise, source code B (if given) is included. The possible conditions are listed in Table 9.1.

Conditional inclusion can be used to include debugging code with something like

```
#ifdef DEBUG
        printf( ... )
#endif
```

To activate such conditional debug statements, you can either add the line

```
#define DEBUG
```

at the beginning of the source code file or compile the source code file with

gcc -DDEBUG *file*

Preventing Multiple Loading of Header Files

In larger C programs, it is common practice to have many source code and header files. The header files often have #include lines to include other headers. This situation often results in the likelihood of certain header files being read more than once during the preprocessing phase. This is not only wasteful, but can also introduce preprocessing errors. To avoid possible multiple inclusion, a header file can be written as a big conditional inclusion construct.

```
/*  A once only header file xyz.h */
#ifndef __xyz_SEEN__
#define __xyz_SEEN__
/* the entire header file*/
        .

        .

        .
#endif /* __xyz_SEEN__ */
```

The symbol __xyz_SEEN__ becomes defined once the file **xyz.h** is read by **cpp** (**Ex:** ex09/gcd.h). This fact prevents it from being read again due to the #ifndef mechanism. This macro uses the underscore prefix and suffix to minimize the chance of conflict with other macros or constant names.

Compilation

The compiling phase takes the output of the preprocessing phase and performs *parsing* and *code generation*. If a -O option is given, then the code generation invokes code optimization routines to improve the efficiency of the generated code. The output of the compilation phase is assembly code.

Assembly

Assembly code is processed by the assembler **as** to produce relocatable object code (.o).

Linking and Loading

Linking/loading produces an executable program (the **a.out** file) by combining user-supplied object files with system-supplied object modules contained in *libraries* (Section 9.5) as well as initialization code needed. GCC uses **collect2** to gather all initialization code from object code files and then calls the loader **ld** to do the actual linking/loading. The **collect2/ld** program treats its command-line arguments in the order given. If the argument is an object file, the object file is relocated and added to the end of the executable binary file under construction. The object file's symbol table is merged with that of the binary file. If the argument is the name of a library, then the library's symbol table is scanned in search of symbols that match undefined names in the binary file's symbol table. Any symbols found lead to object modules in the library to be loaded. Such library object /bin/bash: inking: command not found the same way. Therefore, it is important that a library argument be given after the names of object files that reference symbols defined in the library.

To form an executable, run-time support code (such as crt1.o, crti.o, crtbegin.o, crtend.o in /usr/lib/ or /usr/lib64/) and C library code (such as libgcc.a) must also be loaded. The correct call to **collect2/ld** is generated by **gcc**.

After all object and library arguments have been processed, the binary file's symbol table is sorted, looking for any remaining unresolved references. The final executable module is produced only if no unresolved references remain.

There are a number of options that **collect2/ld** takes. A few important ones are listed:

-l*name*   Loads the library file lib*name*.a, where *name* is a character string. The loader finds library files in standard system directories (normally /lib, /usr/lib, and /usr/local/lib) and additional directories specified by the -L option. The -l option can occur anywhere on the command line, but usually occurs at the end of a **gcc** or **collect2/ld** command. Other options must precede filename arguments.

-L*dir*   Adds the directory *dir* in front of the list of directories to find library files.

-s   Removes the symbol table and relocation bits from the executable file to save space. This is used for code already debugged.

-o *name*   Uses the given *name* for the executable file, instead of a.out.

9.4   The C Library

The C library provides useful functions for many common tasks such as I/O and string handling. Table 9.2 lists frequently used POSIX-compliant libraries. However, library functions do depend on *system calls* (Chapter 10) to obtain operating system kernel services.

TABLE 9.2: Common C Library Functions

Functions	Header	Library File
I/O: **fopen, putc, fprintf, fscanf**, ...	<stdio.h>	*standard*
String: **strcpy, strcmp, strtok**, ...	<string.h>	*standard*
Character: **isupper, tolower**, ...	<ctype.h>	*standard*
Control: **exit, abort, malloc**, ...	<stdlib.h>	*standard*
ASCII conversion: **atoi, atol, atod**, ...	<stdlib.h>	*standard*
Error handling: **perror**, EDOM, errno, ...	<errno.h>	*standard*
Time/Date: **time, clock, ctime**, ...	<time.h>	*standard*
Mathematical: **sin, log, exp**, ...	<math.h>	-lm

An application program may call the library functions or invoke system calls directly to perform tasks. Figure 9.2 shows the relations among the Linux kernel, system calls, library calls, and application programs in C. By using standard library calls as much as possible, a C application program can achieve more *system independence*.

FIGURE 9.2: Library and System Calls

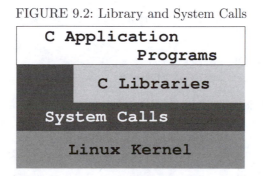

The program in Figure 9.3 implements a command **lowercase**, which copies all characters from standard input to standard output while mapping (a one-to-one transformation) all uppercase characters to lowercase ones. The I/O routines **getchar** and **putchar** are used (**Ex:** ex09/lowercase.c). The C I/O library uses a FILE structure to represent I/O destinations referred to as *C streams*. A C stream contains information about the open file, such as the buffer location, the current character position in the buffer, the mode of access, and so on.

FIGURE 9.3: Source Code File lowercase.c

```
#include <stdlib.h>
#include <stdio.h>

int main()
{   int c;
    while ( (c = getchar()) != EOF )
            putchar( tolower(c) );
    return EXIT_SUCCESS;
}
```

As mentioned before, when a program is started under Linux, three I/O streams are opened automatically. In a C program, these are three *standard C stream* pointers stdin (for standard input from your keyboard), stdout (for standard output to your terminal window), and *stderr* (for standard error to your terminal window). The header file <stdio.h> contains definitions for the identifiers stdin, stdout, and stderr. Output to stdout is buffered until a line is terminated (by \n), but output to stderr is sent directly to the terminal window without buffering. Standard C streams may be redirected to files or pipes. For example,

putc(c, stderr)

writes a character to the standard error. The routines **getchar** and **putchar** can be defined as

```
#define getchar()  getc(stdin)
#define putchar(c) putc(c, stdout)
```

Here is another example that displays the current local date and time (**Ex:** ex09/timenow.c).

```
#include <stdlib.h>
#include <stdio.h>
```

```
#include <time.h>

int main()
{   time_t now=time(NULL);      /* gets current time */
    printf(ctime(&now));        /* displays its string format */
    printf("\n");
    return EXIT_SUCCESS;
}
```

I/O to Files

The I/O library routine **fopen** is used to open a file for subsequent I/O:

FILE *fopen(char *filename, char *access_mode)

This *function prototype* describes the arguments and return value of **fopen**. We will use the prototype notation to introduce C library and Linux system calls.

To open the file passed as the second command-line argument for reading, for example, you would use

FILE *fp = fopen(argv[2], "r");

The allowable access modes are listed in Table 9.3 The file is assumed to be a text file unless the mode letter **b** is given after the initial mode letter (**r**, **w** or **a**) to indicate a binary file. I/O with binary files can be very efficient for certain applications, as we will see in the next section. Now let's explain how to use the *update* modes.

TABLE 9.3: **fopen** Modes

Mode	Opens file for
"r"	reading
"w"	writing, discarding existing contents
"a"	appending at end
"r+"	*updating* (both reading and writing)
"w+"	updating, discarding existing contents
"a+"	updating, writing at end

Because the C stream provides its own buffering, sometimes there is a need to force any output data that remains in the I/O buffer to be sent out without delay. For this the function

int fflush(FILE *stream)

is used. This function is not intended to control input buffering.

File Updating

When the same file is opened for both reading and writing under one of the modes r+, w+, and a+, the file is being updated *in place*; namely, you are modifying the contents of the file. In performing both reading and writing under the update mode, care must be taken when switching from reading to writing and vice versa. Before switching either way, an **fflush** or a file-positioning function (**fseek**, for example) must be called on the stream. These remarks will become clear as we explain how the update modes work.

The r+ mode is most efficient for making one-for-one character substitutions in a file. Under the r+ mode, file contents stay the same if not explicitly modified. Modification is done by moving a *file position indicator* (similar to a cursor in a text editor) to the desired location in the file and writing the revised characters over the existing characters already there. A **lowercase** command based on file updating can be implemented by following the steps (**Ex:** ex09/lower.c):

1. Open the given file with the r+ mode of **fopen**.

2. Read characters until an uppercase letter is encountered.

3. Overwrite the uppercase letter with the lowercase letter.

4. Repeat steps 2 and 3 until end-of-file is reached.

```
/********     lower.c     ********/
#include <stdlib.h>
#include <stdio.h>
#include <ctype.h>
#define SEEK_SET 0

int main(int argc, char *argv[])
{FILE *update;
 int fpos;  /* read or write position in file */
 char c;
 if ((update = fopen(argv[1], "r+")) == NULL)
 { fprintf(stderr, "%s: cannot open %s for updating\n",
                 argv[0], argv[1]);
   exit(EXIT_FAILURE);
 }
 while ((c = fgetc(update)) != EOF)
 { if ( isupper(c) )
   { ungetc(c, update);                   /* back up 1 char  (a) */
     fpos = ftell(update);                /* get current pos (b) */
     fseek(update, fpos, SEEK_SET);       /* pos for writing (c) */
     fputc(tolower(c), update);
     fpos = ftell(update);
```

```
        fseek(update, fpos, SEEK_SET);  /* pos for reading (d) */
    }
}                                       /*                (e) */
fclose(update);
return EXIT_SUCCESS;
}
```

After detecting an uppercase character, the file position is on the next character to read. Thus, we need to reposition the write indicator to the previous character in order to overwrite it. This is done here by backing up one character with **ungetc** (line **a**), recording the current position (line **b**), and setting the write position with **fseek** (line **c**) before putting out the lowercase character. Having done that, we can continue to process the rest of the file. However, we must set the read position with **fseek** (line **d**) before switching back to reading again.

The general form of the file position setting function **fseek** is

```
int fseek(FILE *stream, long offset, int origin)
```

The function normally returns 0, but returns −1 for error. After **fseek**, a subsequent read or write will access data beginning at the new position. For a binary file, the position is set to *offset* bytes from the indicated *origin*, which can be one of the symbolic constants

SEEK_SET   (usually 0) the beginning of the file
SEEK_CUR   (usually 1) the current position
SEEK_END   (usually 2) the end of the file

For a text stream, `offset` must be zero or a value returned by **ftell**, which gives the offset of the current position from the beginning of the file.

After end-of-file is reached, any subsequent output will be appended at the end of the file. Thus, if more output statements were given after (line **e**) in our example, the output would be appended to the file.

The `w+` mode is used for more substantial modifications of a file. A file, opened under `w+`, is read into a memory buffer and then reduced to an empty file. Subsequent read operations read the buffer and write operations add to the empty file. The mode `a+` also gives you the ability to read and write the file, but positions the write position initially at the end of the file.

I/O Redirection

The standard library function **freopen**

```
FILE *freopen(char *file, char *mode, FILE * stream)
```

connects an existing *stream*, such as `stdin`, `stdout`, or `stderr`, to the given `file`. Basically, this is done by opening the given *file* as usual but, instead of creating a new stream, assigning *stream* to it. The original file attached to *stream* is closed. For example, the statement

```
freopen("mydata", "r", stdin);
```

causes your C program to begin reading "mydata" as standard input. A successful **freopen** returns a FILE *.

For example, after the previous **freopen**, the code

```
char c = getc(stdin);
```

reads the next character from the file mydata instead of the keyboard.

A similar library function **fdopen** connects a *file descriptor* (Chapter 10, Section 10.2), rather than a stream, to a file in the same way.

A Linux system provides the Standard C Library, the X Window System library, the networking library, and more. The available library functions are all described in section 3 of the man pages.

9.5   Creating Libraries and Archives

We have mentioned that **collect2/ld** also links in libraries while constructing an executable binary file. Let's take a look at how a library is created and maintained under the Linux system. Although our discussion is oriented toward the C language and C functions, libraries for other languages under Linux are very similar.

A *subroutine library* usually contains the object code versions of functions that are either of general interest or of importance for a specific project. The idea is to avoid reinventing the wheel and to gather code that has already been written, tested, and debugged in a program library, just like books in an actual library, for all to use. Normally, the library code is simply loaded together with other object files to form the final executable program.

On Linux, a library of object files is actually one form of an *archive* file, a collection of several independent files arranged into the *archive file format*. A magic number identifying the archive file format is followed by the constituent files, each preceded by a header. The header contains such information as filename, owner, group, access modes, last modified time, and so on. For an archive of object files (a library), there is also a table of contents in the beginning identifying what symbols are defined in which object files in the archive.

The command **ar** is used to create and maintain libraries and archives. The general form of the **ar** command is

ar *key* [*position*] *archive-name file* ...

Ar will create, modify, display, or extract information from the given *archive-name*, depending on the *key* specified. The name of an archive file normally uses the .a suffix. Some more important keys are listed here.

r   To put the given files into the new or existing archive file, *archive-name*. If a file is already in the archive, it is replaced. New files are appended at the end.

q   To quickly append the given files to the end of a new or existing archive file, *archive-name*, without checking whether a file is already in the archive. This is useful for creating a new archive and to add to a very large archive.

ru   Same as **r**, except existing files in the archive are only replaced if they are older than the corresponding files specified on the command line.

ri or ra   After **ri** or **ra**, a position argument must be supplied, which is the name of a file in the archive. These are the same as **r**, except new files are inserted before (**ri**) or after (**ra**) the *position file* in the archive.

t   To display the table of contents of the archive file.

x   To extract the named files in the archive into the current directory. This, of course, will result in one or several independent files. If no list of names is given, all files will be extracted.

For example, the command (**Ex: ex09/makelibme**)

```
ar qcs libme.a file1.o file2.o file3.o
```

creates the new archive file `libme.a` by combining the given object files. The c modifier tells **ar** to create a new archive and the s modifier causes a table of contents (or index) to be included.

The command

```
ar tv libme.a
```

displays the table of contents of `libme.a`.

```
rw-rw-r-- 0/0    1240 Jul  9 16:18 2009 file1.o
rw-rw-r-- 0/0    1240 Jul  9 16:18 2009 file2.o
rw-rw-r-- 0/0    1240 Jul  9 16:18 2009 file3.o
```

If you do not wish or have permission to locate the `libme.a` file in a system library directory, you can put the library in your own directory and give the library name explicitly to **gcc** for loading. For example,

```
gcc -c myprog.c
gcc myprog.o libme.a
```

Note that `myprog.c` needs to include the header for `libme.a`, say, `me.h`, in order to compile successfully.

9.6   Error Handling in C Programs

An important aspect of system programming is foreseeing and handling errors that may occur during program execution. Many kinds of errors can occur at run time. For example, the program may be invoked with an incorrect number of arguments or unknown options. A program should guard against such errors and display appropriate error messages. Error messages to the user should be written to the **stderr** so that they appear on the terminal even if the stdout stream has been redirected. For example,

fprintf (stderr, "%s: cannot open %s\n", argv[0], argv[i]);

alerts the user that a file supplied on the command line cannot be opened. Note that it is customary to identify the name of the program displaying the error message. After displaying an error message, the program may continue to execute, return a particular value (for example, -1), or elect to abort. To terminate execution, the library routine

exit (*status*) ;

is used, where *status* is of type **int**. For normal termination, *status* should be zero. For abnormal terminal, such as an error, a positive integer *status* (usually 1) is used. The routine **exit** first invokes **fclose** on each open file before executing the system call **_exit**, which causes immediate termination without buffer flushing. A C program may use

_exit (*status*) ;

directly if desired. See Chapter 10, Section 10.14 for a discussion of **_exit**.

Errors from System and Library Calls

A possible source of error is failed system or library calls. A *system call* is an invocation of a routine in the Linux kernel. Linux provides many system calls, and understanding them is a part of learning Linux system programming. When a system or library call fails, the called routine will normally not terminate program execution. Instead, it will return an invalid value or set an external error flag. The error indication returned has to be consistent with the return value type declared for the function. At the same time, the error value must not be anything the function would ever return without failure. For library functions, the standard error values are

- EOF—The error value EOF, usually -1, is used by functions normally returning a non-negative number.

- NULL—The error value NULL, usually 0, is used by functions normally returning a valid pointer (non-zero).

- nonzero—A non-zero error value is used for a function that normally returns zero.

It is up to your program to check for such a returned value and take appropriate actions. The following idiom is in common use:

```
if ( (value = call(...)) == errvalue )
{       /* handle error here            */
        /* output any error message to stderr */
}
```

Failed Linux system calls return similar standard errors -1, 0, and so on.

To properly handle system and library call errors, the header file <errno.h> should be included.

```
#include <errno.h>
```

This header file defines symbolic error numbers and their associated *standard error messages*. For Linux systems, some of these quantities are shown in Table 9.4. You can find all the error constants in the standard C header files, usually under the folder /usr/include.

TABLE 9.4: Basic Linux Error Codes

No.	Name	Message	No.	Name	Message
1	EPERM	Not owner	27	EFBIG	File too large
2	ENOENT	No such file/dir	28	ENOSPC	No space on device
3	ESRCH	No such process	29	ESPIPE	Illegal seek
4	EINTR	Interrupted system call	30	EROFS	Read-only file system
5	EIO	I/O error	31	EMLINK	Too many links
6	ENXIO	No such device/addr	32	EPIPE	Broken pipe
7	E2BIG	Arg list too long			. . .

The external variable **errno** is set to one of these error numbers after a system or library call failure, but it is *not* cleared after a successful call. This variable is available for your program to examine. The system/library call

perror(const char *s)

can be used to display the standard error message. The call **perror(str)** outputs to standard error:

1. The argument string str

2. The COLON (':') character

3. The standard error message associated with the current value of errno

4. A NEWLINE ('\n') character

The string argument given to **perror** is usually argv[0] or that plus the function name detecting the error.

Sometimes it is desirable to display a variant of the standard error message. For this purpose, the error messages can be retrieved through the standard library function

```
char *strerrpr(int n) /* obtain error message string */
```

which returns a pointer to the error string associated with error *n*. Also, there are error and end-of-file indicators associated with each I/O stream. Standard I/O library functions set these indicators when error or end-of-file occurs. These status indicators can be tested or set explicitly in your program with the library functions

int **ferror**(FILE *s)	returns true (non-zero) if error indicator is set
int **feof**(FILE *s)	returns true if eof indicator is set
void **clearerr**(FILE *s)	clears eof and error indicators

Error Indications from Mathematical Functions

The variable **errno** is also used by the standard mathematical functions to indicate *domain* and *range* errors. A domain error occurs if a function is passed an argument whose value is outside the valid interval for the particular function. For example, only positive arguments are valid for the **log** function. A range error occurs when the computed result is so large or small that it cannot be represented as a **double**.

When a domain error happens, **errno** is set to EDOM, a symbolic constant defined in **<errno.h>**, and the returned value is implementation dependent. On the other hand, when a range error takes place, **errno** is set to ERANGE, and either zero (underflow) or HUGE_VAL (overflow) is returned.

9.7 Error Recovery

A run-time error can be treated in one of three ways:

1. Exiting—Display an appropriate error message, and terminate the execution of the program.

2. Returning—Return to the calling function with a well-defined error value.

3. Recovery—Transfer control to a saved state of the program in order to continue execution.

The first two methods are well understood. The third, error recovery, is typified by such programs as **vi**, which returns to its top level when errors occur. Such transfer of control is usually from a point in one function to a point much earlier in the program in a different function. Such *non-local* control transfer cannot be achieved with a goto statement which only works inside a function. The two standard library routines **setjmp** and **longjmp** are provided for non-local jumps. To use these routines, the header file `setjmp.h` must be included.

```
#include <setjmp.h>
```

The routine **setjmp** is declared as

```
int setjmp(jmp_buf env) /* set up longjmp position */
```

which, when called, saves key data defining the current program state in the buffer *env* for possible later use by **longjmp**. The value returned by the initial call to **setjmp** is 0. The routine **longjmp** uses the saved *env* to throw control flow back to the **setjmp** statement.

```
void longjmp(jmp_buf env, int val)
```

FIGURE 9.4: Long Jump

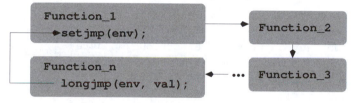

When called with a saved `env` and an integer `val` (must be nonzero), **longjmp** will restore the saved state `env` and cause execution to resume as if the original **setjmp** call has just returned the value `val`. For this *backtracking* to happen correctly, **longjmp** must be called from a function in a sequence of nested function calls leading from the function that invoked **setjmp** (Figure 9.4). In other words, **setjmp** establishes `env` as a non-local goto label, and **longjmp** is used to transfer control back to the point marked by `env`.

After the **longjmp** operation, all accessible global and local data have values as of the time when **longjmp** was called. The ANSI standard states that data values are not saved by the **setjmp** call.

Because of the way it works, **setjmp** can either stand alone or occur in the test condition part of `if`, `switch`, or `while`, and so on. The following is a simple example that shows how to use **setjmp** and **longjmp** (**Ex:** ex09/longjumptest.c).

```
#include <stdio.h>
#include <errno.h>
#include <setjmp.h>
jmp_buf env;

void recover(int n)
```

```
{    /* adjust values of variables if needed */
     longjmp(env, n);
}

void func_2(int j)
{    /* normal processing                            */
     recover(j);
}

void func_1(int i)
{    /* normal processing                            */
     func_2( i * 2);
}

int main()
{    /* initialize  and set up things here       */
     /* then call setjmp                         */
     int err=0;
     if ( (err = setjmp(env)) != 0)
     { /* return spot for longjmp                  */
       /* put any adjustments after longjmp here */
       printf("Called longjmp\n");
       printf("Error No is %d\n", err);
       return err;
     }

     /* proceed with normal processing            */

     printf("After initial setjmp()\n");
     printf("Calling func_1\n");
     func_1(19);
}
```

In this example, the function `main` sets up the eventual **longjmp** called by the function `recover`. Note that `recover` never returns. It is possible to mark several places `env1, env2, ...` with **setjmp** and use **longjmp** to transfer control to one of these marked places.

In addition to error recovery, a non-local jump can also be used to return a value directly from a deeply nested function call. This can be more efficient than a sequence of returns by all the intermediate functions. However, non-local control transfers tend to complicate program structure and should be used only sparingly.

9.8 Debugging with *GDB*

While the C compiler identifies problems at the syntax level, you still need a good tool for debugging at run time. *GDB*, the GNU debugger, is a convenient utility for source-level debugging and controlled execution of programs. Your Linux distribution will usually have it installed. The command is **gdb**.

GDB can be used to debug programs written in many source languages such as C, C++, and **f90**, provided that the object files have been compiled to contain the appropriate symbol information for use by **gdb**. This means that you use the **-g** or better the **-ggdb** option of **gcc** (Section 9.3).

Insight (`sourceware.org/insight/`) is a graphical user interface (GUI) front end for GDB. You can download and install it on your Linux if you prefer a window-menu–oriented environment for using **gdb**.

Other common debuggers include **dbx** and **sdb**. These are generally not as easy to use as **gdb**. We will describe how to use **gdb** to debug C programs. Once learned, **gdb** should be used as a routine tool for debugging programs. It is much more efficient than inserting **fprintf** lines in the source code. The tool can be used in the same way for many other programming languages.

Interactive Debugging

GDB provides an interactive debugging environment and correlates run-time activities to statements in the program source code. This is why it is called a source-level debugger. Debugging is performed by running the target program under the control of the **gdb** tool. The main features of **gdb** are listed below.

1. Source-level tracing—When a part of a program is *traced*, useful information will be displayed whenever that part is executed. If you trace a function, the name of the calling function, the value of the arguments passed, and the return value will be displayed each time the traced function is called. You can also trace specific lines of code and even individual variables. In the latter case, you'll be notified every time the variable value changes.

2. Placing source-level breakpoints—A *breakpoint* in a program causes execution to suspend when that point is reached. At the breakpoint you can interact with **gbx** and use its full set of commands to investigate the situation before resuming execution.

3. Single source line stepping—When you are examining a section of code closely, you can have execution proceed one source line at a time. (Note that one line may consist of several machine instructions.)

4. Displaying source code—You can ask **gbx** to display any part of the program source from any file.

5. Examining values—Values, declarations, and other attributes of identi-fiers can also be displayed.

6. Object-level debugging—Machine instruction-level execution control and displaying of memory contents or register values are also provided.

To debug a C program using **gdb**, make sure each object file has been compiled and the final executable has been produced with **gcc -ggdb**. One simple way to achieve this is to compile all source code (`.c`) files at once using the

gcc -ggdb *source_files*

command. This results in an executable `a.out` file suitable to run under the control of **gdb**. Thus, to use **gdb** on `lowercase.c`, you must first prepare it by

gcc -g `lowercase.c -o lowercase`

Then, to invoke **gdb**, you simply type

gdb `lowercase`

to debug the named executable file. If no file is given, `a.out` is assumed. When you see the prompt

`(gdb)`

the debugger is ready for an interactive session. When you are finished simply type the **gdb** command

quit

to exit from **gdb**. A typical debugging session should follow these steps:

1. Invoke **gdb** on an executable file compiled with the **-ggdb** option.

2. Put in breakpoints.

3. Run the program under **gdb**.

4. Examine debugging output, and display program values at breakpoints.

5. Install new breakpoints to zero in on a bug, deleting old breakpoints as appropriate.

6. Resume or restart execution.

7. Repeat steps 4-7 until satisfied.

Having an idea of what **gdb** can do, we are now ready to look at the actual commands provided by **gdb**.

Basic gdb Commands

As a debugging tool, **gdb** provides a rich set of commands. The most commonly used commands are presented in this section. These should be sufficient for all but the most obscure bugs. The complete set of commands are listed in the **gdb** manual page.

To begin execution of the target program within **gdb**, use

(gdb) **run** [*args*] [< *file1*] [> *file2*]   (start execution in **gdb**)

where *args* are any command-line arguments needed by the binary file. It is also permitted to use > and < for I/O redirection. If **lowercase** is being debugged, then

(gdb) **run** < input_file > output_file

makes sense.

However, before running the program, you may wish to put in breakpoints first. Table 9.5 lists commands for tracing.

TABLE 9.5: Simple GDB Break Commands

Command	Action
break *line*	Stops before execution of *line*
break *function*	Stops before call to *function*
break *address*	Stops at the *address*
display *expr*	Displays the C expression at a break

The **break** command can be abbreviated to **br**. Lines are specified by line numbers which can be displayed by these commands.

list   displays the next 10 lines.
list *line1,line2*   displays the range of lines.
list *function*   displays a few lines before and after *function*

When program execution under **gdb** reaches a breakpoint, the execution is stopped, and you get a (gdb) prompt so you can decide what to do and what values to examine. Commands useful at a breakpoint are in Table 9.6, where the command **bt** is short for **backtrace** which is the same as the command **where**. After reaching a breakpoint you may also single step source lines with **step** (execute the next source line) and **next** (execute up to the next source line). The difference between **step** and **next** is that if the line contains a call to a function, **step** will stop at the beginning of that function block but **next** will not.

As debugging progresses, breakpoints are put in and taken out in an attempt to localize the bug. Commands to put in breakpoints have been given. To disable or remove breakpoints, use

TABLE 9.6: GDB Commands within Breakpoints

Command	Action
bt	Displays all call stack frames
bt *n*	Displays *n* innermost frames
bt -*n*	Displays *n* outermost frames
bt full	Displays all frames and local variable values
print *expr*	Displays the expression *expr*
whatis *name*	Displays the type of *name*
cont	Continues execution
kill	Aborts execution

disable *number* . . .   (disables the given breakpoints)
enable *number* . . .   (enables disabled breakpoints)
delete *number* . . .   (removes the given breakpoints)

Each breakpoint is identified by a sequence *number*. A sequence number is displayed by **gdb** after each **break** command. If you do not remember the numbers, enter

info breakpoints   (displays information on breakpoints)

to display all currently existing breakpoints.

If you use a set of **gdb** commands repeatedly, consider putting them in a file, say, mycmds, and run **gdb** this way

gdb -x mycmds a.out

A Sample Debugging Session with gdb

Let's show a complete debugging session using the source code low.c which is a version of lowercase.c that uses the Linux I/O system calls **read** and **write** (Chapter 10, Section 10.1) to perform I/O (**Ex:** ex09/low.c).

```
#include <unistd.h>
#include <stdlib.h>
#include <stdio.h>
#include <ctype.h>
#define MYBUFSIZ 1024

int main(int argc, char* argv[])
{ char buffer[MYBUFSIZ];
  void lower(char*, int);
  int nc;              /* number of characters */
    while ((nc = read(STDIN_FILENO, buffer, MYBUFSIZ)) > 0)
    {    lower(buffer,nc);
```

```
        nc = write(STDOUT_FILENO, buffer, nc);
        if (nc == -1) break;
    }
    if (nc == -1)   /* read or write failed */
    {    perror(argv[0]);
         exit(EXIT_FAILURE);
    }
    return EXIT_SUCCESS;          /* normal termination */
}


void lower(char *buf, int length)
{    while (length-- > 0)
     {    if ( isupper( *buf ) )
              *buf = tolower( *buf );
          buf++;
     }
}
```

We now show how **gdb** is used to control the execution of this program. User input is shown after the prompt (gdb). Output from **gdb** is indented.

We first compile `lowercase.c` for debugging and invoke **gdb** (**Ex: ex09/debug**).

gcc -ggdb low.c -o low
gdb low

Now we can interact with **gdb**.

```
(gdb) list 10
5
6        int main(int argc, char* argv[])
7        { char buffer[MYBUFSIZ];
8          void lower(char*, int);
9          int nc;               /* number of characters */
10             while ((nc = read(0, buffer, MYBUFSIZ)) > 0)
11             {    lower(buffer,nc);
12                  nc = write(1, buffer, nc);
13                  if (nc == -1) break;
14             }
(gdb) br 10       (line containing system call read)
Breakpoint 1 at 0x400660: file low.c, line 10.
(gdb) br 12       (line containing system call write)
Breakpoint 2 at 0x400671: file low.c, line 12.
(gdb> br lower    (function lower)
Breakpoint 3 at 0x4006ec: file low.c, line 23.
(gdb) run < file1 > file2       (run program)
Starting program: /home/pwang/ex/bug < file1 > file2
```

```
Breakpoint 1, main (argc=1, argv=0x7fff0f4ecfa8) at low.c:10
10              while ((nc = read(0, buffer, MYBUFSIZ)) > 0)
(gdb) whatis nc
type = int
(gdb) cont
Continuing.

Breakpoint 3, lower (buf=0x7fff0f4ecab0 "It Is Time for
       All Good Men\n7", length=28) at low.c:23
23      {      while (length-- > 0)
(gdb) bt
#0  lower (buf=0x7fff0f4ecab0 "It Is Time for All
         Good Men\n7", length=28) at low.c:23
#1  0x0000000000400671 in main (argc=1, argv=0x7fff0f4ecfa8)
         at low.c:11
(gdb) whatis length
type = int
(gdb) cont
Continuing.

Breakpoint 2, main (argc=1, argv=0x7fff0f4ecfa8) at low.c:12
12                  nc = write(1, buffer, nc);
(gdb) bt
#0  main (argc=1, argv=0x7fff0f4ecfa8) at low.c:12
(gdb) cont
Continuing.

Program exited normally.
(gdb) quit
```

GDB offers many commands and ways to debug. When in **gdb**, you can use the
help command to obtain brief descriptions on commands. You can also look
for **gdb** commands matching a regular expression with **apropos** command
inside **gdb**. For example, you can type

```
(gdb) help break        (displays info on break command)
(gdb) help              (explains how to use help)
```

The GUI provided by **insight** can improve the debugging experience. For
one thing, you don't need to memorize the commands because all the available controls at any given time are clearly displayed by the **insight** window
(Figure 9.5).

FIGURE 9.5: Insight in Action

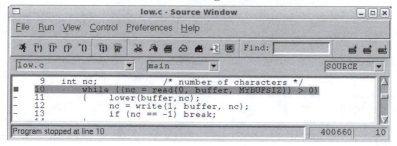

9.9 Examining Core Dumps

In our preceding example (low.c), there were no errors. When your executable program encounters an error, a core dump file is usually produced. This file, named core.*pid*, is a copy of the memory image of your running program, with the process id *pid*, taken right after the error. Examining the core dump is like investigating the scene of a crime; the clues are all there if you can figure out what they mean. A core dump is also produced if a process receives certain signals. For example, you can cause a core dump by hitting the quit character (CTRL+\) on the keyboard.

The creation of a core file may also be controlled by limitations set in your Shell. Typing the Bash command

ulimit -c

will display any limits set for core dumps. A core dump bigger than the limit set will not be produced. In particular,

ulimit -c 0

prevents core dumps all together. To remove any limitation on core dumps use

ulimit -c unlimited

You can use **gdb** to debug an executable with the aid of a core dump by simply giving the core file as an argument.

gdb *executable corefile*

Information provided by the given *corefile* is read in for you to examine. The executable that produced the *corefile* need not have been compiled with the -ggdb flag as long as the executable file passed to **gdb** was compiled with the flag.

Among other things, two pieces of important information are preserved in a core dump: the last line executed and the function call stack at the time of core dump. As it starts, **gdb** displays the call stack at the point of the core dump.

Let's look at an example. Take the following code in file `sample.c` (**Ex: ex09/sample.c**):

```
#include <stdio.h>

int main()
{    int a[10];
     int i=0, j=7;
     while (i <= 10)
           a[i++] = -i*j;
     printf("after while\n");
}
```

If you compile this file and run, you'll find that it takes forever, and the program is most likely stuck in some kind of infinite loop. However, the only loop is the `while` and it does not seem to be obviously wrong. So you hit CTRL+\ to produce a core file and use **gdb** to look into the problem.

gcc –ggdb sample.c -o bad
gdb bad core.12118

and perform a debugging session such as the following:

```
Core was generated by 'bad'.
Program terminated with signal 3, Quit
#0   0x00000000004004ed in main () at sample.c:6
6               while (i <= 10)
(gdb) list
1            #include <stdio.h>
2
3            int main()
4            {    int a[10];
5                 int i=0, j=7;
6                 while (i <= 10)
7                       a[i++] = -i*j;
8                 printf("after while\n");
9            }
(gdb) br 7
Breakpoint 1 at 0x4004d0: file sample.c, line 7.
(gdb) display i
(gdb) run
Starting program: /root/uxlx/source/09/ex/bad

Breakpoint 1, main () at sample.c:7
7                       a[i++] = -i*j;
```

```
1: i = 0
(gdb) c
Continuing.

Breakpoint 1, main () at sample.c:7
7                        a[i++] = -i*j;
1: i = 1
<<< after several more continues >>>
(gdb) c
Continuing.

Breakpoint 1, main () at sample.c:7
7                        a[i++] = -i*j;
1: i = 10                                        (Oops)
(gdb) c
Continuing.

Breakpoint 1, main () at sample.c:7
7                        a[i++] = -i*j;
1: i = -69                                       (Aha!)
```

Clearly, it was looping infinitely, and the execution inside **gdb** had to be stopped by CTRL+C. Tracing the value of the variable i shows that it became -69 after reaching 10. Now we realize that the program goes beyond the last element (a[9]), and the assignment to a[10] actually changes the value of i! The bug is due to the common mistake of going over the declared bounds of the array subscript. The fix is simple: change <= to < on line 6.

When debugging, be on the lookout for any behavior or value that you do not expect based on your program. Find out why it has deviated, and you'll find your bug.

9.10 For More Information

For the official C99 standard, see the document ISO/IEC 9899:1999 from www.iso.org/iso. For C99 features see, for example, this FAQ

www.comeaucomputing.com/techtalk/c99/#getstandard

On Linux, look for the **c99** command to compile Standard C99 programs. Use **man** gcc to display the many options for the GNU C/C++ compiler.

C library functions are documented in section 3 of the Linux manual pages. You can obtain API information for any C library function using the command **man** 3 *function_name*.

For more information on GDB, refer to

- **man** gdb

- www.gnu.org/software/gdb

- sources.redhat.com/gdb/current/onlinedocs/gdb_toc.html

9.11 Summary

The C language is native to Linux and is used to write both application and system programs. Most Linux systems support C with the GCC compiler and the POSIX run-time libraries `glibc` from GNU.

The **gcc** compiler goes through five distinct phases to compile a program: preprocessing, compiling, optimizing (optional), assembly, and linking/loading. GCC calls the preprocessor (**cpp**), the assembler (**as**), and the linker/loader (**collect2/ld**) at different phases and generates the final executable.

The *Standard C Library* is an ISO C99 API for headers and library routines. The GNU `glibc` contains Standard C Library implementations and other POSIX-compliant libraries. In addition, Linux provides many other useful libraries relating to networking, X Windows, etc.

A library is a type of archive file created and maintained using the **ar** command. You can create and maintain your own libraries with **ar**.

Standard header files provide access to system and library calls. Including the correct header files is important for C programs. Library functions, documented in section 3 of the Linux man pages, make application C programs easier to port to different platforms, whereas system calls, documented in section 2 of the man pages, access the Linux kernel directly.

Linux has well-established conventions for command-line arguments and for the reporting and handling of errors from system and library calls. The **gdb** debugger is a powerful tool for interactive run-time, source-level debugging and for analysis of a core dump. The **insight** tool provides a nice GUI for using **gdb**.

9.12 Exercises

1. Modify the **echo** implementation given in Section 9.1 so that using the `-n` option eliminates the carriage return displayed, and using the `-r` option echos the words in reverse order.

2. Write a C program **char_count** that counts the number of characters in `stdin`. Compare your program to the Linux command **wc**.

3. Write a version in C of the Shell script `clean.sh` (Chapter 5, Section 5.20). When is it a good idea to rewrite scripts in C?

4. Implement a basic **tr** command in a C program.

5. Compile several C source files into object (.o) files first. Then use **gcc** to produce the file a.out from the .o files. This should produce a working program. Give the -v option to **gcc** and see what call to the linker/loader is used.

6. Your Linux system may have more than 64 error numbers. To find out, write a C program to access the global external table sys_errlist. Hint: See **man 3 perror**.

7. System header files for C programs are kept in a few system directories. Find out which directories these are on your system.

8. Write four or five C source files containing small routines, and set up some header files that are used by these source files. Establish a library file libme.a of these routines using **ar**. Now write, compile, and run a program that applies a few of these library routines in libme.a. Compile and run your application program.

9. Write an efficient template C program for processing command-line options. The options can be given in any order anywhere on the command line.

10. Revise the lowercase.c program (Section 9.4) so that it takes optional filename arguments:

    lowercase [*infile*] [*outfile*]

    Also provide appropriate error checks.

11. Write a Linux command named **fil**. The usage synopsis is as follows:

    fil [*from*] [*to*]

    to transform text from the named file *from* to the named file *to*. If only one file argument is supplied, it is assumed to be for the *from* file. A hyphen (-) means standard input; a missing *to* means standard output. The **fil** command works as follows:

    - All tabs are replaced by an equivalent number of spaces.
    - All trailing blanks at the end of each line are removed.
    - All lines longer than 80 characters are folded, breaking lines only at spaces.

12. Apply **gdb** to debug your **fil** program.

Chapter 10

I/O and Process Control System Calls

An operating system (OS) provides many tools and facilities to make a computer usable. However, the most basic and fundamental set of services is the *system calls*, specific routines in the operating system kernel that are directly accessible to application programs. There are over 300 system calls in Linux with a kernel-defined number starting from 1. Each system call also has a meaningful *name* and a *symbolic constant* in the form SYS_*name* for its number. With a few exceptions, a system call *name* corresponds to the routine sys_*name* in the Linux kernel source code.

A program under execution is called a *process*. When a process makes a system call at run time, a software-generated interrupt, often known as an *operating system trap*, triggers the process to switch from *user mode* to *kernel mode* and to transfer control to the entry point of the target kernel routine corresponding to the particular system call. A process running in kernel mode can execute instructions that are not available in user mode. Upon system call completion, the process switches back to user mode.

Higher level system facilities are built by writing library programs that use the system calls. Because Linux is implemented in C, its system calls are specified in C syntax and directly called from C programs.

Important Linux system calls are described here. These allow you to perform low-level input/output (I/O), manipulate files and directories, create and control multiple concurrent processes, and manage interrupts. Examples show how system calls are used and how to combine different system calls to achieve specific goals.

Just like library functions, a system call may need one or more associated header files. These header files are clearly indicated with each call described.

The set of system calls and their organization form the C-language interface to the Linux kernel, and this interface is nearly uniform across all major Linux distributions. The reason is because Linux systems closely follow POSIX (Portable Operating System Interface), an open operating system interface standard accepted worldwide. POSIX is produced by IEEE (Institute of Electrical and Electronics Engineers) and recognized by ISO (International Organization for Standardization) and ANSI (American National Standards Intitute). By following POSIX, software becomes easily portable to any POSIX-compliant OS.

Documentation for any system call *name* can be found with

man 2 *name*

in section 2 of the man pages (Section 1.11).

10.1 System-Level I/O

High-level I/O routines such as **putc** and **fopen**, which are provided in the Standard C Library (Chapter 9), are adequate for most I/O needs in C programs. These library functions are built on top of low-level calls provided by the operating system. In Linux, the I/O stream of C (Chapter 9, Section 9.4) is built on top of the *I/O descriptor* mechanism supported by system calls (Figure 10.1).

FIGURE 10.1: I/O Layers

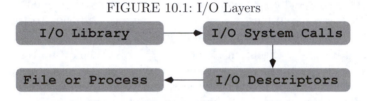

Getting to know the low-level I/O facilities will not only provide insight on how the library functions work, but will also allow you to use I/O in ways not supported by the Standard C Library.

Linux features a uniform interface for I/O to files and devices, such as a terminal window or an optical drive, by representing I/O hardware as *special files*. We shall discuss I/O to files, understanding they apply also to devices, which are nothing but special files. In addition to files, Linux supports I/O between *processes* (concurrently running programs) through abstract structures known as *pipes* and *sockets* (Chapter 11). Although files, pipes, and sockets are different I/O objects, they are supported by many of the same low-level I/O calls explained here.

10.2 I/O Descriptors

Before file I/O can take place, a program must first indicate its intention to Linux. This is done by the **open** system call declared as follows:

```
#include <sys/types.h>
#include <sys/stat.h>
#include <fcntl.h>
int open(const char *filename, int access [, mode_t mode])
```

Arguments to **open** are

filename	character string for the pathname to the file
access	an integer code for the intended access
mode	the protection mode for creating a new file

The call opens **filename**, for reading and/or writing, as specified by **access** and returns an integer *descriptor* for that file. The filename can be given in any of the three valid forms: full pathname, relative pathname, or simple filename. The **open** command is also used to create a new file with the given name. Subsequent I/O operations will refer to this descriptor rather than to the filename. Other system calls return descriptors to I/O objects such as pipes (Section 11.2) and sockets (Section 11.6). A descriptor is actually an index to a per-process *open file table* which contains necessary information for all open files and I/O objects of the process. The **open** call returns the lowest index to a currently unused table entry. Each table entry leads, in turn, to a kernel open file table entry. All processes share the same kernel open file table (Figure 10.2) and it is possible for file descriptors from different processes to refer to the same kernel table entry.

FIGURE 10.2: Open File Tables

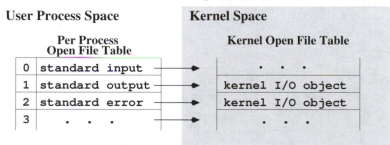

For each process, three file descriptors, STDIN_FILENO (0), STDOUT_FILENO (1), and STDERR_FILENO (2), are automatically opened initially, allowing ready access to standard I/O. The **access** code is formed by the *logical or* (|) of header-supplied single-bit values including

O_RDONLY	to open file for reading only
O_WRONLY	to open file for writing only
O_RDWR	to open file for reading and writing
O_NDELAY	to prevent possible *blocking*
O_APPEND	to open file for appending
O_CREAT	to create file if it does not exist
O_TRUNC	to truncate size to 0
O_EXCL	to produce an error if the O_CREAT bit is on and file exists

Opening a file with O_APPEND instructs each write on the file to be appended to the end. If O_TRUNC is specified and the file exists, the file is truncated to length zero. If access is

(O_EXCL | O_CREAT)

and if the file already exists, **open** returns an error. The purpose is to avoid destroying an existing file.

The third and optional argument to **open** is a file creation mode in case the O_CREAT bit is on. The mode is a bit pattern (of type mode_t from <sys/types.h> with symbolic values from <sys/stat.h>) explained in detail in Section 10.4, where the **creat** system call is described.

If the **open** call fails, a −1 is returned; otherwise, a descriptor is returned. A process may have no more than a maximum number of descriptors open simultaneously. This limit is large enough in Linux to be of no practical concern.

The following example (**Ex:** ex10/open.c) shows a typical usage of the **open** system call. The third argument to **open** is unused because it is not needed for the read-only (O_RDONLY) operation. In this case, any integer can be used as the third argument.

```
/*******    open.c    *******/
#include <stdlib.h>
#include <stdio.h>
#include <fcntl.h>

int main(int argc, char *argv[])
{   int fd;    /* file descriptor */
    /* open argv[1] for reading */
    if ((fd = open(argv[1], O_RDONLY, 0)) == -1)
    {   fprintf(stderr,"%s: cannot open %s\n",
                      argv[0], argv[1]);
        perror("open system call");
        exit(EXIT_FAILURE);
    }

/* other code */
}
```

When a system or library call fails, you can use the code

perror (const char* msg)   (displays system error)

to display the given message msg followed by a standard error message associated with the error (Chapter 9, Section 9.6).

When a descriptor fd is no longer needed in a program, it can be deleted from the per-process open file table using the call

```
int close(int fd)              (closes descriptor)
```

Otherwise, all open file descriptors will be closed when the program terminates.

10.3   Reading and Writing I/O Descriptors

Reading and writing are normally sequential. For each open descriptor, there is a *current position* which points to the next byte to be read or written. After k bytes are read or written, the current position, if movable, is advanced by k bytes. Whether the current position is movable depends on the I/O object. For example, it is movable for an actual file but not for stdin when connected to the keyboard.

The system calls **read** and **write** are declared as

```
#include <unistd.h>
ssize_t read(int fd, void *buf, size_t count);
```

```
ssize_t read(int fd, void *buffer, size_t k)    (reads input from fd)
ssize_t write(int fd, void *buffer, size_t k)   (writes output to fd)
```

where fd is a descriptor to read from or write to, buffer points to an array to receive or supply the bytes, and k is the number of bytes to be read in or written out. Obviously, k must not exceed the length of buffer. **Read** will attempt to read k bytes from the I/O object represented by fd. It returns the number of bytes actually read and deposited in the buffer. The type size_t is usually **unsigned int** (non-negative) and ssize_t is usually **int** (can be negative). If **read** returns less than k bytes, it does not necessarily mean that end-of-file has been reached, but if zero is returned, then the end of the file has been reached.

The **write** call outputs k bytes from the buffer to fd and returns the actual number of bytes written out. Both **read** and **write** return a −1 if they fail.

As an example, we can write a **readline** function with low-level **read** (**Ex:** ex10/readline.c).

```
int readline(char s[], int size)
{   char *tmp = s;
    /* read one character at a time */
    while (0 < --size && read(0, tmp, 1) != 0
                && *tmp++ != '\n');      /* empty loop body   */
    *tmp = '\0';                         /* string terminator */
    return tmp-s; /* number of characters before terminator */
}
```

The `while` loop control is intricate and warrants careful study. The `size` argument is the capacity of the array s. The function returns the number of characters read, not counting the string terminator.

For a complete program, the **lowercase** command (Chapter 9, Figure 9.3) has been rewritten with I/O system calls (**Ex: ex10/lowercase.c**).

```
/********    lowercase.c with I/O system calls    ********/
#include <ctype.h>
#include <stdlib.h>
#include <stdio.h>
#include <unistd.h>

void lower(char *buf, int length)
{   while (length-- > 0)
    {     *buf = tolower( *buf );
          buf++;
    }
}

int main(int argc, char *argv[])
{   char buffer[BUFSIZ];
    ssize nc;          /* number of characters */
    while ((nc = read(STDIN_FILENO, buffer, BUFSIZ)) > 0)
    {     lower(buffer,nc);
          nc = write(STDOUT_FILENO, buffer, nc);
          if (nc == -1) break;
    }
    if (nc == -1)      /* read or write failed */
    {     perror("read/write call");
          exit(EXIT_FAILURE);
    }
    return EXIT_SUCCESS;
}
```

Compared with the version in Chapter 9, Figure 9.3, which uses **putchar**, the program shows the difference between implicit and explicit I/O buffering.

Moving the Current Position

When reading or writing an I/O object that is an actual file, the object can be viewed as a sequence of bytes. The current position is moved by the **read** and **write** operations in a sequential manner. As an alternative to this, the system call **lseek** provides a way to move the current position to any location and therefore allows *random access* to bytes of the file. The standard library function **fseek** (Chapter 9, Section 9.4) is built on top of **lseek**. The call

```
#include <sys/types.h>
```

```
#include <unistd.h>
off_t lseek(int fd,
          off_t offset, int origin)     (moves read/write position)
```

moves the current position associated with the descriptor **fd** to a byte position defined by (**origin** + **offset**). Table 10.1 shows the three possible origins.

TABLE 10.1: The **lseek** Origins

Origin	Position
SEEK_SET	The beginning of a file
SEEK_CUR	The current position
SEEK_END	The end of a file

The offset can be positive or negative. The call **lseek** returns the current position as an integer position measured from the beginning of the file. It returns -1 upon failure. Several calls are illustrated in Table 10.2.

TABLE 10.2: Use of `lseek`

Call	Meaning
lseek(fd, (off_t)0, SEEK_SET)	Puts current pos at first byte.
lseek(fd, (off_t)0, SEEK_END)	Moves current pos to end of the file.
lseek(fd, (off_t)-1, SEEK_END)	Puts current pos at last byte.
lseek(fd, (off_t)-10, SEEK_CUR)	Backs up current pos by 10 bytes.

It is possible to **lseek** beyond the end of file and then **write**. This creates a *hole* in the file which does not occupy file space. Reading a byte in such a hole returns zero.

In some applications, holes are left in the file on purpose to allow easy insertion of additional data later. It is an error to **lseek** a non-movable descriptor such as the STDIN_FILENO. See the example code package (**Ex: ex10/lowerseek.c**) for an implementation of the lowercase program using **lseek** and O_RDWR.

10.4   Operations on Files

System calls are provided for creating and deleting files, accessing file status information, obtaining and modifying protection modes, and other attributes of a file. These will be described in the following subsections.

Creating and Deleting a File

For creating a new file, the **open** system call explained in the previous section can be used. Alternatively, the system call

```
int creat(char *filename, int mode)          (creates a new file)
```

can also be used. If the named file already exists, it is truncated to zero length, and ready to be rewritten. If it does not exist, then a new directory entry is made for it, and **creat** returns a file descriptor for writing this new file. It is equivalent to

open(`filename, (O_CREAT|O_WRONLY|O_TRUNC), mode`)

The lower 9 bits of `mode` (for access protection) are modified by the file creation mask `umask` of the process using the formula

(~umask) & mode

The `mode` is the *logical or* of any of the basic modes shown in Table 10.3.

The initial `umask` value of a process is inherited from the parent process of a running program. We have seen how to set `umask` using the Bash **umask** command (Chapter 2, Section 2.12). The default `umask` is usually 0022, which clears the write permission bits for *group* and *other* (Chapter 6, Section 6.3). A program can set `umask` with the system call

```
#include <sys/types.h>
#include <sys/stat.h>

mode_t umask(mode_t mask);
```

The returned value is the old `umask`. For example,

umask(`0077`);

will force file modes for newly created files to allow file access only for the owner. The value of `umask` is inherited by child processes. After a file is created, it can be read/written with the **read**, **write** calls.

Linking and Renaming Files

For an existing file, alternative names can also be given. The call **link**

```
#include <unistd.h>
int link(const char *file, const char *name)      (a hard link)
int symlink(const char *file, const char *name)   (a symbolic link)
```

establishes another `name` (directory entry) for the existing `file`. The new name is a hard link and can be anywhere within the same filesystem (Chapter 6, Section 6.5). To remove a link, the call

```
int unlink(const char *name)          (deletes file link)
```

is used. When the link removed is the last directory entry pointing to this file, then the file is deleted.

Use a symbolic link (the **symlink** system call) for a directory or a file in a different filesystem.

At the Shell level, renaming a file is done with the **mv** command. At the system call level, use

```
#include <stdio.h>
int rename(const char* old_name, const char* new_name)
```

Both filenames must be within the same filesystem. When renaming a directory, the new_name must not be under old_name.

Accessing File Status

FIGURE 10.3: File Status Structure

```
struct stat
{ dev_t     st_dev;     /* ID containing file    */
  ino_t     st_ino;     /* i-number              */
  mode_t    st_mode;    /* file mode             */
  nlink_t   st_nlink;   /* number of hard links  */
  uid_t     st_uid;     /* user ID of owner      */
  gid_t     st_gid;     /* group ID of owner     */
  dev_t     st_rdev;    /* special file ID       */
  off_t     st_size;    /* total bytes           */
  blksize_t st_blksize; /* filesystem  blocksize */
  blkcnt_t  st_blocks;  /* No. of blocks allocated */
  time_t    st_atime;   /* last access time      */
  time_t    st_mtime;   /* last modification time */
  time_t    st_ctime;   /* last status change time */
};
```

For each file, Linux maintains a set of *status* information such as file type, protection modes, time when last modified and so on. The status information is kept in the i-node (Chapter 6, Section 6.5) of a file. To access file status information from a C program, the following system calls can be used.

```
#include <sys/types.h>
#include <sys/stat.h>
#include <unistd.h>
int stat(const char *file, struct stat *buf)    (of file)
int fstat(int fd, struct stat *buf)             (of descriptor fd)
```

```
int lstat(const char *link, struct stat *buf)    (of the symbolic link)
```

Note that **fstat** is the same as **stat**, except it takes a file descriptor that has been opened already. This parallel exists for many other system calls. The **lstat** is the same as **stat**, except the former does not follow symbolic links. The status information for the given file is retrieved and placed in **buf**. Accessing status information does not require read, write, or execute permission for the file, but all directories listed in the pathname leading to the file (for **stat**) must be reachable.

The **stat** structure (Figure 10.3) has many members. Table 10.3 and Table 10.4 list the symbolic constants for interpreting the value of the **stat** member **st_mode**.

TABLE 10.3: Basic File Modes

Octal Bit Pattern	Symbol	Meaning
00400, 00200, 00100	S_IRUSR, S_IWUSR, S_IXUSR	r, w, or x by u
00040, 00020, 00010	S_IRGRP, S_IWGRP, S_IXGRP	r, w, or x by g
00004, 00002, 00001	S_IROTH, S_IWOTH, S_IXOTH	r, w, or x by o
00700, 00070, 00007	S_IRWXU, S_IRWXG, S_IRWXO	rwx by u, g, or o

There are three *timestamps* kept for each file:

- **st_atime** (last access time)—The time when file was last read or modified. It is affected by the system calls **mknod**, **utimes**, **read**, and **write**. For reasons of efficiency, **st_atime** is not set when a directory is searched.

- **st_mtime** (last modify time)—The time when file was last modified. It is not affected by changes of owner, group, link count, or mode. It is changed by : **mknod**, **utimes**, and **write**.

- **st_ctime** (last status change time)—The time when file status was last changed. It is set both by writing the file and by changing the information contained in the i-node. It is affected by **chmod**, **chown**, **link**, **mknod**, **unlink**, **utimes**, and **write**.

The timestamps are stored as integers, and a larger integer value represents a more recent time. Usually, Linux uses GMT (Greenwich Mean Time). The integer timestamps, however, represent the number of seconds since a fixed point in the past, known as the *POSIX epoch* which is UTC 00:00:00, January 1, 1970. The library routine **ctime** converts such an integer into an ASCII string representing date and time.

The mask **S_IFMT** is useful for determining the file type. For example,

```
if ((buf.st_mode & S_IFMT) == S_IFDIR)
```

TABLE 10.4: File Status Constants

Symbol	Bit Pattern	Meaning
S_IFMT	0170000	File type bit mask
S_IFSOCK	0140000	Socket
S_IFLNK	0120000	Symbolic link
S_IFREG	0100000	Regular file
S_IFBLK	0060000	Block device
S_IFDIR	0040000	Directory
S_IFCHR	0020000	Character device
S_IFIFO	0010000	FIFO
S_ISUID	0004000	Set-UID bit
S_ISGID	0002000	Set-group-ID bit
S_ISVTX	0001000	Sticky bit

determines whether the file is a directory.

As an application, let's consider a function `newer` (**Ex:** ex10/newer.c) which returns 1 if the last modify time of `file1` is more recent than that of `file2` and returns 0 otherwise. Upon failure, `newer` returns -1.

```
/********    newer.c    ********/
#include <sys/types.h>
#include <sys/stat.h>
#include <stdio.h>
#include <stdlib.h>

/* test if file1 is more recent than file2 */
int newer(const char *file1, const char *file2)
{   int mtime(const char *file);
    int t1 = mtime(file1), t2 = mtime(file2); /* timestamps */
    if ( t1 < 0 || t2 < 0) return -1;         /* failed     */
    else if (t1 > t2) return 1;
    else return 0;
}

int mtime(const char *file)     /* last modify time of file  */
{   struct stat stb;
    if (stat(file, &stb) < 0)   /* result returned in stb    */
        return -1;              /* stat failed               */
    return stb.st_mtime;        /* return timestamp          */
}
```

The `stb` structure in the function `mtime` is a return argument supplied to the `stat` system call to collect the status information of a file.

The `newer` function can be used in a `main` program such as

```
int main(int argc, char* argv[])
{   if ( argc == 3 )
    {  if  ( newer(argv[1], argv[2]) )
           return EXIT_SUCCESS;    /* exit status for yes */
       else
           return 1;   /* exit status for no */
    }
    else
    {  fprintf(stderr, "Usage: %s file1 file2\n", argv[0]);
       return -1;
    }
}
```

Note that the correct exit status is returned for logic at the Shell level via the special variable $? (Chapter 5, Section 5.7).

Determining Allowable File Access

It is possible to determine whether an intended read, write or execute access to a file is permissible before initiating such an access. The **access** system call is defined as

```
#include <unistd.h>
int access(const char *file, int a_mode)        (checks access to file)
```

The **access** call checks the permission bits of *file* to see if the intended access given by a_mode is allowable. The intended access mode is a *logical or* of the bits R_OK, W_OK, and X_OK defined by

```
#define R_OK  4     /* test for read permission               */
#define W_OK  2     /* test for write permission              */
#define X_OK  1     /* test for execute (search) permission */
#define F_OK  0     /* test for presence of file              */
```

If the specified access is allowable, the call returns 0; otherwise, it returns −1.

Specifying a_mode as F_OK tests whether the directories leading to the file can be searched and whether the file exists.

10.5   Operations on Directories

Creating and Removing a Directory

In addition to files, it is also possible to establish and remove directories with Linux system calls. The system call **mkdir** creates a new directory.

```
#include <sys/stat.h>
#include <sys/types.h>
int mkdir(const char *dir, mode_t mode)     (makes a new folder)
```

It creates a new directory with the name `dir`. The `mode` works the same way as in the **open** system call. The new directory's owner ID is set to the effective user ID of the process. If the parent directory containing `dir` has the set-group-ID bit on, or if the filesystem is mounted with BSD (Berkeley UNIX) group semantics, the new directory `dir` will inherit the group ID from its parent folder. Otherwise, it will get the effective group ID of the process.

The system call **rmdir**

```
#include <unistd.h>
int rmdir(const char *dir)          (removes a folder)
```

remove the given directory `dir`. The directory must be empty (having no entries other than `.` and `..`). For both **mkdir** and **rmdir**, a 0 returned value indicates success, and a −1 indicates an error. The content of a directory consists mainly of file names (strings) and i-node numbers (i-number). The length limit of a simple fime name depends on the filesystem. Typically, simple file names are limited to a length of 255 characters.

The system call **getdents** can be used to read the contents of a directory file into a character array in a system-independent format. However, a more convenient way to access directory information is to use the directory library functions discussed in the next section.

10.6   Directory Access

In the Linux file system, a directory contains the names and i-numbers of files stored in it. Library functions are available for accessing directories. To use any of them, be sure to include these header files:

```
#include <sys/types.h>
#include <dirent.h>
```

To open a directory, use either

```
DIR *opendir(const char *dir)        (opens directory stream)
```
or
```
DIR *fdopendir(int fd)               (opens directory stream)
```

to obtain a *directory stream* pointer (`DIR *`) for use in subsequent operations. If the named directory cannot be accessed, or if there is not enough memory to hold the contents of the directory, a `NULL` (invalid pointer) is returned.

Once a directory stream is opened, the library function **readdir** is used to sequentially access its entries. The function

```
#include <sys/types.h>
#include <dirent.h>
struct dirent *readdir(DIR *dp)       (returns next dir entry from dp)
```

returns a pointer to the next directory entry. The pointer value becomes NULL on error or reaching the end of the directory.

The directory entry structure **struct dirent** records information for any single file in a directory.

```
struct dirent
{   ino_t            d_ino;       /* i-node number of file    */
    off_t            d_off;       /* offset to the next dirent */
    unsigned short   d_reclen;    /* length of this record    */
    unsigned char    d_type;      /* file type                */
    char             d_name[256]; /* filename                 */
};
```

Each file in a filesystem also has a unique *i-node number* (Chapter 6, Section 6.5). The NAME_MAX constant, usually 255, gives the maxima length of a directory entry name. The data structure returned by **readdir** can be overwritten by a subsequent call to **readdir**.

The function

closedir(DIR *dp)   (closes directory stream)

closes the directory stream **dp** and frees the structure associated with the DIR pointer.

To illustrate the use of these library functions, let's look at a function **searchdir** (Figure 10.4) which searches **dir** for a given **file** and returns 1 or 0 depending on whether the file is found or not (**Ex: ex10/searchdir.c**). Note that the example uses knowledge of the **dirent** structure. Enumeration constants FOUND and NOT_FOUND are used. The **for** loop goes through each entry in **dir** to find **file**. Note the logical not (!) in front of **strcmp**.

Current Working Directory

The library routine

char *get_current_dir_name(void);   (obtains current directory)

returns the full pathname of the current working directory. The system call

int chdir(const char *dir)   (changes directory)

is used to change the current working directory to the named directory. A value 0 is returned if **chdir** is successful; otherwise, a −1 is returned. Because the current directory is a per-process attribute, you will return to the original directory after the program exits.

FIGURE 10.4: Searching a Directory

```
#include <sys/types.h>
#include <sys/dir.h>
#include <string.h>

int searchdir(char *file, char *dir)
{   DIR *dp = opendir(dir);      /* dir pointer        */
    struct dirent *entry;        /* dir entry          */
    enum {NOT_FOUND, FOUND} flag = NOT_FOUND;
  /* go through each entry in dir */
    for (entry=readdir(dp) ;
            entry != NULL ; entry=readdir(dp))
    {   if ( ! strcmp(entry->d_name, file) ) flag = FOUND;
    }
    closedir(dp);
    return flag;
}
```

10.7   An Example: ccp

It is perhaps appropriate to look at a complete example of a Linux command written in C. The command we shall discuss is **ccp** (conditional copy), which is used to copy files from one directory to another (**Ex:** ex10/ccp.c). A particular file is copied or not depending on whether updating is necessary. A version of **ccp** implemented as a Bash script has been discussed in Chapter 5, Section 5.20.

The **ccp** command copies files from a source folder *source* to a destination folder *dest*. The usage is

ccp *source dest* [*file* ...]

The named files or all files (but not directories) are copied from *source* to *dest* subject to the following conditions:

1. If the file is not in *dest*, copy the file.

2. If the file is already in *dest* but the file in *source* is more recent, copy the file.

3. If the file is already in *dest* and the file in *source* is not more recent, do not copy the file.

To check if a file is a directory, we call the `isDir` function (line 1). To compare the recency of two files (line 2), we use the function **newer** presented in Section 10.4.

```
/********    ccp : the conditional copy command    ********/
#include <sys/param.h>
#include <stdio.h>
#include <stdlib.h>
#include <dirent.h>
#include <unistd.h>
#include <string.h>
#include <sys/stat.h>
#include "newer.h"

int isDir(const char *file)
{   struct stat stb;
    if (stat(file, &stb) < 0)  /* result returned in stb */
          return -1;           /* stat failed            */
    return ((stb.st_mode & S_IFMT) == S_IFDIR);
}


void ccp(const char* name, const char* d1, const char* d2)
{ char f1[MAXPATHLEN+1], f2[MAXPATHLEN+1];
  strcpy(f1,d1); strcpy(f2,d2); strcat(f1,"/");
  strcat(f2,"/"); strcat(f1,name); strcat(f2,name);
  if ( isDir(f1)==0 )                               /* (1) */
      if ( access(f2,F_OK) == -1 || newer(f1,f2) )  /* (2) */
          printf("copy(%s,%s)\n", f1, f2);
      else
          printf("no need to copy(%s,%s)\n", f1, f2);
}


int main(int argc,char* argv[])
{ DIR *dirp1;
  struct dirent *dp;
  if (argc < 3)         /* need at least two args */
  { fprintf(stderr, "%s: wrong number of arguments", argv[0]);
    exit(EXIT_FAILURE);
  }
  else if (argc > 3)    /* files specified */
  { int i;
    for (i = 3; i < argc; i++)
        ccp(argv[i],argv[1],argv[2]) ;              /* (3) */
    return EXIT_SUCCESS;
  }
/* now exactly two args */
```

```
if ((dirp1 = opendir(argv[1])) == NULL)
{  fprintf(stderr, "%s: can not open %s", argv[0], argv[1]);
   exit(EXIT_FAILURE);
}
for (dp = readdir(dirp1); dp != NULL;
      dp = readdir(dirp1))                              /* (4) */
    if (strncmp(dp->d_name,".", 1))
          ccp(dp->d_name,argv[1],argv[2]);
return EXIT_SUCCESS;
}
```

If files are given on the command line, we call the function `ccp` on those files (line 3). Otherwise, we go through all files whose names do not begin with a period (line 4). To compile we use

gcc ccp.c newer.c -o ccp

10.8   Shell-Level Commands from C Programs

In the `ccp.c` example, we have not performed any actual file copying. We simply used **printf** to indicate the copying actions needed. To carry out the file copying, it is most convenient to invoke a Shell-level **cp** command from within a C program. Allowing execution of Shell-level commands from within C programs is very useful. With this ability, you can, for example, simply issue a **cp** command to copy a file from a C program rather than writing your own routines. The Linux library call **system** is used for this purpose.

```
#include <stdlib.h>
int system(const char *cmd_str)      /* issues Shell command */
```

The **system** call starts a new Sh process to execute the given string `cmd_str`. The Shell terminates after executing the given command and **system** returns. The returned value represents the exit status of the given command. Thus, to copy *file1* to *file2*, you can use

```
char cmd_string[80];
sprintf(cmd_string, "cp %s %s\n", file1, file2);
system(cmd_string);
```

The string is, of course, interpreted by the Shell before the command is invoked. Any substitution and filename expansion will be done. Also, the Shell locates the executable file (for example, `/bin/cp`) on the command search path for you. Use the full pathname of the command if you do not wish to depend on the `PATH` setting. The **system** call waits until the command is finished before returning.

One shortcoming of the **system** function is that it does not allow you to receive the results produced by the command or to provide input to it. This is remedied by the library function **popen** (Chapter 11, Section 11.1).

10.9   Process Control

A key operating system kernel service is process control. A *process* is a program under execution, and in a multiprogramming system like Linux, there will be multiple processes running concurrently at any given time.

We will look at process address space, states, control structures, creation and termination, executable loading, and inter-process communication here and in later sections.

Virtual Address Space

When created, each individual process has, among other resources, memory space allocated for its exclusive use. This memory space is often referred to as the *virtual address space* (or simply address space) of a process. The address space consists of a *kernel space* which is the Linux kernel shared by all processes and a *user space* which is off limits to other processes. A process executing in *user mode* has no access to the kernel space except through system calls provided by the kernel. Upon a system call, control is transferred to a kernel address through a special signal (Chapter 10, Section 10.16) and the process switches to *kernel mode*. While in kernel mode, the process has access to both user space and kernel space. The process switches back to user mode upon return of the system call.

The process user space is organized into *shared, text, data,* and *stack* regions (Figure 10.5).

FIGURE 10.5: Memory Layout of a Process

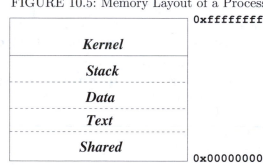

- *Stack*—A last-in-first-out data structure used to manage function calls, returns, parameter passing, and returned values. The memory used for the stack will grow and shrink with the depth of nesting of function calls.

- *Data*—The values of variables, arrays, and structures. Objects allocated at compile time will occupy fixed memory locations in the data area. Room for dynamically allocated space (through **malloc**) is also included in the data area.

- *Text*—The machine instructions that represent the procedures or functions in the program. This part of a process will generally stay unchanged over the lifetime of the process.

- *Shared*—Code from libraries that is not duplicated when shared with other processes.

In addition to the address space, each process is also assigned *system resources* necessary for the kernel to manage the process.

Process Life Cycle

Each process is represented by an entry in the *process table* which is manipulated by the kernel to manage all processes. The kernel schedules the CPU (Central Processing Unit) and switches it from running one process to the next in rapid succession. Thus, the processes appear to make progress concurrently. On a computer with multiple CPUs, a number of processes can actually run simultaneously or in parallel. A process usually goes through a number of *states* before running to completion. Figure 10.6 shows the process life cycle.

FIGURE 10.6: Process Life Cycle

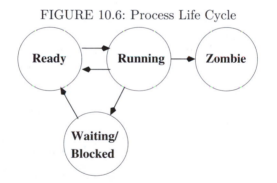

The process states are

- *Running*—The process is executing.

- *Waiting/Blocked*—A process in this state is waiting for an *event* to occur. Such an event could be an I/O completion by a peripheral device, the termination of another process, the availability of data or space in a buffer, the freeing of a system resource, and so on. When a running process has to wait for such an event, it is *blocked* and *waiting* to be unblocked so it can continue to execute. A process blocking creates an opportunity for a *context switch*, shifting the CPU to another process. Later, when the event a blocked process is waiting for occurs, it awakens and becomes *ready* to run.

- *Ready*—A process in this state is then scheduled for CPU service.

- *Zombie*—After termination of execution, a process goes into the *zombie* state. The process no longer exists. The data structure left behind contains its exit status and any timing statistics collected. This is always the last state of a process.

A process may go through the intermediate states many times before it is finished.

From a programming point of view, a Linux process is the entity created by the **fork** system call (Section 10.11). In the beginning, when Linux is booted there is only one process (process 0) which uses the **fork** system call to create process 1, known as the *init* process. The *init* process is the ancestor of all other processes, including your login Shell. Process 0 then becomes the virtual memory *swapper*.

10.10   The Process Table

A system-wide *process table* is maintained in the Linux kernel to control all processes. There is one table entry for each existing process. Each process entry contains all key information needed to manage the process, such as PID (a unique integer process ID), UID (real and effective owner and group ID's of user executing this process), process status, and generally information displayed by the **ps** command. Linux provides a directory under **/proc/** for each existing process, making it easy to access information on individual processes from the Shell level.

The ps Command

You can also obtain various kinds of information on processes with the command

ps   (displays process status)

Because Linux is a multi-user system and because there are many system processes that perform various chores to keep Linux functioning, there are always multiple processes running at any given time. The **ps** command attempts to display a reasonable set of processes that are likely to be of interest to you, and you can give options to control what subset of processes are displayed.

The **ps** command displays information only for your processes. Give the option -a display all interesting processes. Also, **ps** displays in short form unless given the option -f to see a full-format listing. For example,

ps -af

displays, in full format, all interesting processes. Use the option -e (or -A) to display all current processes, including daemon processes (those without a control terminal such as the cron process). See the **ps** man page for quite a few other options.

Information provided for each process includes

PID—The process ID in integer form
PPID—The parent process ID in integer form
S—The single-letter state code from the **ps** man page
STIME or START—The process start time
TIME—CPU time (in seconds) used by the process
TT—Control terminal of the process
COMMAND—The user command which started this process

When you are looking for a particular process, the pipe

ps −e | **grep** *string*

can be handy.

FIGURE 10.7: Process Creation

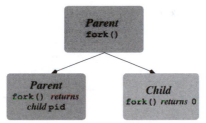

10.11   Process Creation: fork

The **fork** system call is used inside a C program to create another process.

```
#include <sys/types.h>
#include <unistd.h>
pid_t fork();
```

The process which calls **fork** is referred to as the *parent* process, and the newly created process is known as the *child* process. After the **fork** call, the child and the parent run concurrently.

The child process created is a copy of the parent process except for the following:

• The child process has a unique PID.

• The child process has a different PPID (PID of its parent).

The **fork** is called by the parent, but returns in both the parent and the child (Figure 10.7). In the parent, it returns the PID of the child process, whereas in the child, it returns 0. If **fork** fails, no child process is created, and it returns -1. Here is a template for using **fork**.

```
pit_t pid;
if ((pid = fork()) == 0)
{
        /* put code for child here                 */
}
if (pid < 0)
{
        /* fork failed, put error handling here */
}
/* fork successful, put remaining code for parent here */
```

The following simple program (**Ex: ex10/simplefork.c**) serves to illustrate process creation, concurrent execution, and the relationships between the child and the parent across the **fork** call.

```
/********     simplefork.c     ********/
#include <sys/types.h>
#include <unistd.h>
#include <stdlib.h>
#include <stdio.h>

int main()
{  pid_t child_id;
   child_id = fork();                  /* process creation (1)  */
   if ( child_id == 0 )                /* child code begin (2)  */
   {   printf("Child: My pid = %d and my parent pid = %d\n",
             getpid(), getppid());
       _exit(EXIT_SUCCESS);            /* child terminates (3)  */
   }                                   /* child code end        */
   if ( child_id < 0 )                 /* remaining parent code */
   {   fprintf(stderr, "fork failed\n");
       exit(EXIT_FAILURE);
   }
   printf("Parent: My pid = %d, spawned child pid = %d\n",
          getpid(), child_id);
   return EXIT_SUCCESS;
}
```

After calling **fork** (line 1), you suddenly have two processes, the parent and the child, executing the same program starting at the point where **fork** returns.

The child and parent execute different code sections in our example because of the way the program is written. The child only executes the part under if (child_id==0) (line 2). At the end of the child code (line 3), it must terminate execution. Otherwise, the child would continue into the code meant only for the parent. The **_exit** system call is slightly different from library function **exit** and is explained in Section 10.14. Note also that a process can use the system calls **getpid()** and **getppid()** to obtain the process ID of itself and its parent, respectively. The above program produces the following output.

```
Child: My pid = 19603 and my parent pid = 19602
Parent: My pid = 19602, spawned child pid = 19603
```

To further illustrate the use of **fork**, we can write a program where the parent and child run concurrently (**Ex:** ex10/concurrent.c). The child computes the partial sums, and the parent calculates the partial products, of an array of integers.

```
/********    concurrent.c    ********/
#include <sys/types.h>
#include <unistd.h>
#include <stdlib.h>
#include <stdio.h>
#define DIM 8

int main()
{   pid_t pid;
    int i, ans, arr[DIM]={1,2,3,4,5,6,7,8};
    pid = fork();
    if ( pid == 0 )  /* child code begin */
    {  ans = 0;
       for (i = 0 ; i < DIM ; i++)
       {   ans = ans + arr[i];
           printf("Child: sum = %d\n", ans);
           sleep(1); /* 1 sec delay */
       }
       _exit(EXIT_SUCCESS);
    }                   /* child code end   */
    if ( pid < 0 )
    {  fprintf(stderr, "fork failed\n");
       return EXIT_FAILURE;
    }
    ans = 1;
    for (i = 0 ; i < DIM ; i++)
    {   ans = ans * arr[i];
        printf("Parent: product = %d\n", ans);
        sleep(2);   /* 2 sec delay */
    }
```

```
    return EXIT_SUCCESS;
}
```

Both parent and child have access to their own copies of the array **arr**, the variable **ans**, and so on. The fact that both processes are assigning values to **ans** concurrently does not matter because the programs are running in different address spaces. The child delays 1 second after each output line, but the parent delays 2 seconds, giving each other a chance to grab the CPU and run.

Here is one possible set of output by this program.

```
Child: sum = 1
Parent: product = 1
Child: sum = 3
Child: sum = 6
Parent: product = 2
Child: sum = 10
Parent: product = 6
Child: sum = 15
Child: sum = 21
Parent: product = 24
Child: sum = 28
Child: sum = 36
Parent: product = 120
Parent: product = 720
Parent: product = 5040
Parent: product = 40320
```

Depending on the relative speed of execution and other system load factors, the output lines from the parent and the child can be interleaved in a different way.

10.12 Program Execution: exec Routines

A process can load and execute another program by *overlaying* itself with an executable file. The target executable file is read in on top of the address space of the very process that is executing, overwriting it in memory, and execution continues at the entry point defined in the file. The result is that the process begins to execute a new program under the same execution environment as the old program, which is now replaced.

This program overlay can be initiated by any one of the *exec library functions*, including **execl**, **execv**, **execve**, and several others, each a variation of the basic **execv** library function.

```
#include <unistd.h>
extern char **environ;
```

```
int execv(const char *filename, char *const argv[]);
```

where `filename` is the full pathname of an executable file, and `argv` is the command-line arguments, with `argv[0]` being command name.

This **execv** call overlays the calling process with a new executable program. If **execv** returns, an error has occurred. In this case the value returned is −1. The argument `argv` is an array of character pointers to `null`-terminated character strings. These strings constitute the argument list to be made available to the new process. By convention, at least one argument must be present in this array, and the first element of this array should be the name of the executed program (i.e., the last component of `filename`). To the calling program, a successful **execv** never returns.

Other *exec functions* may take different arguments but will work the same way as **execv**. To avoid confusion, we will refer to all of them as an *exec call*.

An *exec call* is often combined with **fork** to produce a new process which runs another program.

1. Process A (the parent process) calls **fork** to produce a child process B.

2. Process B immediately makes an *exec call* to run a new program.

An *exec call* transforms the calling process to run a new program. The new program is loaded from the given `filename` which must be an *executable file*. An executable file is either a binary `a.out` or an *executable text file* containing commands for an interpreter. An executable text file begins with a line of the form

`#!` *interpreter*

When the named file is an executable text file, the system runs the specified *interpreter*, giving it the named file as the first argument followed by the rest of the original arguments. For example, a Bash script may begin with the line

`#!/bin/bash`

and an Sh script with

`#!/bin/sh`

As for an executable binary, Linux has adopted the standard ELF (*Executable and Linking Format*) which basically provides better support for the linking and dynamical loading of shared libraries as compared to the old UNIX `a.out` format. The command

readelf -h a.out

displays the header section of the executable `a.out`. Do a

man 5 elf

to read more about the ELF file format.

The following attributes stay the same after an *exec call*:

- Process ID, parent process ID, and process group ID

- Process owner ID, unless for a set-userid program

- Access groups, unless for a set-groupid program

- Working directory and root directory

- Session ID and control terminal

- Resource usages

- Interval timers

- Resource limits

- File mode mask (`umask`)

- Signal mask

- Environment variable values

Furthermore, descriptors which are open in the calling process usually remain open in the new process. Ignored signals remain ignored across an **exec**, but signals that are caught are reset to their default values. Signal handling will be discussed in Section 10.16.

Example: A Simple Shell

As an example, let's write a program that is a very simple Shell (**Ex:** `ex10/myshell.c`) performing the following tasks:

1. Displaying a prompt

2. Reading a command line from the terminal

3. Starting a background process to execute the command

4. Displaying another prompt and going back to step 1

This cycle is implemented by the main program:

```
/********    myshell.c    ********/
#include <sys/types.h>
#include <unistd.h>
#include <stdio.h>
#include <string.h>
#define MAXLINE 80
```

```
int main()
{   char cmd[MAXLINE];
    void background(char *cmd);
    for (;;)
    {   printf("mysh ready%%");          /* Displays prompt         */
        fgets(cmd, MAXLINE, stdin); /* Reads command          */
        if ( strcmp(cmd,"exit\n") == 0 )
            return EXIT_SUCCESS;
        background(cmd);                 /* Starts background job */
    }
    return EXIT_FAILURE;                 /* Exits abnormally       */
}
```

The function **background** prepares the **argv** array and starts a child process, which then calls **execv** to perform the given **cmd** while **background** returns in the parent process.

```
#define WHITE "\t \n"
#define MAXARG 20

void background(char *cmd)
{   char *argv[MAXARG];
    int id, i = 0;
    /* To fill in argv */
    argv[i++] = strtok(cmd, WHITE);
    while ( i < MAXARG &&
            (argv[i++] = strtok(NULL, WHITE)) != NULL );
    if ((id = fork()) == 0)   /* Child executes background job */
    {   execv(argv[0], argv);
        _exit(EXIT_FAILURE); /* execv failed */
    }
    else if ( id < 0 )
    {   fprintf(stderr, "fork failed\n");
        perror("background:");
    }
}
```

After the program is compiled and named **mysh**, run it and enter a command string as follows:

mysh
mysh ready%   */bin/ls -l*

The directory listing produced this way should match the one obtained in your usual Shell. In fact, virtually any Linux command executed with full pathname will behave the same. Type CTRL-D or the keyboard interrupt character to quit from the **mysh** program.

The **execl** routine is a convenient alternative to **execv** when the filename and the arguments are known and can be given specifically. The general form is

```
int execl(const char *name, const char *arg0, ..., const char
*argn, NULL)
```

For example,

```
execl("/bin/ls","ls","-l",NULL);
```

Since **fork** copies the entire parent process, it is wasteful when used in conjunction with an *exec call* to create a new execution context. In a virtual memory system, the system call

```
int pid; pid = vfork();
```

should be used in conjunction with an **exec call**. Unlike **fork**, **vfork** avoids much of the copying of the address space of the parent process and is therefore much more efficient. However, don't use **vfork** unless it is immediately followed by an *exec call*.

10.13   Synchronization of Parent and Child Processes

After creating a child process by **fork**, the parent process may run independently or elect to wait for the child process to terminate before proceeding further. The system call

```
#include <sys/types.h>
#include <sys/wait.h>
pid_t wait(int *t_status);
```

searches for a terminated child (in *zombie* state) of the calling process. It performs the following steps:

1. If there are no child processes, **wait** returns right away with the value −1 (an error).

2. If one or more child processes are in the zombie state (terminated) already, **wait** selects an arbitrary zombie child, frees its process table slot for reuse, stores its *termination status* (Section 10.14) in *t_status if t_status is not NULL, and returns its process ID.

3. Otherwise, **wait** sleeps until one of the child processes terminates and then goes to step 2.

When **wait** returns after the termination of a child, the variable (*t_status) is set, and it contains information about how the process terminated (normal, error, signal, etc.) You can examine the value of *t_status with predefined macros such as

```
WIFEXITED(*t_status)              (returns true if child exited normally)
WEXITSTATUS(*t_status)            (returns the exit status of child)
```

See **man** 2 `wait` for other macros and for additional forms of **wait**.

A parent process can control the execution of a child process much more closely by using the **ptrace** (process trace) system call. This system call is primarily used for interactive breakpoint debugging such as that supported by the **gdb** command (Chapter 9, Section 9.8). When the child process is *traced* by its parent, the **waitpid** system call is used, which returns when the specific child is stopped (suspended temporarily).

Let's look at a simple example of the **fork** and **wait** system calls (**Ex:** ex10/wait.c). Here the parent process calls **fork** twice and produces two child processes. Each child simply displays its own process ID and terminates. The parent process calls **wait** twice to wait for the termination of the two child processes. After each **wait**, the process ID and the wait status are displayed.

```c
/********     wait.c     ********/
#include <sys/types.h>
#include <sys/wait.h>
#include <unistd.h>
#include <stdio.h>
#include <stdlib.h>

int main()
{   pid_t pid1, pid2, pid;
    int status;
    if ((pid1 = fork()) == 0)    /* child one */
    {   printf("child pid=%d\n", getpid());
        _exit(EXIT_SUCCESS);
    }
    printf("forking again\n");
    if ((pid2 = fork()) == 0)    /* child two */
    {   printf("child pid=%d\n", getpid());
        _exit(EXIT_FAILURE);
    }
    printf("first wait\n");
    pid = wait(&status);
    printf("pid=%d, status=%d\n", pid, WEXITSTATUS(status));
    printf("2nd wait\n");
    pid = wait(&status);
    printf("pid=%d, status=%d\n", pid, WEXITSTATUS(status));
    return EXIT_SUCCESS;
}
```

Note that the second child in this example returns an exit status 1 on purpose.

10.14   Process Termination

Every running program eventually comes to an end. A process may terminate execution in three different ways:

1. The program runs to completion and the function **main** returns.

2. The program calls the library routine **exit** or the system call **_exit**.

3. The program encounters an execution error or receives an interrupt signal, causing its premature termination.

The argument to **_exit/exit** is the process *exit status* and is part of the termination status of the process. Conventionally, a zero exit status indicates normal termination and non-zero indicates abnormal termination. The system call

```
void _exit(int status)
```

terminates the calling process with the following consequences:

1. All of the open I/O descriptors in the process are now closed.

2. If the parent process of the terminating process is executing a **wait**, then it is notified of the termination and provided with the child termination status.

3. If the terminating process has child processes yet unfinished, the PPIDs of all existing children are set to 1 (the init process). Thus, the new orphan processes are adopted by the init process.

Most C programs call the library routine **exit** which performs clean-up actions on I/O buffers before calling **_exit**. The **_exit** system call is used by a child process to avoid possible interference with I/O buffers shared by parent and child processes.

10.15   The User Environment of a Process

The parameters **argc** and **argv** of a C program reference the explicit arguments given on the command line. Every time a process begins, another array of strings, representing the *user environment*, called the *environment list*, is also passed to the process. This provides another way through which to control the behavior of a process. If the function *main* is declared as

```
int main(int argc, char* argv[], char* arge[])
```

then **arge** receives additional values for the environment list which is always available for a process in the global array **environ**:

```
extern char **environ
```

Each environment string is in the form

name=value

Although direct access to **environ** is possible in a C program, it is simpler to access environment values in a C program with the library routine **getenv**:

```
#include <stdlib.h>
char* getenv(const char* name)
```

This routine searches the environment list for a string, of the form *name=value*, that matches the given **name** and returns a pointer to *value* or **NULL** if no match for **name** is found.

With **getenv** we can write a simple test program (**Ex: ex10/envtest.c**).

```
/********     envtest.c    ********/
#include <stdlib.h>
#include <stdio.h>

int main(int argc, char* argv[], char* arge[])
{   char *s;
    s = getenv("PATH");
    printf("PATH=%s\n", s);
    return EXIT_SUCCESS;
}
```

You can set environment values at the Shell level. With Bash, a variable is exported to the environment as shown in Chapter 2, Section 2.10. Environment variables and their values are contained in the environment list. Frequently used environment variables include PATH, HOME, TERM, USER, SHELL, DISPLAY, and so on (Chapter 2, Section 2.10).

In Bash, we can also pass additional environmental values to any single command by simply listing them before the command. For example,

gcc envtest.c -o envtest
foo=bar ... ./envtest

At the C level, the **execl** and **execv** library calls pass to the invoked program their current environment. The system call

```
#include <unistd.h>
int execve(const char *file, char *const argv[], char *const
envp[]);
```

can be used to pass an environment array **envp** containing additional environmental values to the new program (**Ex: ex10/execve.c**).

```
/*  passing environment with execve  */
#include <unistd.h>
#include <stdlib.h>
char* envp[3];

int main(int argc, char* argv[])
{   envp[0]="first=foo";
    envp[1]="second=bar";
    envp[2]=NULL;
    execve("target-program", argv, envp);
    exit(EXIT_FAILURE);       /* execve failed */
}
```

Example: Command Search

The **which** command

which *cmdname* ...

locates the given command names (or aliases) on the command search path
defined by the environment variable PATH (Chapter 2, Section 2.10). It displays
the full pathname of each command when located or an error message. To
illustrate the use of system and library calls further, a simplified version of
the **which** command is implemented here.

The program **mywhich** that follows is the same as the **which** command,
except it takes only one command and no aliases (**Ex: ex10/mywhich.c**). The
appropriate header files are included at the beginning:

```
/*      File: mywhich.c
 *      Usage:  mywhich cmdname
 */
#include <stdio.h>
#include <sys/param.h>       /* for MAXPATHLEN */
#include <unistd.h>          /* for access    */
#include <string.h>          /* for strncpy   */
#include <stdlib.h>          /* for getenv    */

int has_command(char* name, char* dir)
{   int ans=-1;
    char wd[MAXPATHLEN+1];
    getcwd(wd, MAXPATHLEN+1);                        /* 1 */
    if ( chdir(dir)==0 )                             /* 2 */
    {   ans = access(name, F_OK | X_OK);             /* 3 */
        chdir(wd);                                   /* 4 */
    }
    return ans==0;
}
```

Before changing, the current working directory is saved (line 1). Note that **getcwd** is a library function and not a system call. If the directory is accessible (line 2), the existence of an executable file, not directory, is tested (line 3). The working directory is restored (line 4). The function `has_command` returns 1 if the command is found; otherwise, it returns 0 .

The main program extracts individual directories on the environment variable PATH and calls `has_command` to locate the given command:

```
int main(int argc, char* argv[])
{   char* path=getenv("PATH");                    /* 5 */
    char  dir[MAXPATHLEN+1];
    int   dir_len;
    char* pt=path;
    while ( dir_len=strcspn(path, ":") )          /* 6 */
    {   strncpy(dir, path, dir_len);              /* 7 */
        dir[dir_len]='\0';                        /* 8 */
        if ( has_command(argv[1],dir) )
        {   printf("%s/%s\n", dir, argv[1]);
            return EXIT_SUCCESS;
        }
        path += dir_len+1;                        /* 9 */
    }
    printf("%s not found on\n%s\n", argv[1], pt);
    return EXIT_FAILURE;
}
```

The main program initializes **path** with the value of the environment variable PATH (line 5). The first directory on **path** is copied as a string into the variable **dir** (line 6-8) and is used in a call to `has_command`. If the command is not found in this directory, **path** is advanced to the next directory (line 9) and the iteration continues.

10.10 Interrupts and Signals

Basic Concepts

We already know that a program executes as an independent process. Yet, events outside a process can affect its execution. The moment when such an event would occur is not predictable. Thus, they are called *asynchronous* events. Examples of such events include I/O blocking, I/O ready, keyboard and mouse events, expiration of a time slice, as well as interrupts issued interactively by the user. Asynchronous events are treated in Linux using the *signal* mechanism. Linux sends a certain signal to a process to signify the occurrence of a particular event. After receiving a signal, a process will react to it in a well-defined manner. This action is referred to as the *signal disposition*. For example, the process may be terminated or suspended for later resumption. There is a system-defined default disposition associated with each signal. A process normally reacts to a signal by following the default action. However, a program also has the ability to redefine its disposition to any signal by specifying its own handling routine for the signal.

TABLE 10.5: Some Linux Signals

Symbol	Default action	Meaning
SIGHUP	Terminate	Hangup (for example, terminal window closed)
SIGINT	Terminate	Interrupt (for example, CTRL+C from keyboard)
SIGQUIT	Core dump	Quit (for example, CTRL+\ from keyboard)
SIGILL	Core dump	Illegal instruction
SIGTRAP	Core dump	Trace/breakpoint trap
SIGABRT	Terminate	Abort (**abort()**)
SIGBUS	Core dump	Memory bus error
SIGFPE	Core dump	Floating point exception
SIGKILL	Terminate	Force terminate
SIGSEGV	Core dump	Invalid memory reference
SIGALRM	Terminate	Time signal (**alarm()**)
SIGPROF	Terminate	Profiling timer alarm
SIGSYS	Core dump	Bad argument to system call
SIGCONT	Resume	Continue if stopped
SIGSTOP	Suspend	Suspends process
SIGTSTP	Suspend	Stop (for example, CTRL+Z) from terminal

There are many different signals. For instance, typing CTRL+\ on the keyboard usually generates a signal known as *quit*. Sending the quit signal to a process makes it terminate and produces a *core image* file for debugging. Each kind of signal has a unique integer number, a symbolic name, and a default action defined by Linux. Table 10.5 shows some of the many signals Linux handles. A complete list of all signals can be found with **man 7 signal**.

Sending Signals

You may send signals to processes connected to your terminal window by typing certain control characters such as CTRL+\, CTRL+C, and CTRL+Z typed at the Shell level. These signals and their effects are summarized below.

CTRL+C	SIGINT	terminates execution of foreground process
CTRL+\	SIGQUIT	terminates foreground process and dumps core
CTRL+Z	SIGTSTP	suspends foreground process for later resumption

In addition to these special characters, you can use the Shell-level command **kill** to send a specific signal to a given process. The general form of the **kill** command is

kill [−*sig_no*] *process*

where *process* is a process number (or Shell jobid). The optional argument specifies a signal number *sig_no*. If no signal is specified, `SIGTERM` is assumed which causes the target process to terminate. Recall that we used **kill** in Chapter 2, Section 2.6 where we discussed job control.

In a C program, the standard library function

int **raise**(int sig_no)   (sends `sig_no` to the process itself)

is used by a process to send the signal `sig_no` to itself, and the system call

int **kill**(pid_t pid, int sig_no)   (sends `sig_no` to process `pid`)

is used to send a specified signal to a process identified by the given numerical `pid`.

Signal Delivery and Processing

When a signal is sent to a process, the signal is added to a set of signals pending delivery to that process. Signals are delivered to a process in a manner similar to hardware interrupts. If the signal is not currently *blocked* (temporarily ignored) by the process, it is delivered to the process by the following steps:

1. Block further occurrences of the same signal during the delivery and handling of this occurrence.

2. Temporarily suspend the execution of the process and call the handler function associated with this signal.

3. If the handler function returns, then unblock the signal and resume normal execution of the process from the point of interrupt.

There is a default handler function for each signal. The default action is usually exiting or core dump (Table 10.5). A process can replace a signal handler with a handler function of its own. This allows the process to *trap* a signal and deal with it in its own way. The `SIGKILL` and `SIGSTOP` signals, however, cannot be trapped.

Signal Trapping

After receiving a signal, a process normally (by the default signal handling function) either exits (terminated) or stops (suspended). In some situations, it is desirable to react to specific signals differently. For instance, a process may ignore the signal, delete temporary files before terminating, or handle the situation with a **longjmp**.

The system call **sigaction** is used to trap or catch signals.

```
#include <signal.h>
int sigaction(int signum,
        const struct sigaction *new,
        struct sigaction *old);
```

where **signum** is the number or name of a signal to trap. The **new** (**old**) structure contains the new (old) handler function and other settings. The handling action for **signum** is now specified by **new**, and the old action is placed in **old**, if it is not **NULL**, for possible later reinstatment.

The **struct sigaction** can be found with **man 2 sigaction**, but you basically can use it in the following way:

```
struct sigaction new;
new.sa_handler=handler_function;
new.sa_flags=0;
```

The *handler_function* can be a routine you write or one that is defined by the system. If *handler_function* is **SIG_IGN**, the signal is subsequently ignored. If it is **SIG_DFL**, then the default action is restored. The new handler normally remains until changed by another call to **sigaction**. Default actions of some signals are indicated in Table 10.5. The **sa_flags** control the behavior of the signal handling. For example, **sa_flags=SA_RESETHAND** automatically resets to the default handler after the new signal handler is called once.

We now give a simple example that uses the **sigaction** system call to trap the **SIGINT** (interrupt from terminal) signal and adds one to a counter for each such signal received (**Ex: ex10/sigcountaction.c**). To terminate the program type CTRL+\ or use **kill -9**.

```
#include <signal.h>
#include <stdio.h>

void cnt(int sig)
{   static int count=0;
    printf("Interrupt=%d, count=%d\n", sig, ++count);
}

int main()
{   struct sigaction new;
    struct sigaction old;
    new.sa_handler=cnt;
    new.sa_flags=0;
    sigaction(SIGINT, &new, &old);
    printf("Begin counting INTERRUPTs\n");
    for(;;);   /* infinite loop */
}
```

If the signal handler function, such as **cnt** here, is defined to take an **int** argument (for example, `sig`), then it will automatically be called with the signal number that caused a trap to this function. Of course, counting the number of signals received is of limited application. A more practical example, `cleanup.c`, has to do with closing and deleting a temporary file used by a process before terminating due to a user interrupt (**Ex: ex10/cleanup.c**).

```c
#include <stdio.h>
#include <signal.h>
#include <stdlib.h>
FILE *tempfile=NULL;
char filename[32];

void onintr()
{   extern FILE* tempfile;
    if ( tempfile != NULL )
    {   printf("closing and deleting %s\n", filename);
        fclose(tempfile);   unlink(filename);
    }
    exit(EXIT_FAILURE);
}

/* Installs onintr() handler, if SIGINT is not being ignored */
void sigtrap(int sig)
{   struct sigaction new;
    struct sigaction old;
    new.sa_handler=SIG_IGN;
    new.sa_flags=0;
    sigaction(SIGINT, &new, &old);
    if ( old.sa_handler != SIG_IGN )
    {    new.sa_handler=onintr;
        sigaction(sig, &new, &old);
    }
}

int main()
{   extern char filename[32];
    extern FILE* tempfile;
    sigtrap(SIGINT);                            /* trap SIGINT    */
    sprintf(filename, "/tmp/%d", getpid()); /* temp file name */
    /* open temporary stream for reading and writing */
    tempfile = fopen(filename, "w+");
    /* other code of the program  */
    for(;;) sleep(3);
    /* remove temporary file before termination */
    fclose(tempfile);   unlink(filename);
    return EXIT_SUCCESS;
```

```
}
```

In this example, trapping of SIGINT is done only if it is not being ignored. If a process runs with its signal environment already set to ignore certain signals, then those signals should continue to be ignored instead of trapped. For example, the Sh arranges a background process to ignore SIGINT generated from the keyboard. If a process proceeds to trap SIGINT without checking to see if it is being ignored, the arrangement made by the Shell would be defeated.

Furthermore, as with interactive utilities such as the **vi** editor, it is often desirable to use the keyboard interrupt to *abort to the top level* within a program. This can be easily done by combining signal trapping with the **longjmp** mechanism (Chapter 9, Section 9.7).

Generally, when the signal handler function returns or when a process resumes after being stopped by CTRL+Z (SIGTTSP), a process resumes at the exact point at which it was interrupted. For interrupted system calls, the external **errno** is set to EINTR, and the system call returns -1. If interrupted while reading input from the keyboard, a process may lose a partially typed line just before the interrupt.

10.17   For More Information

For a list of Linux system calls, see the HTML version of the man page for **syscall**, which is a system call used to make all system calls. You can find the man page from the `resources` page on the book's companion website. The example code package for this book has an example (**Ex:** `ex10/sysopen.c`) demonstrating how to use **syscall**.

The POSIX standard documentation can be purchased from IEEE.

10.18   Summary

All open I/O channels are represented by *I/O descriptors*. With I/O descriptors, the Linux kernel treats file, device, and inter-process I/O uniformly. This uniformity provides great flexibility and ease in I/O programming. For I/O, a C program may use the low-level system calls or the higher level standard I/O library routines. I/O descriptors are identified by small integers. Three pre-opened descriptors 0, 1, and 2 give each process access to the standard input, output, and error output, respectively.

In addition to a complete set of file manipulation calls, Linux also offers a set of library functions for accessing directories. File- and directory-related system calls are summarized in Table 10.6.

Linux supports multiprogramming. Processes are created with **fork**, terminated with **_exit**, overlaid with another executable program with **exec**, and synchronized with **wait**. Interrupt signals can be sent from one process

TABLE 10.6: File and Directory System Calls

Call	Action
int **open**(const char *file,int a,mode_t mode)	Returns descriptor to file
ssize_t **read**(int fd,void *b,size_t k)	Reads up to k bytes into b
ssize_t **write**(int fd,const void *b,size_t k)	Writes k bytes from b to fd
int **close**(int fd)	Closes descriptor fd
off_t **lseek**(int fd,off_t offset,int pos)	Moves r/w position of fd
int **access**(const char *name,int a_mode)	Tests access permission
int **chdir**(const char *dir_name)	Changes working directory
int **link**(const char *file,const char *name)	Creates link
int **unlink**(const char *name)	Removes link
int **mkdir**(const char *name,mode_t mode)	Creates new directory
int **rmdir**(const char *dir_name)	Removes directory
int **stat**(const char *name,struct stat *buf)	Accesses file status
int **fstat**(int fd,struct stat *buf)	Accesses file status
mode_t **umask**(mode_t newmask)	Sets file permission mask

to another by **kill** and trapped by **sigaction**. The *environ[] array contains string-valued environment variables for a process which can be consulted with **getenv**.

10.19 Exercises

1. What is the difference between a file descriptor and a C file stream? Please explain.

2. Explain the effect of the umask values 077 and 022.

3. Do **cat /proc/sys/fs/file-max** to see the limit on the maximum number of open files for your system.

4. The Linux command **pwd** displays the current working directory. Write your own version of this command.

5. Write a Linux command **testaccess** that takes an access flag (-r, -w, and so on) and a filename as command-line arguments and returns an exit status of 0 or 1 depending on whether the specified access is permitted or not.

6. Write a Linux command **rmold** that takes a date string and removes all files older than the given date in the current directory. If the command is invoked with the -i flag, then the program will go into interactive mode and asks the user at the terminal for approval before actually deleting a file.

7. Write your own version of a simple **cp** program (file to file) using low-level I/O.

8. Write a program which will print out the information given by the **stat** system call for each file given as its argument.

9. How is a child process produced? How does a parent process obtain the PID of a child process? How does a child obtain the PID of its parent? How does the parent process learn about the termination of a child process?

10. What is the difference between the C **exit**() function and the _**exit**() system call? Where should each be used?

11. Consider the simple Shell in Section 10.12. Add a **wait** call to the program so that the Shell waits until the child process has finished before displaying the next prompt.

12. Modify the simple Shell in the previous exercise so that it uses the command search path.

13. Write your own version of the **system** library call.

14. Write a program that prints the value of the environment variables PATH, HOME, USER, and TERM and other variables specified as arguments on the command line.

15. Write a program **nls** which is similar to the **ls** command but which, by default, displays regular files and directories separately.

16. Write a program, using a mixture of C and Shell commands if you wish, to provide a facility which takes a C source program as input and generates a list of correctly formatted **include** statements for system header files.

17. Linux provides the **flock** system call to aid the management of mutually exclusive operations. Find out how this works and how it is used to achieve mutual exclusion.

18. The Linux system calls **semctl**, **semget**, and **semop** support *semaphores*. Find out how semaphores work and how they can be used to achieve mutual exclusion.

Chapter 11

Inter-process and Network Communication

The many applications discussed in Chapter 7 clearly illustrate the convenience and the enormous potential networking can bring. Here we will describe how to write C programs for networking and illustrate how some of the Linux networking commands are actually implemented.

As mentioned before, a networking application usually involves a client process and a server process, residing on different hosts or on the same host. At the C program level, networking simply means communication between such independent processes.

We consider two types of *inter-process communication* (ipc): ipc between related processes and ipc between unrelated processes. For processes related by **fork**, ipc can be arranged with I/O redirection and the **pipe** system call. Between unrelated processes, ipc is usually performed through the *socket* mechanism.

A processes communicates through its own socket with another socket attached to a different process. Sockets belong to different *address families*, and only sockets within the same address family can communicate with one another. Within the same address family, different types of sockets support different networking protocols. Familiarity with sockets is essential to network programming. The topic is presented in detail, and many code examples help illustrate how clients and servers work together.

11.1　Opening a Process for I/O

In the previous two chapters, we became familiar with I/O to/from files using either C streams or Linux kernel file descriptors, but I/O between processes is not very different. The simplest ipc involves a parent process and a child process. The parent initiates the child to run some program and sends input to or receives output from the child. The Standard C Library function **popen**

```
#include <stdio.h>
FILE *popen(const char *cmd_string, char *mode)
```

creates a child process to execute

sh -c *cmd_string*

and establishes a read or write stream (FILE *) to the child. The stream established is either for reading the standard output or writing the standard input of the given command, depending on whether *mode* is "r" or "w".

Once opened, the stream can be used with any of the Standard C I/O Library functions. Finally, the stream created by **popen** can be shut down by

int **pclose**(FILE *stream)

As an application of **popen**, let's write a simple program that is a version of **ls**, but lists only the names of subdirectories in a given directory (**Ex:** ex11/lsdir.c):

```
/********   lsdir.c   ********/
#include <stdio.h>
#include <stdlib.h>

int main(int argc, char* argv[])
{   int i, count, total = 0;
    size_t len=1024;
    char* line=malloc(len);
    if ( argc > 1 )  chdir(argv[1]);
    /* reads output of ls cmd  */
    FILE *in = popen("/bin/ls -ldF *\n", "r");
    while( getline(&line, &len, in) > 0 )
    {       /* reads one line of input */
       /* if a dir, displays line   */
       if ( line[0] == 'd' ) printf(line);
    }
    pclose(in);   /* closes stream  */
    free(line) ;
    return EXIT_SUCCESS;
}
```

The program uses the Linux command **ls** with the option -ldF to list the current working directory. The output is read, one line at a time, using the standard library function **getline**. If a line begins with the character d (a directory), then it is displayed by the parent process. Otherwise, we ignore the line and move on to the next. Here is a sample output.

```
drwx------ 2 pwang faculty  4096 2009-08-07 16:49 Art/
drwx------ 2 pwang faculty  4096 2009-08-08 20:31 ex/
drwx------ 2 pwang faculty  4096 2009-08-07 16:49 info/
```

The **popen** function relies on the basic *pipe* mechanism which is our next topic.

11.2 IPC with pipe

A *pipe* is a direct (in memory) I/O channel between processes. It is often used together with the system calls **fork**, **exec**, **wait**, and **_exit** to make multiple processes cooperate and perform parts of the same task. A pipe is a flexible tool to arrange ipc among **fork**-related processes.

At the Shell level, you can connect commands into a pipeline. The pipe can be thought of as a first-in-first-out character buffer (Figure 11.1) with a *read* descriptor pointing to one end and a *write* descriptor pointing to the other end. To create a pipe, the system call

```
#include <unistd.h>
int pipe(int fildes[2])
```

is used which establishes a buffer and two descriptors:

```
fildes[0]                (for reading the pipe)
fildes[1]                (for writing the pipe)
```

FIGURE 11.1: Pipe between Processes

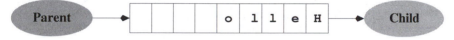

The **pipe** system call is used in conjunction with subsequent **fork** calls to establish multiple processes having access to the same pipe, thereby allowing them to communicate directly (Figure 11.2).

FIGURE 11.2: Pipe after fork()

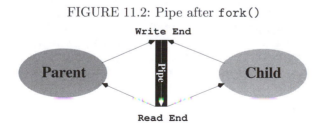

The **pipe** call returns 0 for success or −1 for failure. Consider the following piece of code:

```
int fildes[2];
pipe(fildes);            /* setting up the pipe        */
if (fork() == 0)
{   close(fildes[1]); /* child will read fildes[0]  */
    .
    .
    .
```

```
    _exit(0);
}
close(fildes[0]);        /* parent will write fildes[1] */
```

After the **fork**, both parent and child have their copies of `fildes[0]` and `fildes[1]` referring to the same pipe buffer. The child closes its write descriptor and the parent closes its read descriptor because they are not needed in this case. Now the child process can read what the parent writes into the pipe.

To perform I/O through a pipe, you use the **read** and **write** system calls on the pipe file descriptors. The call **read** removes characters from the buffer, whereas **write** adds them. The capacity of the pipe buffer is usually 4096 characters, but the buffer size is system dependent. Writing into a full pipe buffer causes the process to be blocked until more space is available in the buffer. Reading more characters than there are in the buffer results in one of the following:

1. Returning end-of-file (0) if the buffer is empty and the write end of the pipe has been closed

2. Returning what is left in the pipe if the buffer is not empty and the write end of the pipe has been closed

3. Blocking the reading process to await the arrival of additional characters if at least one file descriptor to the write end of the pipe remains open

The example (**Ex: ex11/p2cpipe.c**) below shows a parent process writing the message `"Hello there, from me."` to a child process through a pipe (Figure 11.1).

```
/********    p2cpipe.c    ********/
#include <unistd.h>
#include <stdio.h>
#include <stdlib.h>
#include <string.h>
#include <sys/wait.h>

int main(int argc, char *argv[])
{   int p[2];
    int i, status;
    pid_t pid;
    char buffer[20];
    pipe(p);                            /* setting up the pipe    */
    if ((pid = fork()) == 0)
  /* in child */
    {   close(p[1]);                    /* child closes p[1]      */
        while ((i = read(p[0], buffer, 6)) != 0)
```

```
        { buffer[i] = '\0';              /* string terminator      */
          printf("%d chars %s received by child\n", i, buffer);
        }
        _exit(EXIT_SUCCESS);             /* child terminates       */
    }
  /* in parent */
    close(p[0]);                         /* parent writes p[1]     */
    write(p[1], "Hello there,", sizeof("Hello there,")-1);
    write(p[1], " from me.", sizeof(" from me.")-1);
    close(p[1]);                         /* finished writing p[1] */
    while (wait(&status)!=pid);          /* waiting for pid        */
    if (status == 0)  printf("child finished\n");
    else printf("child failed\n");
    return EXIT_SUCCESS;
}
```

After the **fork**, both parent and child have the file descriptors p[0] and p[1]. In order to establish the parent as the sender and the child as the receiver of characters through the pipe, the child closes its own p[1] and the parent closes its own p[0]. The parent process writes to the pipe "Hello there" and " from me." in two separate **write** calls and closes its write descriptor (p[1]). In the meantime, the child reads the pipe and displays what it gets, six characters at a time (just to show multiple read operations). The following output is produced by this program:

```
6 chars :Hello : received by child
6 chars :there,: received by child
6 chars : from : received by child
3 chars :me.: received by child
child finished
```

By closing its p[1], the parent causes the pipe's write end to be completely closed—no processes can write to the pipe any more. This condition causes the final successful read in the child process to return with the last 3 characters. The next read by the child returns 0, indicating end of file.

Pipe between Two Commands

Now let's show how a Shell may establish a pipe between two arbitrary programs by combining **pipe**, **fork**, and **exec**.

A command **mypipeline** takes as arguments two command strings separated by a %. It sends the standard output of the first command to the standard input of the second command. Thus,

mypipeline /bin/ls -l % /bin/grep pwang

should work as expected (same as **ls -l | grep pwang**). Of course, we shall use a pipe between the two processes; one executing the first command and the other the second. The key in this example is connecting `stdout` in the first process to the write end of the pipe and connecting `stdin` in the second process to the read end of the pipe. This can be accomplished by the **dup2** system call (Figure 11.3).

FIGURE 11.3: Pipe and I/O Redirection

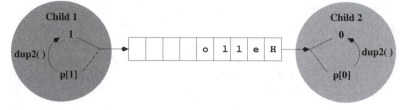

`int dup2(int fd, int copyfd)`

Dup2 duplicates an existing I/O descriptor, `fd`, which is a small non-negative integer index in the per-process descriptor table. The duplicate entry is made in the descriptor table at an entry specified by the index `copyfd`. If the descriptor `copyfd` is already in use, it is first deallocated as if a **close**(`copyfd`) had been done first. The value returned is `copyfd` if the call succeeded; otherwise, the error value returned is −1.

After **dup2**, both `fd` and `copyfd` reference the same I/O channel. In the following program (**Ex: ex11/mypipeline.c**), **dup2** is used to identify descriptor 1 (in child one) with the write end of a pipe and descriptor 0 (in child two) with the read end of the same pipe.

```
/********    mypipeline.c    ********/
#include <unistd.h>
#include <stdio.h>
#include <stdlib.h>
#include <string.h>

int main(int argc, char *argv[])
{   int p[2];
    int i,pid1,pid2, status;
    argv++;                                 /* lose argv[0] */
    for (i = 1; i <= argc ; i++)
        if (strcmp(argv[i],"%") == 0)
        {   argv[i] = '\0';          /* break into two commands */
            break;
        }
    pipe(p);                            /* setting up the pipe */
    if ((pid2 = fork ()) == 0)              /* child one       */
```

```
{   close(p[0]);
    dup2(p[1],1);        /* 1 becomes a duplicate of p[1] */
    close(p[1]);
    execv(argv[0],argv);         /* this writes the pipe */
    _exit(EXIT_FAILURE);         /* bad error execv failed */
}
if ((pid1 = fork ()) == 0)                       /* child two */
{   close(p[1]);
    dup2(p[0],0);        /* 0 becomes a duplicate of p[0] */
    close(p[0]);
    execv(argv[i+1], &argv[i+1]); /* this reads the pipe */
    _exit(EXIT_FAILURE);         /* bad error execl failed */
}
/* parent does not use pipe */
close(p[0]);    close(p[1]);
while (wait(&status)!=pid2);              /* waiting for pid2 */
if (status == 0) printf("child two terminated\n");
    else printf("child two failed\n");
return EXIT_SUCCESS;
}
```

Because open I/O descriptors are unchanged after an **exec** call, the respective programs in the two stages of the pipeline execute as usual, reading standard input and writing standard output, not knowing that these descriptors have been diverted to a pipe. The same principles are used by the Shell to establish a pipeline.

After compilation into **mypipeline**, we can run the command

./mypipeline /bin/ls -l % /bin/fgrep '.c'

and it should be entirely equivalent to

ls -l | fgrep '.c'

11.3   Connecting a File Descriptor to a File Stream

The **dup2** system call redirects I/O at the file descriptor level. At the file **stream** level, we have seen (Chapter 9, Section 9.4) the Standard C Library function **freopen**, which reconnects an existing file **stream** to another file.

In addition to these two mechanisms, there is also the standard library function **fdopen**, which establishes a **stream** that connects to an existing file descriptor.

FILE * **fdopen**(int fd, char *mode)

The function **fdopen** establishes a file **stream** with the given file descriptor **fd**. The **mode** must be compatible with that of the descriptor **fd**.

The **fdopen** call is useful when converting an **fd** into a **stream** for use with Standard C I/O Library functions. For instance, a pipe descriptor can be connected to a **stream** in this way.

11.4   Two-Way Pipe Connections

As an application, let's see how a parent process can pass some input to a child process and then receive the results produced. To the parent, the child process simply produces a well-defined result based on the input given. The desired ipc can be achieved by establishing a two-way pipe, an outgoing and an incoming pipe, between the parent and child processes (Figure 11.4).

FIGURE 11.4: A Two-Way Pipe

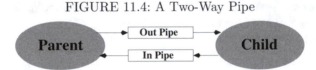

The outgoing pipe is used by the parent to send input to the child and the incoming pipe is used to receive results returned by the child. The function **pipe_2way** (**Ex: ex11/pipe2way.c**) is defined for this purpose. Given the command strings **cmd**, **pipe_2way** will establish a process to run the command and return the quantities **piped[0]** and **piped[1]**, the read end of the incoming pipe and the write end of the outgoing pipe, respectively.

```
int pipe_2way(char *cmd[], int piped[])
{    int pid, wt[2], rd[2];
     pipe(rd);                        /* incoming pipe: read by parent */
     pipe(wt);                        /* outgoing pipe: write to child */
     if ((pid=vfork()) == 0)
   /* in child */
     { close(wt[1]);
       dup2(wt[0],0);                 /* 0 identified with wt[0] */
       close(wt[0]); close(rd[0]);
       dup2(rd[1], 1);                /* 1 identified with rd[1] */
       close(rd[1]);
       execv(cmd[0],cmd);             /* execute given command   */
       perror("execv failed");        /* normally not reached     */
       _exit(EXIT_FAILURE);
     }
   /* in parent */
     close(wt[0]); piped[1] = wt[1];
```

```
        close(rd[1]); piped[0] = rd[0];
        return 0;
}
```

The return parameter, `piped`, is filled with the two proper descriptors before the function returns. To test `pipe_2way`, let's write a program that sends characters to the command **lowercase** and receives the transformed string back. The latter is performed by the `readl` function

```
int readl(int fd, char s[], int size)
{   char *tmp = s;
    while (0 < --size && read(fd, tmp, 1)!=0 && *tmp++ !='\n');
    *tmp = '\0';                /* string terminator */
    return(tmp - s);
}
```

Now the main program to test `pipe_2way` is

```
/********    pipe2way.c    ********/
/* headers, readl, and pipe_2way functions */
#define SIZE 256

int main()
{   int pd[2];
    char *str[2];
    char test_string[] = "IPC WITH TWO-WAY PIPE.\n";
    char buf[SIZE];
    char *tmp = buf;
    str[0] = "./lowercase";
    str[1] = NULL;
    pipe_2way(str, pd);
    /* write to lowercase process  */
    write(pd[1], test_string, strlen(test_string));
    readl(pd[0], buf, SIZE);    /* read lowercase process */
    printf("Received from lowercase process:\n%s", buf);
    return EXIT_SUCCESS;
}
```

If you compile and run this program,

gcc lowercase.c -o lowercase
gcc pipe2way.c
./a.out

you'll see the display

```
Received from lowercase process:
ipc with two-way pipe.
```

11.5   Network Communication

Inter-process communication so far works for processes related by **fork**. Extending ipc to unrelated processes executing on different hosts achieves true networking. For network communication, independent processes must be able to initiate and/or accept communication requests in an asynchronous manner, whether the communicating processes are on the same computer or on different hosts in a network. The standard *Linux ipc* today was first introduced by Berkeley UNIX in the 1980s. The scheme is centered on the *socket* mechanism and supports the Internet protocols well. Its wide use contributed to the explosive growth of the Internet.

Linux ipc provides access to a set of communication *domains* characterized by their protocol family. Important ipc domains are

1. The *Local domain* uses the Linux socket-type file and the pipe mechanism for communication between processes within the local Linux system.

2. The *Internet domains* IPv4 and IPv6 use the corresponding Internet protocols for local-remote communications.

Other domains, for example, the *ATMPVC domain* (Asynchronous Transfer Mode Permanent Virtual Connection), exist.

The ipc communication domains are characterized by such properties as addressing scheme, protocols, and underlying communications facilities. The central mechanism is the *socket*. A socket is an endpoint of communication within a specific communication domain. A socket may be assigned a name (that is, an address) that allows others to refer to it. A process communicates (exchanges data) through its own socket with another socket in the same domain, belonging to a different process. Thus, communication is conducted through a pair of cooperating sockets, each known as the *peer* of the other. In the Local domain, sockets are named with file system pathnames, for example, `/tmp/soc`. In the Internet domain, a socket address is more complicated. It consists of an *address family*, an *IP address*, and a transport layer *port number*. In the same domain, different types of sockets use different communications protocols. Processes communicate through sockets of the same type.

Processes connected by sockets can be on very different computers that may use different data representations. For example, an `int` is 32 bits on some systems but 64 bits on others. Even when the data sizes agree, systems may still use either the high or the low byte to store the most significant part of a number. In this *heterogeneous environment*, data are sent and received, at the socket level, as a sequence of bytes. Thus, a sequence of ASCII characters can usually be sent and received directly through sockets. Other types of data need to be *serialized* into a sequence of bytes before sending and to be *deserialized* from a byte sequence into the local data type at the receiving end.

Client and Server

As stated in Chapter 7, a network service usually involves a *server* and a *client*. A server process provides a specific service accessible through the network communications mechanism. A client process provides user access to a particular network service. A well-defined set of conventions must exist to govern how services are located, requested, accepted, delivered, and terminated. This set of conventions comprises a protocol that must be followed by both server and client.

Most Internet services use protocols sitting on top of the basic transport layer protocol TCP/IP or UDP/IP. For example, HTTP (the Web protocol) sits on top of TCP. Internet domain sockets support TCP and UDP.

11.6 Sockets

A socket is an abstraction that serves as an endpoint of communication within a networking domain. A program accesses ipc through the socket. In other words, the socket is the ipc mechanism's interface to application programs. Each socket potentially can exchange data with any other socket within the same domain. Each socket is assigned a *type* property. Different types of sockets use different protocols. The following types of sockets are generally supported:

- *stream* socket—Supports the bidirectional, reliable, sequenced, and unduplicated flow of data without record boundaries. When put to use, a stream socket is connected to another stream socket, and the connected pair forms a two-way pipe across the network. Each socket in the pair is called the *peer* of the other. Aside from the bidirectionality of data flow, a pair of connected stream sockets provides an interface nearly identical to that of a pipe. Within the Local domain, a pair of connected sockets is used to implement a pipe. Stream sockets in the Internet domain use the *Transmission Control Protocol* (TCP/IP).

- *datagram* socket—Provides bidirectional flow of data packets called *messages*. The communications channel is not promised to be sequenced, reliable, or unduplicated. That is, a process receiving messages on a datagram socket may find messages duplicated and, possibly, not in the order in which they were sent. A datagram socket does not have to be connected to a peer. A message is sent to a datagram socket by specifying its address. Datagram sockets closely model the facilities of packet-switched networks. Datagram sockets in the Internet domain use the *User Datagram Protocol* (UDP/IP).

- *raw* socket—Gives access to the underlying communication protocols that support socket abstractions. These sockets are normally datagram oriented, although their exact characteristics are dependent on the interface provided by the protocol. Raw sockets are not intended for the general user, but for those interested in developing new communication protocols or for gaining access to esoteric facilities of an existing protocol. Raw sockets in the Internet domain give direct access to the *Internet Protocol* (IP).

The domains and standard socket types are defined in the header file <sys/socket.h>. Some defined constants for sockets are given in Table 11.1.

TABLE 11.1: Socket Constants

Symbol	Meaning
PF_UNIX, PF_LOCAL	Local domain
PF_INET	IPv4 domain
PF_INET6	IPv6 domain
SOCK_STREAM	Stream socket type
SOCK_DGRAM	Datagram socket type
SOCK_SEQPACKET	Sequenced two-way datagram type
SOCK_RAW	Raw socket type

Creating Sockets

The **socket** system call

```
#include <sys/types.h>
#include <sys/socket.h>
int socket(int domain, int type, int protocol)
```

is used to create a socket of the indicated *type* in the given *domain*. It returns a descriptor that is used to reference the socket in other socket operations. Defined constants (Table 11.1) are used to specify the arguments. If the protocol is left unspecified (with a 0 value), an appropriate protocol in the domain that supports the requested socket type will be selected by the system. For example,

```
s = socket(PF_LOCAL, SOCK_DGRAM, 0);
```

creates a datagram socket for use within the Local domain supported by UDP, whereas the call

```
s = socket(PF_INET, SOCK_STREAM, 0);
```

creates an Internet stream socket supported by TCP.

Socket Address

Typically, a process that provides a specific network service first creates a socket in an appropriate domain and of the appropriate type. Then an address is assigned to the socket so that other processes can refer to it. The socket address is important because a client process must specify the address of a socket to send a message or make a connection. Therefore,

1. A server process must assign its socket an address and make it known to all potential clients.

2. A client process must be able to obtain the correct socket address of any server on any host.

Linux supports many different networking protocols and address families. Here we will focus on local ipc and the Internet.

Local and Internet Socket Addresses

A local socket address is just a pathname for a socket-type file in the local file system. An Internet socket address combines a host IP address (Chapter 7, Section 7.16) and a transport layer *port number*. Standard network services are assigned the same port numbers on each host. The file **/etc/services** contains a list of services and their port numbers. It lists one line for each service with four fields:

- An official name of the service

- A unique transport layer port number

- The protocol to use

- Any aliases (other names for the service)

For example, the entry

```
ssh 22/tcp
```

specifies that the Secure Shell service is at port 22 and uses the TCP protocol.

Sixteen bits (two bytes) are used for representing a port number. Standard ports (below 1024) are *privileged* and their access restricted to widely used server programs with the right privilege. Port numbers 1024 and higher are referred to as non-privileged ports and are used for other applications. For socket programs written by regular users, we need to find a port that is not privileged and not used by other well-known services as listed in **/etc/services**. The Shell level command

/sbin/sysctl net.ipv4.ip_local_port_range

FIGURE 11.5: Local Domain Socket Address Structure

```
#define UNIX_PATH_MAX     108
struct  sockaddr_un
{  sa_family_t sun_family;          /* AF_LOCAL  */
   char sun_path[UNIX_PATH_MAX]; /* pathname  */
};
```

displays local port numbers that you can use in socket programming exercises.

Data structures used for socket addresses are

- In the Local domain, a socket address is stored in a `sockaddr_un` structure usually defined in `<sys/un.h>` (Figure 11.5).

- In the Internet domain, a socket address is declared by the `sockaddr_in` structure contained in `<netinet/in.h>` (Figure 11.6).

FIGURE 11.6: Internet Socket Address Structure

```
struct sockaddr_in
{  sa_family_t      sin_family;  /* AF_INET    */
   in_port_t        sin_port;    /* port no.   */
   struct in_addr   sin_addr;    /* IPv4 address */
   char             sin_zero[8];
};
```

In practice, Internet socket addresses are often used in very specific ways.

- A client must construct a *destination socket* address to be used either in making a connection (**connect()**) to the server or in sending (**sendto()**) and receiving (**recvfrom()**) datagrams without making a connection. Here is a typical code sequence (minus error checking) for building an Internet destination socket address.

  1. `struct sockaddr_in d`—Creates socket addr structure d

  2. `memset(&d, 0, sizeof(d))`—Zeros out the structure

3. `d.sin_family = AF_INET`—Sets IP address family

4. `struct hostent* hep=gethostbyname(host)`—Obtains host entry structure

5. `memcpy(&d.sin_addr, hep->h_addr, hep->h_length)`—Copies IP address into `d`

6. `d.sin_port=getservbyname(service,transport)->s_port`—Sets standard port number

The IP address of a target host is usually obtained by consulting the domain name server (Chapter 7, Section 7.16) via the **gethostbyname** call. The standard service port is retrieved with the **getservbyname** call (Section 11.11). To use a non-standard port, set `sin_port` to **htons**(*port_number*).

- A server, on the other hand, must construct a *service socket address* and bind it to a socket for the server to receive incoming connections or datagrams. The typical code sequence for building an Internet service socket address is

  1. `struct sockaddr_in s`—Creates Internet socket addr structure `s`

  2. `memset(&s, 0, sizeof(s))`—Zeros out the structure

  3. `s.sin_family = AF_INET`—Sets IP address family

  4. `s.sin_port=getservbyname(service,transport)->s_port`—Sets port to standard port number

  5. `s.sin_addr.s_addr = INADDR_ANY`—Sets server addr to any local host IP address

The constant `INADDR_ANY` gets you the IP address of the local host.

To bind a socket address to a socket, the system call

bind(`int soc, struct sockaddr *addr, int addrlen`)

is used, where `soc` is a socket descriptor, `addr` is a pointer to the appropriate address structure, and `addrlen` is the size of the address. The parameter `addr` can receive pointers of type `struct sockaddr_un *` or `struct sockaddr_in *`.

Let's look at an example demonstrating Internet stream socket usage in a client program.

11.7   A TCP Echo Client

The standard Internet *echo service* is useful in testing sockets. The echo
server can receive messages from any client connected to it and then sends
that same message back to where it came from. The echo service normally
uses TCP and port number 7.

The program `tcp_echo.c` is a client program that connects to the echo
server on any particular host and sends it a message. You might say that this
is our *Hello World* example of socket programming. The program is used in
the following way:

gcc `tcp_echo.c -o tcpEcho`
./tcpEcho *host* "*Any Message*"

The program starts with the necessary header files and a helper function
for exiting on error (**Ex:** ex11/tcp_echo).

```
/********    tcp_echo.c    ********/
#include <stdio.h>
#include <stdlib.h>
#include <sys/socket.h>
#include <netinet/in.h>
#include <netdb.h>
#include <string.h>
#define B_SIZE 1024

void Quit(const char *err)
{  perror(err);
   exit(EXIT_FAILURE);
}
```

The **main** program first checks for correct command-line arguments and declares variables.

```
int main(int argc, char* argv[])
{   if (argc != 3)
    {  fprintf(stderr, "Usage: %s host \"message\"\n", argv[0]);
       exit(EXIT_FAILURE);
    }
    int soc;                      /* socket descriptor          */
    char buf[B_SIZE];
    struct sockaddr_in cl;     /* client socket addr (local) */
    memset(&cl, 0, sizeof(cl));
    struct sockaddr_in sr;     /* server socket addr (remote) */
```

Then, it fills each field in the server socket address structure **sr** by first ze-
roing out the structure (line **A**), assigning the address family (**AF_INET** for
IPv4, line **B**), finding and setting the standard port number (line **C**) via the

getservbyname library call, and copying the host Internet address obtained by **gethostbyname** (line D) into the `sin_addr` field of the socket address structure (line E). See Section 11.11 for information on the library calls.

```
memset(&sr, 0, sizeof(sr));                         /* (A) */
sr.sin_family=AF_INET;                              /* (B) */
sr.sin_port=getservbyname("echo","tcp")->s_port;   /* (C) */
hostent *hp = gethostbyname(argv[1]);              /* (D) */
if ( hp == NULL )
{  sprintf(buf, "%s: %s unknown host\n", argv[0], argv[1]);
   Quit(buf);
}
memcpy(&sr.sin_addr, hp->h_addr, hp->h_length);    /* (E) */
```

With the target remote server address completed, the program can now create a local client-side socket (line F) in the `PF_INET` protocol family using the TCP protocol and connect (line G) it to the server socket identified by the socket address `sr` which was just filled in (lines A–E).

```
/* creates socket */
    if ( (soc=socket(PF_INET, SOCK_STREAM,         /* (F) */
                   IPPROTO_TCP)) < 0 )
    {  Quit("Problem creating socket");   }
/* requests connection to server */
    if (connect(soc, (struct sockaddr*)&sr,        /* (G) */
                  sizeof(sr)) == -1)
    {  close(soc);
       Quit("client:connect\n");
    }
```

FIGURE 11.7: TCP/IP Socket Connection

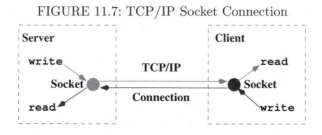

After successful connection of the local socket to the server socket, the program can begin to read/write the local socket as a file descriptor (lines H and I). Data written to the socket gets sent to the remote socket, and data sent by the remote socket can be read from the local socket. Because we are connected to the standard *echo* service, the program should read back whatever it had sent on to the server in the first place.

```
    write(soc, argv[2], strlen(argv[2]));              /* (H) */
    read(soc, buf, sizeof(buf));                       /* (I) */
    printf("SERVER ECHOED: %s\n", buf);
    close(soc); return EXIT_SUCCESS;
}
```

We can use this program to access the *echo* service on an actual host.

./tcpEcho monkey.cs.kent.edu "Here is looking at you, kid."
SERVER ECHOED: Here is looking at you, kid.

Refer to the example code package for the complete `tcp_echo.c` Internet client program.

11.8   Using Datagram Sockets

To further illustrate socket communication, let's look at a simple example involving a sender process and a receiver process using Internet datagram sockets. The receiver is a server ready and waiting to receive datagrams from any sender client on the Internet (Figure 11.8).

FIGURE 11.8: Datagram Socket Communication

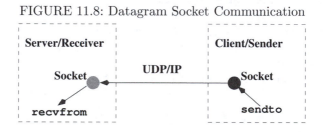

The receiver first creates a blank **sender** socket address. Then it builds its own socket address **self** (line a) using port 8080 (line b) and the IP address of the server host (INADDR_ANY line c). To run this server yourself, please find a usable UDP port on your host and modify line b accordingly (**Ex:** ex11/ireceiver.c).

```
/********     ireceiver.c    ********/
/** Same headers and Quit() helper function **/
#define B_SIZE 1024

int main()
{   struct sockaddr_in sender;
    memset(&sender, 0, sizeof(sender));
    struct sockaddr_in self;                           /* (a) */
    memset(&self, 0, sizeof(self));
```

```
self.sin_family=AF_INET;
self.sin_port=htons(8080);                        /* (b) */
self.sin_addr.s_addr = htonl(INADDR_ANY);         /* (c) */
```

Now we can create a socket to receive datagrams (line d) and bind the address self to it (line e).

```
soc = socket(PF_INET, SOCK_DGRAM, IPPROTO_UDP); /* (d) */
n = bind(soc, (struct sockaddr *)&self,          /* (e) */
         sizeof(self));
if ( n < 0 ) Quit("bind failed\n");
```

In a loop, the receiver calls **recvfrom** (line f and Section 11.9) to wait for the next incoming datagram. When it arrives, the message is received in buf, and the sender socket address is stored in the **sender** structure. The **recvfrom** call blocks until an incoming message is received. It returns the actual length of the message or a negative number if something goes wrong. In case the buffer space is too small for the incoming message, the rest of the message may be discarded by **recvfrom**. To use it as a string, we place a string terminator at the end of the message received (line g).

```
int soc, n, len=0;
char buf[B_SIZE], client[INET_ADDRSTRLEN];
while(1)
{   n = recvfrom(soc, buf, sizeof(buf)-1,          /* (f) */
            0, (struct sockaddr *)&sender, &len);
    if ( n < 0 )
    {   close(soc);
        Quit("recvfrom failed\n");
    }
    buf[n] = '\0';                                 /* (g) */
    inet_ntop(AF_INET, &(sender.sin_addr),         /* (h) */
                    client, INET_ADDRSTRLEN);
    printf("Received from %d %s %d chars= %s\n", /* (i) */
                    sender.sin_addr, client,  n, buf);
    if ( strncmp(buf, "Stop", 4)==0 ) break;       /* (j) */
}
close(soc);
return EXIT_SUCCESS;
}
```

In this receiver example, we used the **inet_ntop** library function to convert the sender IP address to a quad notation string in the character buffer client (line h). The receiver displays the information received to standard output (line i). In our example, if the message received starts with "Stop", the receiver will terminate execution (line j).

We can compile and run the receiver on a selected server host, say,

dragon.cs.kent.edu, and experiment with it by sending messages to it using
the **nc** command (Chapter 7, Section 7.19):

gcc ireceiver.c -o ireceiver   (on **dragon**)
./ireceiver

nc -u dragon.cs.kent.edu 8080   (on any other host)
Here is a test message.
Here is another test message.
Stop
CTRL+C

The display by the receiver looks like

Received from 1141709121 65.25.13.68 23
chars= Here is a test message.

As another experiment, we can write a client program (isender.c) that
uses the **sendto** call (Section 11.9) to send datagrams to the receiver. Make
sure the receiver is running, on **dragon**, say, and then experiment with the
sender as follows.

gcc isender.c -o isender
./isender dragon.cs.kent.edu 8080

Let's look at the program isender.c (**Ex:** ex11/isender.c).

```
/********    isender.c    ********/
/** headers and the Quit() helper functions **/

int main(int argc, char* argv[])
{   if (argc != 3)
    {  fprintf(stderr, "Usage: %s host port\n", argv[0]);
       exit(EXIT_FAILURE);
    }
    char buf[] = "Hello there, it is me.";
    char end[] = "Stop.";
    struct sockaddr_in receiver;
    memset(&receiver, 0, sizeof(receiver));        /* (1) */
    receiver.sin_family=AF_INET;                   /* (2) */
    receiver.sin_port=htons(atoi(argv[2]));        /* (3) */
    struct hostent *hp = gethostbyname(argv[1]);
    if ( hp == NULL )
    {  sprintf(buf, "%s: %s unknown host\n", argv[0], argv[1]);
       Quit(buf);
    }
    memcpy(&receiver.sin_addr, hp->h_addr,         /* (4) */
           hp->h_length);
```

After checking the command-line arguments, the server socket address structure `receiver` is built (lines 1–4).

An Internet datagaram socket is created (line 5) and used to send the message in `buf` to the `receiver` socket address (line 6).

```
int soc = socket(PF_INET, SOCK_DGRAM, 0);   /* (5) */
int n = sendto(soc, buf, strlen(buf), 0,    /* (6) */
      (struct sockaddr *)&receiver,
      sizeof(receiver));
if ( n < 0 ) {   Quit("sendto failed"); }
printf("Sender: %d chars sent!\n", n);
n = sendto(soc, end, strlen(end), 0,
      (struct sockaddr *)&receiver,
      sizeof(receiver));
close(soc);
return EXIT_SUCCESS;
}
```

11.9   Socket I/O System Calls

For connected sockets, the basic **read** and **write** calls can be used for sending and receiving data:

```
read(soc, buffer, sizeof(buffer));
write(soc, buffer, sizeof(buffer));
```

Each process reads and writes its own socket, resulting in a bidirectional data flow between the connected peers. The socket I/O calls

```
recv(soc, buffer, sizeof(buffer), opt);
send(soc, buffer, sizeof(buffer), opt);
```

are exclusively for stream sockets. If the argument *opt* is zero, then they are the same as the **write** and **read**. If opt has the `MSG_PEEK` bit turned on, then **recv** returns data without removing it so a later **recv** or **read** will return the same data previously previewed.

The **sendto** and **recvfrom** system calls send and receive messages on sockets, respectively. They work with any type of socket, but are normally used with datagram sockets.

```
int sendto(int soc, char *buf, int k, int opt,
           struct sockaddr *to, int tosize)
```

sends, via the socket `soc`, k bytes from the buffer `buf` to a receiving socket specified by the address `to`. The size of `to` is also given. The `to` is a pointer to any valid socket address, in particular, `struct sockaddr_un` or `struct sockaddr_in`. Most current implementations of `struct sockaddr` limit the length of the active address to 14 bytes. The **opt** parameter specifies different options for **sendto/recvfrom** and works just like the **opt** argument for **send/recv**. The **sendto** call returns the number of bytes sent or −1 to indicate an error.

On the receiving end, the call

```
int recvfrom(int soc, char *buf, int bufsize, int opt,
                struct sockaddr *from, int *fromsize)
```

receives, into the given buffer **buf** of size **bufsize**, a message coming from another socket. If no messages are available, the call waits unless the socket is non-blocking (set via the **fcntl** system call). The peer's address structure is returned in ***from** and its size in ***fromsize**. The argument **from** is a result parameter that is filled with the address of the sending socket. The **fromsize** is a *value-result parameter*; it initially should contain the amount of space in ***from**. On return, ***fromsize** contains the actual size (in bytes) of the address ***from**. The number of bytes received is the return value of **recvfrom**.

Shutting Down Sockets

The **close** system call can, of course, be used on a socket descriptor:

```
int close(int soc)
```

The read and write halves of a socket can also be independently closed with the **shutdown** system call.

```
int shutdown(int soc, int flag)
```

closes the read portion if **flag** is 0, the write portion if **flag** is 1, and both the read and the write if **flag** is 2. When **shutdown** is combined with the **socketpair** call, which creates two connected sockets in the Local domain, the **pipe** system call can be emulated exactly.

11.10   TCP-Based Servers

We have seen in Section 11.7 a TCP client that accesses the standard **Echo** service.

TCP-based servers use stream sockets. A stream socket is connected with its peer to form a two-way pipe between a client and a server. A client process uses its socket to initiate a connection to a socket of a server process, and a server process arranges to listen for connection requests and accepts a connection. After a connection is made, data communication can take place using the **read**, **write**, **recv**, and **send** I/O system calls. Figure 11.7 illustrates server and client stream socket connections.

A server process binds a published address to a socket. To initiate a connection, a client process needs to

1. Find the correct address of the desired server socket.

2. Initiate a connection to the server socket.

as we have seen in Section 11.7.

Accepting a Connection

A server process with a stream socket takes the following steps to get ready to accept a connection:

1. Creates a socket in the appropriate domain of type SOCK_STREAM.

2. Constructs the correct server socket address, and binds it to the socket.

3. Indicates a willingness to accept connection requests by executing the **listen** system call.

4. Uses the **accept** call to wait for a connection request from any client and to establish a connection (Figure 11.9).

FIGURE 11.9: Stream Socket Connections

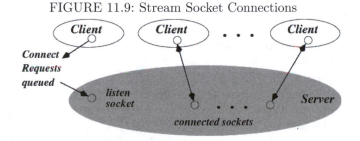

The call

int **listen**(int *soc*, int *n*)

initializes the socket *soc* for receiving incoming connection requests and sets the maximum number of pending connections to *n*. After the **listen** call, the **accept** call

int **accept**(int *soc*, struct sockaddr **addr*, socklen_t **addrlen*)

accepts connections on the stream socket *soc* on which a **listen** has been executed. If there are pending connections, **accept** extracts the first connection request on the queue, creates a new socket (say, **ns**) with the same properties as *soc*, connects the new socket with the requesting peer, and returns the descriptor of this new socket. The connection listening socket *soc* remains ready to receive connection requests. If no pending connections are present on the queue and the socket is not marked as non-blocking (say, with the **fcntl** system call), **accept** blocks until a connection request arrives. If the socket is marked as non-blocking and no pending connections are present on the queue, **accept** will return an error instead of blocking.

The accepted socket, **ns**, is used to communicate with its peer and may not be used to accept additional connections. The argument *addr* is filled with the address of the connected peer. Again, the *addrlen* is a value-result parameter.

An Example TCP/IP Server

Let's look at an example server (**Ex: ex11/inetserver.c**) that uses TCP/IP and forks child processes to take care of clients while the parent process continues to monitor incoming connection requests.

The program begins by checking command-line arguments and preparing the **peer** and **self** socket address structures (lines up to I).

```
int main(int argc, char* argv[])
{   if (argc != 2)
    {   fprintf(stderr, "Usage: %s port \n", argv[0]);
        exit(EXIT_FAILURE);
    }
    int soc, ns;
    struct sockaddr_in peer;
    int peer_len=sizeof(peer);
    memset(&peer, 0, sizeof(peer));
    peer.sin_family=AF_INET;
    struct sockaddr_in self;
    memset(&self, 0, sizeof(self));
    self.sin_family=AF_INET;
    self.sin_addr.s_addr = htonl(INADDR_ANY);
    self.sin_port=htons(atoi(argv[1]));            /* (I)   */
/* set up listening socket soc */
    if ( (soc=socket(PF_INET, SOCK_STREAM, 0)) < 0 )
    {   Quit("server:socket");   }
    if (bind(soc, (struct sockaddr*)&self, sizeof(self)) == -1)
    {   close(soc); Quit("server:bind"); }        /* (II)  */
    listen(soc, 1);                               /* (III) */
/* accept connection request */
    int pid;
    while ( (ns = accept(soc, (struct sockaddr*)  /* (IV)  */
                    &peer, &peer_len)) >= 0 )
    {   if ( (pid=fork()) == 0 )                   /* (V)   */
        action(ns, &peer);
    }
    close(soc);
    Quit("server:accept");
}
```

After creating the server socket **soc** and binding the local address to it (line **II**), we begin listening (line **III**) and accepting incoming connections (line **IV**) on **soc**.

When **accept** returns, we fork a child process to perform the service (line **V**), defined entirely by the **action** function. The parent calls **accept** again for the next connection.

The **action** function repeatedly reads the incoming data, echos it back, and displays the data received (line **VI**). When the child is done, it calls **_exit** (line **VII**).

```
/* Performs service */
int action(int ns, struct sockaddr_in* peer)
{   int k;
    char buf[256];
    char* client[INET_ADDRSTRLEN];
    inet_ntop(AF_INET, &(peer->sin_addr),
                client, INET_ADDRSTRLEN);
    while ( (k=read(ns, buf, sizeof(buf)-1)) > 0 ) /* (VI)   */
    {   buf[k]='\0';
        printf("SERVER id=%d RECEIVED FROM %s: %s\n",
                getpid(), client, buf);
        write(ns, buf, k);
    }
    printf("Child %d Done.\n", getpid());
    close(ns);
    _exit(EXIT_SUCCESS);                              /* (VII)  */
}
```

Run this program, say, on port 4900, by

gcc inetserver.c -o myecho
./myecho 4900

and connect to it with

nc localhost 4900
nc *host* 4900

The example code package contains the complete **inetserver.c** program.

11.11   Network Library Routines

Linus provides a set of standard routines in the *Internet networking library* to support network address mapping. These routines, with the help of the DNS and data files such as **/etc/services** and **/etc/hosts**, return C structures containing the needed information. Routines are provided for mapping domain names to IP addresses, service names to port numbers and protocols, network names to network numbers, and so on. We have seen some use of these already. Now we will describe these routines in more detail.

The header file `<netdb.h>` must be included in any file that uses these networking library routines. For instance, the library function

```
#include <netdb.h>
struct hostent *?gethostbyname(const char *host)
```

consults the DNS and returns a pointer to a `hostent` structure for the *host* as follows:

```
struct  hostent
{  char *h_name;        /* official name of host       */
   char **h_aliases;    /* aliases                     */
   int  h_addrtype;     /* address type: PF_INET       */
   int  h_length;       /* length of address           */
   char **h_addr_list;  /* IP addresses (from name server) */
};
```

A `NULL` pointer is returned for error. The *host* argument can be given either as a domain name or as an IP address. In the latter case, no DNS query is necessary.

For example, to obtain the IP address of a host with the name `monkey.cs.kent.edu.`, use

```
struct hostent *hp;
hp = gethostbyname("monkey.cs.kent.edu.");
```

and the numerical IP address is in

```
hp->h_addr_list[0]            /* IP address */
```

which can be copied into the `sin_addr` field of a `sockaddr_in` structure for a target socket. If a partial domain name such as `monkey` is given, then it is interpreted relative to the Local domain. The IP address is stored as bytes in *network byte order*: byte 0 is the most significant and byte 4 is the least significant. This order is commonly known as *big endian*. The network byte order may or may not be the same as the *host byte order* used to store `longs`, `ints`, and `shorts` on a particular computer system. There are big endian and *little endian* CPUs. The library routine **htonl** (**htons**) is used to transform an `unsigned int` (`unsigned short`) from host to network order. The routine **ntohl** (**ntohs**) does the opposite.

To determine the port number for standard network services, use

```
struct servent *
getservbyname(const char *service, const char *proto)?
```

which returns the port number of the given service with the given protocol in a `servent` structure:

```
struct servent
{   char  *s_name;      /* official name of service    */
    char **s_aliases;  /* alias list                  */
    int    s_port;      /* port no in network short byte order */
    char  *s_proto;    /* protocol used               */
};
```

A NULL pointer is returned for error. For example,

```
struct servent* sp;
sp = getservbyname("ssh", "tcp");
```

gets `sp->s_port` to be 22 (after conversion by **ntohs**), the designated port for the SSH over TCP service.

Similar sets of library functions are provided to access the network and protocol databases. Examples are **getnetbyname** and **getprotobyname**.

11.12   On-Demand Internet Services

Any program that uses standard I/O can be made into an Internet TCP server without changing its code. Such servers are deployed under the control of the *Internet server daemon* **xinetd**[1] (Figure 11.10), which monitors all designated ports and invokes registered server programs on-demand.

If your distribution does not have **xinetd**, install it easily (Section 8.24) with either

yum `install xinetd`

or

sudo apt-get `install xinetd`

and start the **xinetd** running either with the **system-config-services** tool or with

service `xinetd start`

Server programs under the control of **xinetd** do not need to deal with sockets, protocols, or Internet addresses. In fact, such servers work as a filter reading input (incoming data from a client) from `stdin` and producing output (outgoing data to a client) to `stdout`. Thus, many Linux commands can easily be servers without change. New servers can also be written easily in any source code language, including scripting languages such as Bash. The

---

[1]**Xinetd** is a successor of the original **inetd**, with better security and ease of configuration.

FIGURE 11.10: Server under **xinetd** Control

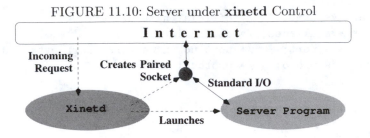

xinetd reduces the processing load on the server host because we can run fewer daemon processes by making many services on-demand.

Let's see how we can place the **fortune** program under **xinetd** to serve the Internet.

1. Edit /etc/services. Go to the end of the file and add the line

   ```
   fortune            55000/tcp    # a test on-demand server
   ```

   to make **fortune** a network service. Note that we have selected a *local port* which does not conflict with any standard or well-known services. You can find the local port range with

   sysctl net.ipv4.ip_local_port_range

2. Make **fortune** an on-demand service by registering it under **xinetd**. This is done by adding a file /etc/xinetd.d/fortune as follows:

   ```
   ## Our test on-demand server
   service fortune
   {        socket_type              = stream
            protocol                 = tcp
            wait                     = no
            user                     = root
            group                    = yes
            server                   = /usr/bin/fortune
            server_args              =
            disable                  = no
   }
   ```

   Restart your **xinetd**.

Your fortune server is ready! Try it with

nc localhost 55000

If it works you should see a display from **fortune**.

Each file in the `/etc/xinetd.d` folder represents a different on-demand service made possible under **xinetd**. Some of the available configuration settings are

- `service`—The service name, usually to match a service listed in the /etc/services file

- `socket_type`—The network socket type (`stream` or `dgram`)

- `wait`–For a single-threaded server (`yes`) or multithreaded server (`no`).

- `user`—Effective user ID of server

- `server`–The server executable pathname

- `disable`–Service is active or not

All on-demand services can be controlled via the `On Demand Services` tab of the GUI tool **system-config-services**.

11.13   Daemon Processes

On Linux, there are many hidden processes that work quietly in the background to perform a variety of tasks as though by magic. These are the so-called *daemon* processes, and they run concurrently with other active user processes. For example,

- The **cron** daemon (usually `/usr/sbin/crond`) executes commands at specified dates and times scheduled through the **crontab** command (see Chapter 7 Exercises).

- The **httpd** Web server (usually `/usr/sbin/httpd`) is a daemon that handles HTTP requests (Chapter 8).

- Several daemons, including **rpc.nfsd**, **rpc.lockd**, **rpc.statd**, and **rpc.mountd** provide the Network Filesystem (NFS) service (Section 6.9).

- The **named** (usually `/usr/sbin/named`) is the Internet DNS server (Section 7.16).

- The **sendmail** daemon (usually `/usr/sbin/sendmail -bd`) is the Internet email server.

- The **sshd** daemon (usually `/usr/sbin/sshd`) is the secure Shell login server.

- The *Internet Services* daemon (usually `/usr/sbin/xinetd`) monitors specific transport layer ports and launches appropriate on-demand servers (for example, `rsyncd`).

Many other network servers not listed here run as daemons, but there are also servers, such as the X Window server, that are not considered daemons. Newer workstations have multiple hardware processors to execute several processes in parallel, resulting in greatly increased system speed.

The `xinetd`, on the other hand, reduces the number of idling servers waiting for action by monitoring many different service ports and invoking the appropriate server when a request arrives.

Programming a Daemon

Daemon programs such as `sshd`, `xinetd`, and `sendmail -bd` have these four important characteristics:

1. A daemon never exits.

2. A daemon has no control terminal window.

3. A daemon does not use standard I/O.

4. A system daemon is normally started at boot time, is controlled by the init process (process 1), and can be restarted if it dies for some reason.

In Chapter 8, Section 8.6 we presented how a Linux is configured to start the Apache Web server at boot time. Follow the same procedure for other servers.

A process can disassociate itself from its control terminal window with the system call **setsid()**.

```
#include <unistd.h>
pid_t setsid(void);
```

The call creates a new *session* and a new *process group*. It sets the calling process as the session leader and the process group leader. No control terminal is assigned yet. The calling process is the only process in the new process group and the only process in the new session.

Thus, a daemon process often executes the sequence in Figure 11.11 to disassociate itself from the control terminal and the parent process. Once orphaned, the daemon process is controlled by the init process.

11.14   Input/Output Multiplexing

Programs such as the `inetd` and the X Window server require the capability to monitor or multiplex a number of I/O descriptors at once. On-line

FIGURE 11.11: Disassociating from Control Terminal Window

```
setsid():
close(0); close(1); close(2);
if ( vfork() == 0 )
      perform_duty();   /* infinite loop  */
exit(0);                /* child orphaned */
```

chat programs are good examples. They need to deal with many I/O channels simultaneously.

The **select** system call provides a general synchronous multiplexing scheme.

```
#include <sys/select.h>
int select(int nfds, fd_set* readfds, fd_set* writefds,
     fd_set* exceptfds, struct timeval *timeout)
```

The **select** call monitors the I/O descriptors specified by the bit masks `*readfds`, `*writefds`, and `*exceptfds`. It checks if any of the `*readfds` is ready for reading; if any of the `*writefds` is ready for writing; and if any of the `*exceptfds` has an exceptional condition pending. Each mast has bit 0 through `nfds-1`. The nth bit of a mask represents the I/O descriptor n. That is, if bit n of a mask is 1, then file descriptor n is monitored. For example, if `*readfds` has the value 1 (a 1 in bit position 0), then I/O descriptor 0 is monitored for data available for reading. The call returns when it finds at least one descriptor ready. When **select** returns, the bit masks are modified to indicate (in the same manner) the I/O descriptors that are ready. The integer value returned by **select** is the total number of ready descriptors.

The parameter `timeout` is a non-zero pointer specifying a maximum time interval to wait before **select** is to complete. To affect a poll, the `timeout` argument should be non-zero, pointing to a zero valued timeval structure. If `timeout` is a zero pointer, **select** returns only when it finds at least one ready descriptor. The code fragment in Figure 11.12 is an example where **select** monitors using a two-second timeout. The `int` masks can accommodate descriptors 0 through 31. Different methods are used to handle a larger number of descriptors. One is to use several `int`s for a mask. Linux systems may not work in the same way in this regard.

Let's look at a server that monitors a stream and a datagram socket with **select** (**Ex:** ex11/selectExample.c).

```
#include <stdlib.h>
#include <sys/types.h>
#include <sys/socket.h>
```

FIGURE 11.12: I/O Multiplexing

```
#include <sys/select.h>

    struct timeval wait;
    int fd1, fd2, read_mask, nready;
    wait.tv_sec = 2
    wait.tv_usec = 0;
    ...
    read_mask = (1 << fd1) | (1 << fd2)
    ...
    nready = select(32, (fd_set*)&read_mask, 0, 0, &wait);
```

```
#include <sys/select.h>
#include <netinet/in.h>       /* Internet domain header */

#define SERVER_PORT0   3900
#define SERVER_PORT1   3901

int main()
{   int soc_s, soc_d, s_mask, d_mask, read_mask, nready;
/* set up listening socket soc */
    struct sockaddr_in addr0 = {AF_INET};
    addr0.sin_addr.s_addr = htons(SERVER_PORT0);
    struct sockaddr_in addr1 = {AF_INET};
    addr0.sin_addr.s_addr = htons(SERVER_PORT1);
    soc_s = socket(AF_INET, SOCK_STREAM, 0);              /* A */
    soc_d = socket(AF_INET, SOCK_DGRAM, 0);
    if (soc_s < 0 || soc_d < 0)
    {   perror("server:socket"); exit(EXIT_FAILURE); }
    if (bind(soc_s, (struct sockaddr *)&addr0,
            sizeof(addr0))==-1 ||
        bind(soc_d, (struct sockaddr *)&addr1,
            sizeof(addr1))==-1)
    {   perror("server:bind"); exit(EXIT_FAILURE);  }
    listen(soc_s, 3);                                     /* B */
/* monitor sockets */
    s_mask= 1 << soc_s; d_mask= 1 << soc_d;               /* C */
    for (;;)
    {   read_mask =  s_mask | d_mask;                     /* D */
```

```
nready = select(2, (fd_set*)&read_mask, 0, 0, 0); /* E */
while ( nready )                                   /* F */
{  if ( read_mask & s_mask )
   {  nready--; do_stream(soc_s);                  /* G */
   }
   else if ( read_mask & d_mask )
   {  nready--; do_dgram(soc_d);                   /* H */
   }
} /* end of while */
} /* end of for */
}
```

The stream socket `soc_s` and the datagram socket `soc_d` are created, bound to correct addresses, and made ready to receive input (lines A – B). After the bit masks are set correctly by bit shifting operations (line C), the program goes into an infinite loop to monitor these two sockets (line D). When **select** (line E) returns, each of the ready descriptors is treated in a **while** loop (line F) and monitoring is resumed.

The functions `do_stream` (line G) and `do_dgram` (line H) each handle a different kind of ready socket.

A similar system call **pselect** is also available, which allows you to block signals while multiplexing I/O.

11.15   TCP Out-of-Band Data

TCP/IP sockets support two independent logical data channels. Normal data are sent/received *in-band*, but *urgent messages* can be communicated *out-of-band* (oob). If an abnormal condition occurs while a process is sending a long stream of data to a remote process, it is useful to be able to alert the other process with an urgent message. The oob facility is designed for this purpose.

Out-of-band data are sent outside of the normal data stream and received independently of in-band data. TCP supports the reliable delivery of only one out-of-band message at a time. The message can be a maximum of one byte long. When an oob message is delivered to a socket, a `SIGURG` signal is also sent to the receiving process so it can treat the urgent message as soon as possible. The system calls,

send(soc, buffer, sizeof(buffer), opt);
recv(soc, buffer, sizeof(buffer), opt);

with the `MSG_OOB` bit of `opt` turned on, send and receive out-of-band data. For example, a TCP/IP client program can use the code

send(soc, "B", 1, MSG_OOB);

to send the one-character urgent message B to a peer socket.

To treat oob data, a receiving process traps the `SIGURG` signal (Chapter 10, Section 10.16) and supplies a handler function that reads the out-of-band data and takes appropriate action. For example, the following code defines a function `oob_handler` which reads the oob data.

```
int oobsoc;

void oob_handler()
{   char buf[1];
    ssize_t k;
    k = recv(oobsoc, buf, sizeof(buf), MSG_OOB);
    if ( k > 0 )
    {   /* process urgent msg */
    }
}
```

To treat signals sent via oob, for example, this handler function can check the received message to see which oob byte is received and use

kill(*SIGXYZ*, getpid());

to send some signal to itself (**Ex:** ex11/oob.c).

The `SIGURG` signal, indicating pending oob data, is trapped with

```
#include <signal.h>
#include <fcntl.h>

    struct sigaction new;
    struct sigaction old;
    oobsoc = ns;    /* ns is Internet stream socket */
    new.sa_handler=oob_handler;
    new.sa_flags=0;
    sigaction(SIGURG, &new, &old);
```

To ensure that the process is notified the moment urgent oob data arrives, the following codes should also be executed:

```
#include <unistd.h>
#include <fcntl.h>
    if (fcntl(ns, F_SETOWN, getpid()) < 0)
    {   perror("fcntl F_SETOWN:");
        _exit(EXIT_FAILURE);
    }
    ...
```

The code requests that when a `SIGURG` associated with the socket `ns` arises, it is sent to the process itself. The **fcntl** file control call sets the process to receive `SIGIO` and `SIGURG` signals for the file descriptor `ns`.

You'll find a program (**Ex: ex11/inetserverOOB.c**) in the example code package which adds the out-of-band data capability to the `inetserver.c` program.

11.16   For More Information

Consult section 7 of the Linux man pages for all supported socket address families. For `AF_INET` see `ip(7)`, for `AF_INET6` see `ipv6(7)`, for `AF_UNIX` (same as `AF_LOCAL`) see `unix(7)`, for `AF_APPLETALK` see `ddp(7)`, for `AF_PACKET` see `packet(7)`, for `AF_X25` see `x25(7)`, and for `AF_NETLINK` see `netlink(7)`. For Linux kernel socket support see `socket(7)`.

For networking and network protocols see *Computer Networking: Internet Protocols in Action* by Jeanna Matthews (Wiley). For Networking on Linux see *Advanced Guide to Linux Networking and Security* by Ed Sawicki (Course Technology).

11.17   Summary

Linux supports networking applications by providing a set of system-level facilities for ipc among distributed processes. Network services often use a client and server model where server processes provide specific services accessed by client programs that act as user or application interfaces. Different socket types support different networking protocols. Clients access servers by locating the server's socket address and initiating a request.

The ipc hinges on the socket mechanism, which serves as endpoints for communication within any specific communication domain. The *Local domain* and the *Internet domain* are usually supported on Linux. The former is used for communication within the local Linux system. The latter supports the various Internet protocols that exist in the Internet protocol family, including IP, TCP, and UDP.

There are several types of sockets. *Stream sockets* are connected in pairs to support a bidirectional communications channel, which can be likened to a two-way pipe. *Datagram sockets* may or may not be connected and can send/receive messages similar to data packets. *Raw sockets* give access to the underlying communication protocols that support socket abstractions. Raw sockets are not intended for the general programmer. A process uses its own socket to communicate across the network with a socket belonging to a remote process (the peer). The two sockets must be of the same type. The DNS and a set of networking system calls combine to retrieve network addresses and service ports. Library routines make it straightforward to find and construct socket addresses in a program.

Network server programs may run as *daemon processes*, divorced from control terminal windows and standard I/O, to run constantly but quietly in the background. In contrast, on-demand services do not run as daemons, but get started by the **xinetd** super server only when a client request arrives.

Monitoring I/O with **select** or **pselect** enables the multiplexing concurrent I/O. Out-of-band data, supported by Internet stream sockets, can be used to send urgent messages such as interrupts to peer sockets.

11.18   Exercises

1. The **system** or **popen** call executes an *sh* command. How would you get such a call to execute a command string for the Bash Shell?

2. Is it possible for a parent process to send data to the standard input of its child? How? Is it possible for a parent process to receive output from the standard output of a child process? How?

3. Refer to the `Hello there` pipe example in Section 11.2. What would happen if the child did not close its descriptor `p[1]`? What would happen if the parent did not close its descriptor `p[1]`?

4. Write a C function `pipe_std("Shell-command-string")` which creates a child process to execute any given regular Linux command. Furthermore, it connects the file descriptors 0 and 1 of the calling (parent) process to the corresponding descriptors of the child process.

   The usage of the `pipe_std` function is as follows:

   - In the parent process, a call to `pipe_std` is made with a specific command string. This sets up the two-way pipe between the parent process and the child process. Then, `pipe_std` returns.

   - Now in the parent process, file descriptor 0 reads the standard output of the child process, and output to file descriptor 1 is read as standard input by the child process. This allows the parent process to feed input to the child process and collect the child's output.

   - After interaction with the child process is over, the parent process calls

     ```
     end_pipe_write();
     end_pipe_read();
     ```

     two additional functions associated with `pipe_std`, to restore the parent's file descriptors 0 and 1.

   - Since the parent process and the child process can form a circular producer-consumer relationship, the danger of deadlock is always there. It is the parent program's responsibility (not that of `pipe_std`) to guard against deadlock.

5. What different system calls can be used to read/write a socket? What are their differences? Include calls not covered in the text.

6. Write a lowercase server that takes messages from a client and turns all uppercase characters into lowercase before echoing the message back to the client. Implement the service using an Internet datagram socket.

7. Do the previous problem with an Internet stream socket.

8. Add code to your lowercase server that checks the address and port number of the client socket and only accepts requests from "allowable" clients.

9. Use the out-of-band mechanism of Internet stream sockets to send Linux signals to a remote process.

10. Write a command **service** that takes a service name, such as `ftp` and a host name, such as `monkey.cs.kent.edu`, and displays the the IP address and port number.

11. Maxima is a powerful program for mathematical computations. Install the `maxima` package if your Linux does not already have it, and then make it into an on-demand Internet server (Section 11.12).

12. Write a *chat* application where multiple people can join in the same chat session on different hosts. This problem requires a clear overview of the problem and a careful design before implementation.

Chapter 12

GUI Programming with Ruby/GTK2

A *Graphical User Interface* (GUI) program displays windows, buttons, menus, and icons on the screen. These are the on-screen representations of *windowing objects*, or *widgets*, inside the GUI program. Various types of widgets support a variety of GUI features and functionalities.

GUI programs are *event driven* and perform tasks in response to user-initiated *events* such as a mouse move, a mouse click, a key press, a button icon click, or a menu selection. The occurrence of an event can trigger a call to its *event handler*, which is a function programmed to perform the desired task. Thus, GUI applications are usually easier to learn and more intuitive to use.

GTK+, an object-oriented version of the GTK (GIMP Toolkit), is a C library for GUI programming. GTK+ works mostly on top of X Windows in Linux/UNIX systems, but is supported also on other platforms such as Microsoft Windows and Mac OS X. GTK+ is part of the GNU Project and therefore freely available. In addition to C and C++, GTK+ can be used from Perl, PHP, Ruby, and some other programming languages.

Ruby is a scripting language that was first developed in the mid-1990s. Ruby combines features from several other languages, including Perl, Smalltalk, Eiffel, Ada, and Lisp. Ruby/GTK2 provides a GTK+ binding for Ruby and works well under Linux, especially if you use the Gnome desktop. The `ruby-gtk2` package is part of the Ruby-GNOME2 bundle (`http://ruby-gnome2.sourceforge.jp`). This chapter introduces GUI programming using Ruby/GTK2.

To run the examples in this chapter, you'll need to have Ruby and Ruby/GTK2 installed on your Linux. It is likely that your Linux already has these installed. If not, you can easily install them with (Section 8.24)

sudo apt-get install libgtk2-ruby   (Ubuntu/Debian)
yum install ruby-gtk2   (CentOS/Fedora)

12.1 Getting Started with Ruby/GTK2

As our first example, we will look at a Ruby/GTK2 program (**Ex: ex12/click.rb**) which displays a button that you can click. After each click, the button shows the new click count (Figure 12.1).

FIGURE 12.1: First Ruby/GTK2 Example

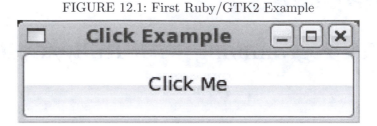

The program starts with the standard Linux executable text file line which indicates the interpreter, **/usr/bin/ruby** in this case, for the program.

The **=begin** and **=end** lines bracket any number of comment lines. The **require** statement loads and executes another Ruby file. The argument to **require** can be the name of a library (located in a known library folder) or a pathname to any Ruby source file. For a Ruby/GTK2 program, we will always need the **gtk2** library (located at **/usr/lib/ruby/site_ruby/1.8/gtk2.rb**, for example).

```
#!/usr/bin/ruby
=begin
 file:   click.rb
=end
require "gtk2"
```

The program starts by setting a variable **count** to zero (line 1). Then a new **Button** instance (a widget) is created, with the button label set to the string **Click Me** (line 2). The event **clicked** for the **Click Me** button, which is caused by a left mouse click on the **Click Me** button, is connected to the *event handling* code (line 3). Event handling actions are to increment **count** and to set the button label to **count** converted to a string (line 4).

```
count = 0                               ## (1)
button = Gtk::Button.new("Click Me")    ## (2)
button.signal_connect("clicked") {      ## (3)
    count += 1
    button.label = count.to_s()         ## (4)
}
```

Having set up the widget **button**, we now proceed to create an instance of **Window**, with title **Click Example**, as our top-level container widget, one that can contain child widgets (line 5). The width and height of **win** is set to 250 x 50 pixels (line 6). The **add** method of the **win** widget places **button** in **win** (line 7).

```
win = Gtk::Window.new("Click Example")   ## (5)
win.set_default_size(250, 50)            ## (6)
```

```
win.add(button)                         ## (7)
win.signal_connect("destroy") {         ## (8)
  Gtk.main_quit
}


win.show_all                            ## (9)
Gtk.main                                ## (10)
```

The `destroy` event caused by the user clicking on the *close-window icon* is connected to the standard `Gtk` class method `main_quit` which causes the displayed window to disappear and the program to terminate (line 8).

Finally, the `show_all` method of `win` is called to make visible `win` and all its child widgets (line 9), and the `Gtk` class method `main` is invoked to draw the display and start event monitoring (line 10).

This simple example provides a template for writing other, more complicated, Ruby/GTK+ programs.

12.2   GTK+ Event Handling Basics

GUI programs are event driven to allow easy user interactions. An event-driven program normally does nothing until an event triggers some preprogrammed action. There are different kinds of events, including mouse button events (clicks, presses, releases, and double clicks), key events, mouse move events, and so on.

A GUI program begins monitoring events after being initialized. Part of the initialization sets up the monitoring and handling of specific events. When such an event occurs, the GUI program reacts to the event, handles it quickly, and goes back to doing nothing—being ready for the next event.

Basically, writing an event-driven program involves indicating which events are monitored by which widgets and specifying actions in response to these events when they occur. When an event takes places, a *signal* is sent to the widget on which the event occurred. By setting up a *signal handler* (also known as an *event handler*), you tell a widget to catch a specific signal and then execute the handler code.

In our `click.rb` example (Section 12.1), we connected an event handler to the `clicked` signal on a `Button` widget `button`:

```
button.signal_connect("clicked") {
    count += 1
    button.label = count.to_s()
}
```

The event handler is given by a code block (`{ ... }` or `do ... end`). Here, the variables `count` and `button` in the code block refer to the variables defined in its *enclosure* (calling context).

Depending on the event type, the enclosure may also pass one or more parameters to the event handler code block. For most events, the first parameter is *self*, namely, the widget itself that is receiving the signal. The second parameter is an *event object* containing information about the event. To receive such parameters, you would write the event handling code as

```
widget.signal_connect("signal_name") {|w|      ## (w is self)
    ...
}
widget.signal_connect("signal_name") {|w,e|  ## (e an event obj)+
    ...
}
```

For example, we can rewrite our click handler above as

```
button.signal_connect("clicked") {|b|
    count += 1
    b.label = count.to_s()
}
```

where the local variable b gets passed the `button` widget when this handler is called.

An event handler may return `true` or `false`. The `true` return value stops the further propagation of the signal. The `false` return value allows further propagation. We will discuss more about events in Section 12.5.

12.3   A Ruby Primer

Ruby/GTK2 is Ruby with GTK+ binding. Thus, it enables us to write GUI programs in an object-oriented scripting language. We will describe a select set of Ruby constructs to get us started. Complete documentation for Ruby can be found at `ruby-doc.org`.

A Ruby program is a sequence of expressions. Each expression is terminated with a semicolon (;) which can be omitted if it is at the end of a line. Variable names may involve letters, digits, and the _ (UNDERSCORE), but must not begin with a digit.

Strings

Ruby string constants can be given between single or double quotes. In double quotes, backslash characters such as \n and expression substitution (#{*exp*}) are allowed. You may check string equality with == and compare strings with *s1*<=>*s2* which returns 0 (strings are equal), 1 (*s1* is greater), −1 (*s2* is greater). Strings can be concatenated with the plus operator, *s1*+*s2*. Here are some useful string expressions (**Ex: ex12/str.rb**).

```
str = "Green is good.\n"
str[0]                # Ascii code in octal for G: 71
str[0].chr            # The character G (Ruby character literal ?G)
str[0,3]              # substring "Gree"
str[-3,-1]            # substring "d.\n"
str.length            # 15
str.index(?G)         # index of character G: 0 (nil if not found)
str.index("good")     # index of substring: 9 (nil if not found)
```

The method `to_i` (`to_f`) converts a string to an integer (floating point). The method `to_s` converts a number to a string. Refer to the Ruby `String` class for more string operations.

Arrays

Arrays in Ruby are given as a list of elements inside square brackets (`[ ]`). Elements are accessed by zero-based indexing. Here are some examples (**Ex: ex12/arr.rb**).

```
arr=[1,2,?A,"ok"] # Array with 2 integers, a char, and a string
arr[2]              # The 3rd element, ?A
arr[1..3]           # Sub array [2, ?A, "ok"]
arr[-2..-1]         # Sub array [?A, "ok"]
arr[0,3]            # Sub array [1,2, ?A] (start and length)
arr[1]=9            # Setting 2nd element, [1, 9, ?A, "ok"]
arr.length          # 4
str=arr.join(";")   # String of elements, "1;2;A;ok"
str.split(";")      # The array ["1", "2", "A", "ok"]
arr.delete_at(1)    # Deletes element, arr now [1, ?A, "ok"]
arr.delete(x)       # Deletes all x from arr
new_arr=arr+arr2    # Array concatenation
new_arr=arr&arr2    # Array intersection
new_arr=arr * 4     # Array repetition
new_arr=arr-[2,3]   # Copy of arr with given elements removed
```

You can also create an empty array with

```
arr = Array.new
```

and then append elements to it with the notation `arr << element`.

Associative arrays are formed with curly braces and can use symbolic indices (**Ex: ex12/asso.rb**).

```
asso = {"a"=>3, "b"=>5,  0=>"ok", 7=>"done"}
asso["b"]              # is 5
asso[7]                # is "done"
asso.delete("a")       # {"b"=>5,  0=>"ok", 7=>"done"}
asso["new"]=9          # {"a"=>3,"b"=>5,0=>"ok",7=>"done","new"=>9}
```

To conveniently iterate over all elements of an array, use

```
arr.each { |item|
    ## one or more expressions
    ## to process item
}
```

With each iteration, the loop-control variable `item` takes on the value of the next element in `arr`.

Functions

Let's now show how to define a function with an example (**Ex:** `ex12/factorial.rb`).

```
def factorial(n)
    if ( n < 0 )
        return nil               ## logical false
    end ## of if
    if (n == 0 || n == 1 )
        return 1
    end ## of if
    return n*factorial(n-1)    ## or just n*factorial(n-1)
end ## of def
```

When a Ruby function runs to completion through its last statement, the value of the last statement becomes the return value. Hence, the word `return` can be ommitted from the last statement of a function. Also, a function may return multiple values. For example, if `some_fn` uses `return(a,b,c)`, then you can call it with `x,y,z=some_fn()`.

Function parameters may have default values (**Ex:** `ex12/testFn.rb`):

```
def test_fn(x=1, y=2, z=x+y)
    puts "#{x}, #{y}, #{z}"
end
```

```
test_fn          ##=>  1, 2, 3
test_fn 7        ##=>  7, 2, 9
test_fn(3,5)     ##=>  3, 5, 8
```

Ruby has many useful built-in functions.[1] You can find them easily on the Web.

The `ARGV` built-in array holds command-line arguments passed to Ruby programs. For example, we can call `factorial` with an argument passed on the command line:

---

[1] Actually, methods of the `Object` class.

```
ans=factorial(ARGV[0].to_i())    ## ARGV is built-in array
```

To implement `factorial` with a `while` loop (**Ex:** ex12/while.rb), we can use

```
def factorial(n)
    if ( n < 0 ); return nil; end
    ans=1;
    while  n > 1
        ans = ans*n
        n=n-1
    end ## of while
    ans
end ## of def
```

We can replace the `while` loop with a `for` loop (**Ex:** ex12/for.rb).

```
for i in 2..n do
        ans = ans*i
end ## of for each
```

Ruby *regular expressions* are similar to those used by the Linux **grep** command. A regular expression is enclosed in / / or %r{ }. To test if a string contains a regular expression, *re*, you can use

```
if ( /re/ =~ str )



    ...



end
```

More will be said about *object-oriented programming* in Ruby (Section 12.6). Complete documentation for Ruby can be found at `ruby-doc.org`.

12.4   GTK+ Widgets

In GUI programming, *windowing objects* (widgets) play a central role. A GUI program employs and displays various widgets to provide visual control/operation for a program. An *atomic widget* is one that corresponds to a single GUI feature such as a button or label. A *container widget*, on the other hand, is a GUI component that can contain/manage other widgets. A GUI application usually has a top-level widget (the root window) that contains and manages other widgets in the program.

A GUI program works by responding to specific events from the user. Reactions to an event are written into the GUI program. The GUI program execution environment usually supplies event monitoring, input focusing, window rendering, graphics drawing, and parent-child window coordination.

GTK+ provides a large number of widgets to make GUI easy to build. A set of important GTK+ widgets are described briefly here to get you started.

Each GTK+ widget belongs to a *class*, say, `GTK::`*Xyz*, that defines how that widget works. The code

```
x = GTK::Xyz.new( ... )
```

creates a new *instance* of class `GTK::`*Xyz* and assigns it to the variable `x`. After being created, you can use the *methods* of the widget `x` with the notation `x.`*method_name*`(...)`.

The Top-Level `Window`

In GTK+, you use a `Window` instance as the top-level container widget to enclose and build your GUI. The code we have seen,

```
win = Gtk::Window.new("Click Example")
```

creates a new `Window` object and gives it a window title. The default size of a top-level window can be set with

```
win.set_default_size(width, height)          ## in pixels
```

A `Window` widget can contain one child widget at any given time. To display multiple widgets, you simply add a *layout container* (Section 12.4) to house the desired widgets.

Layout Containers

GTK+ *layout containers* are invisible widget containers that help us position child widgets to achieve a desired layout. GTK+ provides many different layout containers, including

- `Alignment`—Controls the alignment and size of a child widget
- `AspectFrame`—Constrains its child to a particular aspect ratio
- `HBox`—Provides a horizontal container box
- `VBox`—Provides a vertical container box
- `HButtonBox`—Arranges child buttons horizontally
- `VButtonBox`—Arranges child buttons vertically
- `Fixed`—Places child widgets at specified coordinates
- `HPaned`—Provides left-right panes

- VPaned—Provides up-down panes

- Layout—Gives an infinitely scrollable area for child widgets and/or custom drawing

- Notebook—Displays child widgets in a tabbed notebook

- Table—Aligns child widgets in a row-column grid

- Expander—Hides and reveals a child widget

You can find demos together with sample source code for these layout containers and other GTK+ widgets in the Ruby/GTK2 documentation, found, for example, under **/usr/share/doc/ruby-gtk2-***version***/sample**.

Among the various layout containers, the HBox and the VBox are the most basic and common. Widgets packed into an HBox (a VBox) are displayed in a horizontal row (vertical column).

To demonstrate layout with boxes, let's look at an example (Figure 12.2) that makes accessing the API (application programming interface) documentation of GTK+ widgets quick and easy (**Ex: ex12/api.rb**). By entering any widget class name, you'll arrive directly at the API Web page.

FIGURE 12.2: Ruby/GTK2 Widget API Example

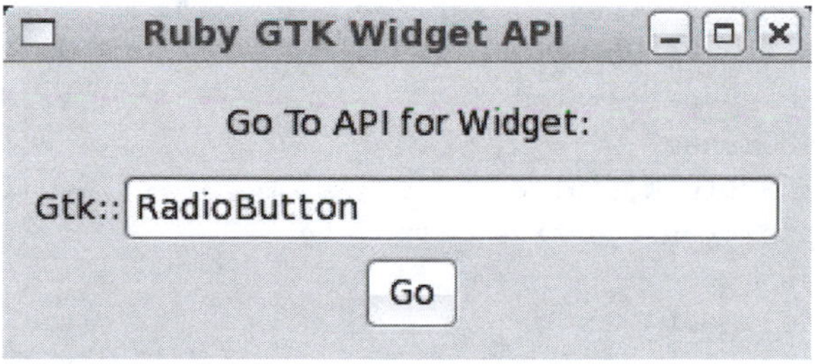

The api.rb program starts by setting the Web URL for the Ruby GTK+ Widget API documentation and the the Web browser to use (specified via the environmental variable $BROWSER which is available in Ruby as ENV["BROWSER"]).

```
#!/usr/bin/env ruby
=begin
Example:  Accessing Ruby GTK Widget API documentation
=end
require "gtk2"

### Ruby Gnome2 documentation retrieval URL
url="http://ruby-gnome2.sourceforge.jp/hiki.cgi?Gtk::"
browser=ENV["BROWSER"]    ## Web browser to use
```

Next, `api.rb` constructs the GUI. The layout involves a `Label` (title on line 2) and two `HBox`es packed (lines 8-10) into a `VBox` (lines 1). When creating a new `VBox` or `HBox`, two arguments can be specified:

```
Gtk::HBox.new(homogeneous = false, spacing = nil)
Gtk::VBox.new(homogeneous = false, spacing = nil)
```

If *homogeneous* (default `false`) is `true`, then all child widgets are of the same width. The number of pixels between child widgets is specified by *spacing* (default `nil` means unspecified).

The `Label` gtk and a single-line text entry widget `entry` (lines 3-4) are packed into the `HBox` enBox (lines 5-6). The `pack_start` method packs successive child widgets from left/top to right/bottom. The `pack_end` method, on the other hand, packs successive child widgets from right/bottom to left/top.

The arguments are *child*, *expand*, *fill*, and *padding*. If *expand* is `true`, the box will have the full width/height allocated to it. If *expand* is `false`, the box will have just enough width/height to contain the child widget. The *fill* setting is only effective when *expand* is `true`. Setting *fill* to `true` stretches the children to fill the horizontal/vertical space (line 6). Setting *fill* to `false` prevents such stretching. You may also specify extra *padding* around the *child* in pixels.

```
vbox = Gtk::VBox.new(true, 5)                      # (1)
title=Gtk::Label.new("Go To API for Widget:")      # (2)
gtk=Gtk::Label.new("Gtk::")                         # (3)
entry=Gtk::Entry.new()                              # (4)
enBox=Gtk::HBox.new
enBox.pack_start(gtk, false, false, 0)              # (5)
enBox.pack_start(entry, true, true, 0)              # (6)
goBox=Gtk::HBox.new
go=Gtk::Button.new(" Go ")
goBox.pack_start(go, true, false, 0)                # (7)
vbox.pack_start(title, false, true, 0)              # (8)
vbox.pack_start(enBox, false, true, 0)              # (9)
vbox.pack_start(goBox, false, true, 0)              # (10)
```

After entering the target widget class name in the `Entry` field, a user can press the ENTER key (line 12) or click the go button (line 13) to invoke the function `api` (line 11) which invokes the designated browser on the correctly constructed URL.

In Ruby, calling the `system` function is one way of executing a Shell command. Another is with BACKQUOTES (` `cmdString` ` as in Bash Shell). Yet another is with the Ruby notation `#[cmdString]`. In either case, the `#{exp}` substitution is allowed in *cmdString*.

```
def api(browser, url, className)                   # (11)
  cmd=browser + " " + url + className
```

```
    system cmd
end

entry.signal_connect("activate") {              # (12)
  api(browser, url, entry.text)
}

go.signal_connect("clicked") {                  # (13)
  api(browser, url, entry.text)
}

window = Gtk::Window.new("Ruby GTK Widget API")
window.border_width = 10
window.add(vbox)                                # (14)
window.signal_connect("delete_event") {
  Gtk.main_quit
}

window.set_size_request(320, -1)
window.show_all
Gtk.main
```

The rest of the `api.rb` program simply establishes the top-level window, adds `vbox`, and displays the GUI as usual.

The `api.rb` shows how to use `HBox` and `VBox` for layout. `Table` is another layout container for placing child widgets in neatly aligned rows and colums. See the `tz.rb` program in Section 12.4 for an example using `Table` layout.

The `Expander` layout container allows you to pull down and pull up contents displayed (Figure 12.3). It is useful for displaying multiple items in a limited space.

FIGURE 12.3: Ruby/GTK2 `Expander` Example

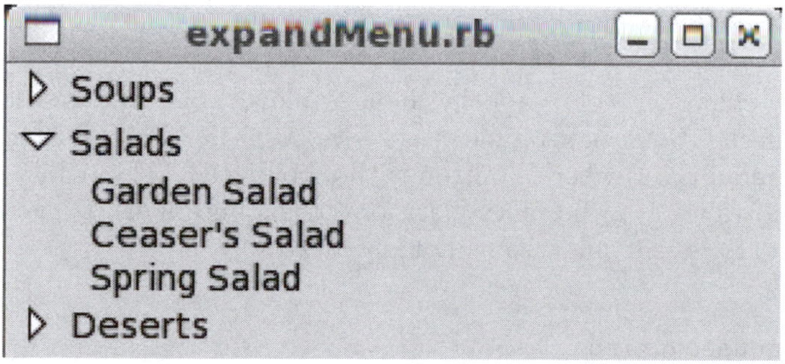

In the example code package, you can find an example (**Ex:** `ex12/ExpandMenu.rb`) that applies an `Expander` to display a menu.

Atomic Widgets

Ruby/GTK2 provides *atomic widgets* to support many common GUI operations. Atomic widgets are basic building blocks of GUI programs. They are laid out in containers to create a visual display. Atomic widgets include

- Label—For displaying a small amount of text with possible color and font settings

- Button—For making a child (usually a Label but can be any widget) clickable (via the clicked event)

- CheckButton—For user choice (via the toggled event)

- RadioButton—For one in many choice (via the toggled event)

- Entry—For single-line text input (via the activate event)

- HScale/VScale—For a horizontal/vertical sliding scale with a user-movable slider (via the value-changed event)

- SpinButton—For increasing/decreasing values (via the value-changed event)

- ComboBox—For providing a pull-down list of choices (via the changed event)

- ComboBoxEntry—Same as ComboBox except the selected entry can be edited by the user

- ColorButton/FontButton—For choosing color or font (via the color-set or font-set event)

Becoming familiar with atomic widgets, their purposes, and their event handling can make writing GUI code much easier. Let's do this by working with examples.

Button with Rollover

A widget changing color as the mouse moves over it is the familiar *rollover* effect GUI designers and users love. Let's look at an example (**Ex:** ex12/mouseover.rb) where a button with a boldface label in white over a red background (lines 1-2) becomes black on green (lines 3-4), triggered by the enter-notify-event and leave-notify-event.

```
=begin
 file:   mouseover.rb
=end
require "gtk2"
```

```
red=Gdk::Color.new(45500,0,0)  # deep red
green=Gdk::Color.new(0,45500,0)  # deep green
white=Gdk::Color.new(45500,45500,45500)  # white
black=Gdk::Color.new(0,0,0)


la = Gtk::Button.new                         ## (1)
lb=Gtk::Label.new
lb.set_markup("<b>Mouse over me</b>")
lb.modify_fg(Gtk::STATE_NORMAL, white)
la.add(lb)
la.modify_bg(Gtk::STATE_NORMAL, red)         ## (2)

la.signal_connect("enter-notify-event") {    ## (3)
    la.modify_bg(Gtk::STATE_NORMAL, green)
    lb.modify_fg(Gtk::STATE_NORMAL, black)
}
la.signal_connect("leave-notify-event") {    ## (4)
    la.modify_bg(Gtk::STATE_NORMAL, red)
    lb.modify_fg(Gtk::STATE_NORMAL, white)
}

win = Gtk::Window.new("MouseOver Example")
win.set_default_size(250, 50)
win.add(la)
win.signal_connect("destroy") {
  Gtk.main_quit
}

win.show_all
Gtk.main
```

Refer to the Gtk::Color API for more information on color settings.

A GUI for tar

Our next example (**Ex: ex12/tz.rb**) uses RadioButtons and Table layout to build a GUI for the Linux **tar**, a command that creates and extracts *archive files* (Chapter 6, Section 6.12).

The program begins by setting some global variables.

```
#!/usr/bin/ruby
=begin
  example:  tz.rb
=end
require "gtk2"
$operation="x"    ## "x" for extract, or "c" for create
```

FIGURE 12.4: A GUI for **tar**

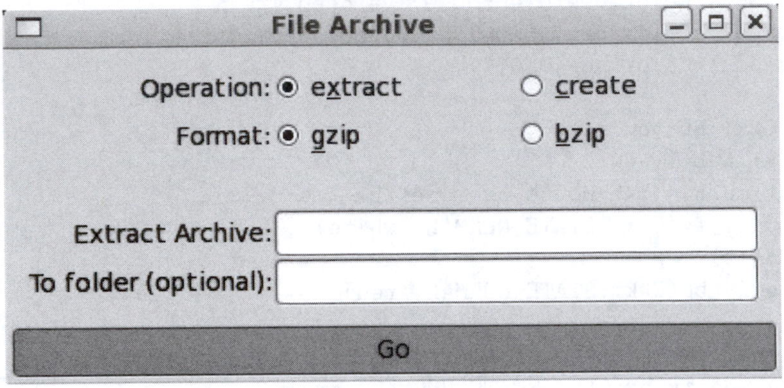

```
$format="z"        ## "z" for gzip, or "j"  for bzip2
$suffix=".tgz"     ## or "tbz", archive file name suffix
### Text for labels
$ac_text="Create Archive:"
$fc_text="From files and folders:"
$ax_text="Extract Archive:"
$fx_text="To folder (optional):"
```

Two groups of `RadioButtons` are used. Group one selects the operation (creating or extracting an archive), and group two decides the data compression format (`gzip` or `bzip2`).

```
## radio buttons group one
extract = Gtk::RadioButton.new("e_xtract")
create = Gtk::RadioButton.new(extract, "_create")

## radio buttons group two
gzip = Gtk::RadioButton.new("_gzip")
bzip = Gtk::RadioButton.new(gzip, "_bzip")
```

The UNDERSCORE character used in `RadioButton` names is significant. It designates the next character as the *keyboard mnemonic* for selecting the button (see Figure 12.4).

A 5-row by 3-column layout table `tb` displays the **Operation** selection (row one), **Format** selection (row two), an empty spacer (row 3), the archive name entry field (row 4), and the source archive file (or destination file/folder) entry field (row 5).

The third argument to `Table.new` indicates whether to make all table cells the same size. The `attach` method indicates the placement of a child widget in the layout table. Specifically,

`tb.attach(`*widget, c0, c1, r0, r1*`)`

places the *widget* between column *c0* and *c1* and between rows *r0* and *r1* (Figure 12.5).

FIGURE 12.5: Table Layout

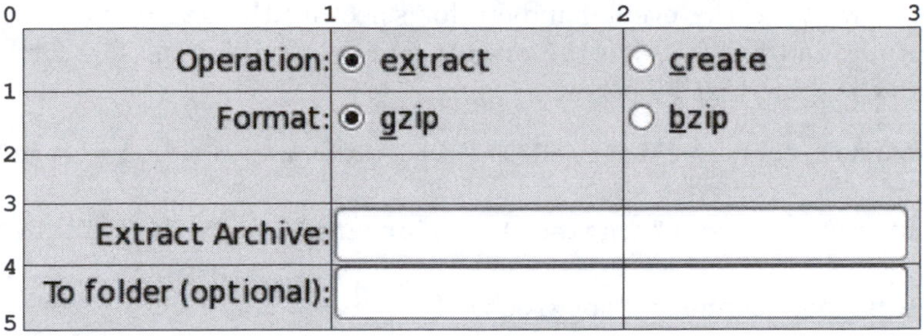

```
tb=Gtk::Table.new(5,3,true)               ## layout table

op=Gtk::Label.new("Operation:")
op.set_xalign(1)                          ## align to the right
tb.attach(op,0,1,0,1)                     ## row 1
tb.attach(extract,1,2,0,1)
tb.attach(create,2,3,0,1)

fm=Gtk::Label.new("Format:")
fm.set_xalign(1)
tb.attach(fm,0,1,1,2)                     ## row 2
tb.attach(gzip,1,2,1,2)
tb.attach(bzip,2,3,1,2)

### text entry fields
$a_label=Gtk::Label.new($ax_text)
$a_label.set_xalign(1)
$f_label=Gtk::Label.new($fx_text)
$f_label.set_xalign(1)
$folder=Gtk::Entry.new()
$ar=Gtk::Entry.new()

tb.attach(Gtk::Label.new(" "),0,1,2,3)    ## row 3, spacer
tb.attach($a_label,0,1,3,4)               ## row 4 archive name
tb.attach($ar,1,3,3,4)
tb.attach($f_label,0,1,4,5)               ## row 5 file/folder
tb.attach($folder,1,3,4,5)
```

The layout table tb is then placed in a VBox (box2), and box2 together with the go Button is placed in another VBox (box1). The visual layout is now complete.

```
box2 = Gtk::VBox.new(false, 10)
box2.border_width = 10; box2.add(tb)
box1 = Gtk::VBox.new; box1.add(box2)
go = Gtk::Button.new(" Go "); box1.add(go)
```

Next, we have the event handlers for selecting the `gzip/bzip` compression format and for choosing the `create/extract` operation. The latter also involves changing text labels.

```
gzip.signal_connect("toggled") { $format="z"; $suffix=".tgz"; }

bzip.signal_connect("toggled") { $format="j"; $suffix=".tbz"; }

create.signal_connect("toggled") {
      $operation="c"
      $a_label.text=$ac_text    ## relabeling
      $f_label.text=$fc_text    ## relabeling
}

extract.signal_connect("toggled") {
      $operation="x"
      $a_label.text=$ax_text;    ## relabeling
      $f_label.text=$fx_text;    ## relabeling
}
```

Two functions are defined: `tarCommand` constructs the correct **tar** command to invoke, and `tarResult` displays a confirmation dialog or an error dialog (see API for `Gtk::Dialog`).

```
def tarCommand()
   aName = $ar.text;    tail = ""
   if ( $operation=="x" && $folder.text != "" )
      tail = "-C " + $folder.text
   end
   if ( $operation=="c" )
      if ( $folder.text == "" )
         ### error no files to archive
      else
         tail = $folder.text
         aName = $ar.text + $suffix
      end
   end
   cmd="tar "+$format+$operation+"pf "+aName+" "+tail
   return cmd
end

def tarResult(parent, msg, type)
```

```
dialog = Gtk::MessageDialog.new( parent, Gtk::Dialog::MODAL,
                    type, Gtk::MessageDialog::BUTTONS_OK, msg)
  dialog.title = "Tar Result"
  dialog.run { };  dialog.destroy
end
```

Clicking the **go** button causes a call to `tarCommand()` (line A) and execution of the resulting **tar** command (line B). Success/failure is then reported by calling `tarResult` (lines C and D).

```
window = Gtk::Window.new("File Archive")
go.signal_connect("clicked") {
   cmd=tarCommand()                              # (A)
   if ( system(cmd) )                            # (B)
      tarResult(window, cmd + " Successful.",    # (C)
               Gtk::MessageDialog::INFO)
   else
      tarResult(window, cmd + " Failed.",        # (D)
               Gtk::MessageDialog::WARNING)
   end
}
```

The usual top-level window code follows.

```
window.add(box1)
window.signal_connect("destroy"){Gtk.main_quit}
window.show_all
Gtk.main
```

Decorator Containers

Ruby/GTK2 also provides a variety of single-child containers, known as *Decorators*, to add a certain functionality or visual effect to a child. For example,

- `Frame`—Adds a frame box with optional in-frame caption for the child

- `MenuItem/ToolItem`—Enables a child to be placed on a Menu or Tool-Bar

- `ScrolledWindow`—Provides scrolling to any child added by the `add_with_viewport` or `add` method

Now let's look at another example (**Ex: ex12/volume.rb**) that applies the `Frame` and the `HScale` to create a GUI for controlling the sound volume setting of your Linux box (Figure 12.6).

The volume `control` is a horizontal scale with minimum value 0.0, maximum value 1.0, and an adjustment step size of 0.1 (line I).

FIGURE 12.6: Volume Control

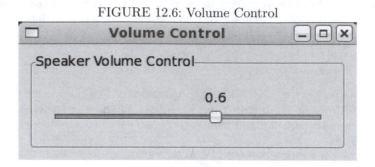

The initial setting of `control` is based on the current volume setting of the *front speakers* (assuming left and right front speakers are in lock step). In the function `setLevel`, the **amixer** command's `get` operation retrieves the volume information, and the level (between 0 and 31) is converted to a value between 0.0 and 1.0 for the `HScale` control (lines II-III).

Depending on the sound system used on your own Linux, you may need to use some command other than **amixer**.

```ruby
#!/usr/bin/ruby
=begin
     volume.rb---Volume control example
=end
require "gtk2"

##   min, max,   step
control = Gtk::HScale.new(0.0, 1.0,   0.1)        ## (I)

def setLevel(f)                                   ## (II)
   level=`/usr/bin/amixer get Front |
         grep "Front Left:" | cut -d " " -f 6`
   f.set_value(level.to_f() / 31.0)
end

setLevel(control)                                 ## (III)
```

The `value-changed` event of `control` sets the front speaker volume with the **amizer** operation `sset` which can take a percentage value (line IV).

```ruby
control.signal_connect("value-changed") {         ## (IV)
   v=control.value*100
   v=v.to_i();          ## float to int conversion
   cmd="/usr/bin/amixer sset Front #{v}% >/dev/null"
   system cmd
}
```

Now, for the layout, the horizontal scale is placed in an **HBox** to center it with some padding (line **V**). We then surround the **HBox** with a captioned **Frame** (line **VI**).

```
hbox=Gtk::HBox.new(true, 0)
hbox.pack_start(control, false, true, 20)        ## (V)
cf = Gtk::Frame.new("Speaker Volume Control")
cf.add(hbox)                                      ## (VI)
```

The usual top-level window code follows.

```
window = Gtk::Window.new(Gtk::Window::TOPLEVEL)
window.set_title  "Volume Control"
window.border_width = 10
window.signal_connect("delete_event") { Gtk.main_quit }
window.set_size_request(350, 120)
window.add(cf)
window.show_all
Gtk.main
```

It is true that your Linux desktop panel will most likely already supply a GUI volume control. Our example simply shows how you can also construct a similar GUI program easily with Ruby/GTK2.

12.5   More about Events

In Ruby/GTK2, events fall into two broad categories: *low-level events* supported by **Gdk** and higher level *semantic events* defined by **Gtk** atomic, decorator, and container widgets. The API documentation of each **Gtk** widget describes, under the *signal* section, the events accepted by that widget. For example, **Button** accepts the **clicked** event and **Window** accepts the **destroy** event, as we have seen in prior examples. Of course, each widget also accepts signals allowed by ancestors in its object hierarchy.

Being the ultimate ancestor of Gtk widgets, a **Gtk::Widget** object accepts many low-level events including these common events:

- Mouse-button events: **button-press-event**, **button-release-event**

- Input focus events: **focus-in-event**, **focus-out-event**

- Key events: **key-press-event**, **key-release-event**

We will illustrate their use with examples. Our first example (**Ex: ex12/mousebutton.rb**) detects which mouse button is clicked (Figure 12.7).

The **Click Me** button catches the **button-press-event** and checks the button number in the event object (of type **Gdk::EventButton**) to determine which mouse button was pressed.

FIGURE 12.7: Which Mouse Button

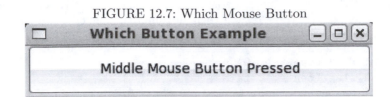

```
button = Gtk::Button.new("Click Me")
button.signal_connect("button-press-event") {|w,e|
    ## e is event object of type Gdk::EventButton
    if ( e.button == 1 )
       button.label="Left Mouse Button Pressed"
    elsif ( e.button == 2 )
       button.label="Middle Mouse Button Pressed"
    elsif ( e.button == 3 )
       button.label="Right Mouse Button Pressed"
    else
       button.label="Mouse Button " + e.button.to_s()
                     + " Pressed"
    end
    true
}
```

FIGURE 12.8: Mouse Button Events

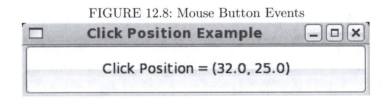

The event object also carries the mouse press location information (x and y coordinates relative to the upper-left corner of the widget) as well as double/triple-clicking information. Figure 12.8 shows the display made by the following event handling code (**Ex: ex12/position.rb**):

```
button.signal_connect("button-press-event") {|w,e|
    button.label="Click Position = (#{e.x}, #{e.y})"
    true
}
```

If we catch the **key-press-event**, then we get an event object of type Gdk::EventKey, which has a keyval property indicating which key has been pressed. Figure 12.9 shows the keyval of the *Left Shift Key* produced by the following event handling code (**Ex: ex12/key.rb**):

FIGURE 12.9: Key Events

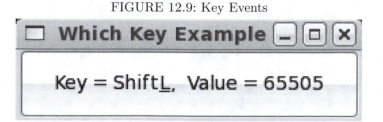

```
button.signal_connect("key-press-event") {|w,e|
    kv=e.keyval
    button.label="Key = " + Gdk::Keyval.to_name(kv) +
                    ",  Value = #{kv}"
    true
}
```

Gdk provides constants for keys in the form Gdk_x, where *x* is the name of the key and a utility function Gdk::Keyval.to_name that can turn a keyval into a key name.

Our next example (**Ex: ex12/convert.rb**) is an inch-centimeter converter program (Figure 12.10).

FIGURE 12.10: Focus Events

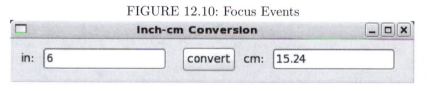

By entering a number into one of the entry fields and clicking the convert button, the equivalent value shows up in the other entry field.

First, we create an inch and a cm Entry widget, each 16-chars wide (lines 1-2). Two text labels and a convert button are also created (line 3-4). These are then packed into an HBox for visual presentation (lines 5-6).

```
inch=Gtk::Entry.new                        ## (1)
cm=Gtk::Entry.new
inch.width_chars=16; cm.width_chars=16     ## (2)
inlabel=Gtk::Label.new("in:")              ## (3)
cmlabel=Gtk::Label.new("cm:")
convert=Gtk::Button.new("convert")         ## (4)
hbox=Gtk::HBox.new                         ## (5)
hbox.pack_start(inlabel, false, false, 0)
hbox.pack_start(inch, false, false, 10)
hbox.pack_start(convert, false, false, 10)
hbox.pack_start(cmlabel, false, false, 0)
```

```
hbox.pack_start(cm, false, false, 10)
convert.height_request=30                         ## (6)
```

Now let's look at event handling. The `convert` button `clicked` signal triggers a call to `doConvert`, a function that performs the actual conversion. Each `Entry` widget has its `focus-in-event` connected to a call to the `reset` function which blanks out both entry fields. So they are ready for the next conversion.

```
convert.signal_connect("clicked") { doConvert(inch, cm) }
cm.signal_connect("focus-in-event") { reset(cm, inch) }
inch.signal_connect("focus-in-event") { reset(cm, inch) }

def reset(cm, inch)
   cm.text=""; inch.text=""
end

def doConvert(inch, cm)
   i=inch.text; c=cm.text
   if ( i == "" )
     inch.text=(c.to_f/2.54).to_s
   else
     cm.text=(i.to_f*2.54).to_s
   end
end
```

The rest of `convert.rb` is the usual top-level window code.

```
win = Gtk::Window.new("Inch-cm Conversion")
win.set_default_size(350, 50)
win.border_width=10; win.add(hbox)
win.signal_connect("destroy") { Gtk.main_quit }
win.show_all; Gtk.main
```

See the API for `Gtk::Widget` for a complete list of signals available to all widgets. Each signal description specifies the arguments passed to event handlers and the event object type. In Ruby/GTK2, event objects are descendants of the `Gdk::Event` class.

12.6   OOP with Ruby/GTK2

Ruby supports *Object-Oriented Programming* (OOP), and GTK+ is a class library for GUI programming. An OO program builds objects and has them interact with one another at run time to achieve desired tasks. With OOP, a *class* is defined as a template to build *objects* that are *instances* of the class. Thus, a class is a blueprint for creating new objects belonging to that class. The general form of a class in Ruby is

```
class NameOfClass
    def initialize(...)
        ...
    end  ## of initialize

    def methodName(...)
        ...
    end  ## of method

    ...


end  ## of class definition
```

A class has *attributes* (variables defined in the class) and *methods* (functions defined in the class). The `initialize` is a special method known as a *constructor*. The code *NameOfClass*.`new(...)` creates a new object instance and passes any arguments to `initialize` for setting up that object.

An *instance method* is a method within an individual object. A method is an *instance method* unless its method name uses the *NameOfClass.* prefix, in which case the method becomes a *class method*. You invoke an instance method *i_method* through an object.

an_object.*i_method*(...)

whereas you invoke a class method *c_method* with the class name.

NameOfClass.*c_method*(...)

When no arguments are passed, the entire (...) is often omitted.

An *instance attribute* is a local variable in an object. An instance attribute name must use the `@` prefix. A *class attribute* is a variable local to the class itself. A class attribute name must use the `@@` prefix. A global variable, one that is accessible anywhere, uses the `$` prefix.

In Ruby, the period (.) operator is used for method calls **but not for attribute access**. In fact, an attribute is *private* (not accessible from outside the object or class) unless the class provides *accessor methods* for it.

You can easily arrange the conventional *reader* and *writer* methods for any instance attribute `@var`:

```
def var; @var; end          (Reader for @var, usage: foo=obj.var
def var=(x); @var=x; end     (Writer for @var, usage: obj.var=value
```

Ruby makes coding this even easier. With the abbreviation notations

```
attri_reader :var1, :var2, ...
attri_writer :var1, :var2, ...
attri_accessor :var1, :var2, ...
```

you can create readers, writers, and *accessors* (both readers and writers) for any list of instance attributes of your choice (see Section 12.7 for an example).

After this brief introduction to OOP in Ruby, we will demonstrate how to write OO code in Ruby/GTK2 with examples. Let's rewrite our `click.rb` example using OOP as follows (**Ex: ex12/ooclick.rb**):

```
#!/usr/bin/ruby
=begin
 file:   ooclick.rb
=end
require "gtk2"

class ClickMe < Gtk::Window                     # (1)
  def initialize(title) ## constructor
      super(title)                              # (2)
      @count=0                                  # (3)
      @button = Gtk::Button.new("Click Me")     # (4)
      add(@button)                              # (5)
      @button.signal_connect("clicked"){        # (6)
            clickHandler }
      set_default_size(250, 50)                 # (7)
      signal_connect("destroy") {Gtk.main_quit} # (8)
  end  ## of initialize

  def clickHandler    ## method                 # (9)
      @count+=1;   @button.label=@count.to_s()
  end  ## of clickHandler
end ## of class ClickMe

win = ClickMe.new("OO Click Example")           # (10)
win.show_all; Gtk.main
```

A great advantage of OOP is the ability to write new classes by modifying existing classes through *inheritance*. Here our `ClickMe` class inherits from `Gtk::Window` (with the < notation on line 1), meaning that `ClickMe` gets all the methods and attributes in `Gtk::Window`, the *superclass* of `ClickMe`. As a result, `ClickMe` becomes a specialized version (a *subclass*) of `Gtk::Window` and can be used as such.

The `initialize` method of `ClickMe` passes the `title` argument (line 2) to the superclass constructor (`initialize` in `Gtk::Window`), sets up a `Button` object `@button` and an integer `@count` (lines 3-4), adds `@button` as a child to itself (line 5), and connects the method `clickHandler`, defined on line 9, to the `@button clicked` event (line 6).

Finally, it sets its own default size and `destroy` event handler (lines 7-8).

After the class `ClickMe` is defined, we simply create an instance (line 10), show it, and call `Gtk.main`.

12.7 A Tic-Tac-Toe Game

Our next example (**Ex: ex12/tic/tictactoe.rb**) demonstrates OOP and the use of images in a realistic application (Figure 12.11).

FIGURE 12.11: Tic-Tac-Toe

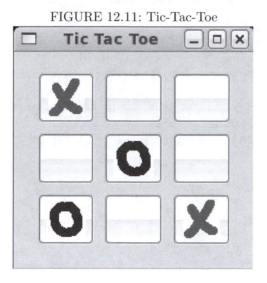

First, we define a PlayButton class as a subclass of Gtk::Button. PlayButton objects will form the nine clickable positions on the Tic-Tac-Toe game board.

```
class PlayButton < Gtk::Button
  attr_reader :r, :c, :open, :lb
  attr_writer :open
  def initialize(r,c)
      super()    ## initialize the super-class object
      @lb=Gtk::Label.new("  ")
      add(@lb)
      @r=r; @c=c; @open=true
  end
end
```

The rows and columns of our game board are indexed 0 through 2. The PlayButton row (@r) and column (@c) attributes record its board position. The @open attribute indicates if the position is occupied or not. A PlayButton initially contains a blank label. When the position is played, it will display either an X image or an O image.

The TicBoard, a subclass of Gtk::Window, implements the game board display and game logic.

```
class TicBoard < Gtk::Window
```

```
def initialize(x_file, o_file)  ## constructor
    super("Tic-Tac-Toe")
    @x=x_file; @o=o_file; @xTurn=true           ## (a)
    set_default_size(220, 200)
    signal_connect("destroy") {Gtk.main_quit}
    reset()
end  ## of initialize
```

A new `TicBoard` gets the X and O token image files and saves them in instance attributes `@x` and `@o`, respectively. The `@xTurn` attribute keeps track of which token to play (line a).

The constructor calls `reset` to prepare values for a new game. It creates a 3x3 layout `Table` (`@grid`, lines b-c) and attaches a new `PlayButton` to each of the nine positions with a double loop (line e).

```
def reset
    @gameEnd=false; @moveCount=0
    @mv=Array.new(3){Array.new(3, nil)}
    @grid=Gtk::Table.new(3,3,true)              ## (b)
    @grid.row_spacings=10
    @grid.column_spacings=10
    @grid.border_width=20                       ## (c)
    for r in 0..2 do                            ## (e)
        for c in 0..2 do
            b=PlayButton.new(r,c)
            b.set_size_request(37,39)
            @grid.attach(b,r,r+1,c,c+1)
            b.signal_connect("clicked"){|pb|     ## (f)
                makeMove(pb) }
        end  ## of for c loop
    end  ## of for r loop
    add(@grid)
end  ## of reset
```

The `reset` method can be called at any time to begin a new game.

The `clicked` event of each `PlayButton` is connected (line f) to the event handler `makeMove`, which works as follows:

1. If the game has ended, starts a new game by removing and destroying the `@grid` child, calling `reset`, and causing re-display (line g)

2. If the play position is not open, returns; otherwise, marks it occupied and continues (line h)

3. Increments the `@moveCount`, replaces the blank label child with the appropriate game token image, records the move in the `@mv` array and calls the `winner` method to see if the game has ended (lines i-j)

4. Toggles the @xTurn attribute before finishing (line k)

```
def makeMove(pb)
    if ( @gameEnd )                              ## (g)
        remove(@grid); @grid.destroy
        reset(); show_all; return(true)
    end
    if (! pb.open); return(true)                 ## (h)
    else; pb.open=false; end
    @moveCount=@moveCount+1                       ## (i)
    r=pb.r; c=pb.c
    pb.remove(pb.lb); pb.lb.destroy
    if ( @xTurn )
        token=Gtk::Image.new(@x)
        pb.add(token); token.show
        @mv[r][c]="X";
        if (winner("X")); @gameEnd=true; end
    else
        token=Gtk::Image.new(@o)
        pb.add(token); token.show
        @mv[r][c]="O"
        if (winner("O")); @gameEnd=true; end
    end                                          ## (j)
    @xTurn=!@xTurn                               ## (k)
    true
end  ## of makeMove
```

To determine if token t has won, the **winner** method checks the rows, columns, and diagonals only if the @moveCount is 5 or greater.

```
def winner(t)
    if ( @moveCount > 4 ); return(
        (@mv[0][0]==t && @mv[0][1]==t && @mv[0][2]==t) ||
        (@mv[1][0]==t && @mv[1][1]==t && @mv[1][2]==t) ||
        (@mv[2][0]==t && @mv[2][1]==t && @mv[2][2]==t) ||
        (@mv[0][0]==t && @mv[1][0]==t && @mv[2][0]==t) ||
        (@mv[0][1]==t && @mv[1][1]==t && @mv[2][1]==t) ||
        (@mv[0][2]==t && @mv[1][2]==t && @mv[2][2]==t) ||
        (@mv[0][0]==t && @mv[1][1]==t && @mv[2][2]==t) ||
        (@mv[0][2]==t && @mv[1][1]==t && @mv[2][0]==t)  )
    end
    return(false)
end
end ## of class TicBoard
```

With the **TicBoard** class defined, we only need three lines of code to get the game going.

```
$win = TicBoard.new("x.png", "o.png")
$win.show_all;   Gtk.main
```

The complete, ready-to-run code for this and other examples in this chapter
can be found in the example code package.

12.8   Menu Bar

A game GUI, even one as simple as that for Tic-Tac-Toe, will usually have
options for the game and for players. Such options are typically placed on a
menu bar for easy access.

Let's see how to add a menu bar to the Tic-Tac-Toe game given in Section 12.7.

FIGURE 12.12: Tic-Tac-Toe with Menu Bar

In Figure 12.12, the menu bar contains two menu items, and each pulls
down a menu (list of options). In this example, the Player menu provides the
option of X or O going first. The Game menu allows players to start a new
game or take back a move.

The createBar method creates and returns a MenuBar (line A). After obtaining a new MenuBar, topbar, we set its background (to orange, line B); pass
two options, op1 and op2, to the constructor of a MyMenuItem named Game
(linesC-D); and place the menu item on topbar (line E). The Player menu
item is constructed similarly.

```ruby
def createBar
    topbar=Gtk::MenuBar.new                       ## (A)
    topbar.modify_bg(Gtk::STATE_NORMAL,           ## (B)
            Gdk::Color.parse("#FFFFAAAA0000"))
    op1=Gtk::MenuItem.new("New Game")             ## (C)
    op1.signal_connect("activate"){
        @vbox.remove(@grid); @grid.destroy
        reset(); show_all
    }
    op2=Gtk::MenuItem.new("Unmove")
    op2.signal_connect("activate"){
        puts "Take back a move!"
    }
    it1=MyMenuItem.new("Game", op1, op2)          ## (D)
    topbar.append(it1)

    op1=Gtk::MenuItem.new("X First")
    op1.signal_connect("activate"){ @xTurn=true }
    op2=Gtk::MenuItem.new("O First")
    op2.signal_connect("activate"){ @xTurn=false }
    it1=MyMenuItem.new("Player", op1, op2)
    topbar.append(it1)
    return topbar
end
```

The `MyMenuItem` class is a subclass of `Gtk::MenuItem` created with any given title and number of options passed to its constructor. Note that the `*options` parameter gets all arguments after the first one (line **F**).

```ruby
class MyMenuItem < Gtk::MenuItem
    def initialize(title, *options)              ## (F)
        super(title)
        submenu = Gtk::Menu.new
        options.each {|op| submenu.append(op) }
        set_submenu(submenu)
    end
end
```

The full program (**Ex:** ex12/tic/ticmenu.rb) can be found in the example package.

12.9   Drag and Drop

A particular advantage of GUI is the ability to provide drag-and-drop (*dnd*) operations. Here, we will demonstrate Ruby/GTK2 support for dnd by a program that can receive files via dnd from the Gnome desktop, the Nautilus file manager, or other similar applications and upload the files to another computer (Figure 12.13).

FIGURE 12.13: Drag-And-Drop File Upload

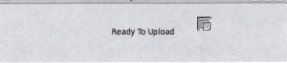

With dnd, you drag from a *source window* and drop over a *destination window*. The data items transferred from source to destination are known as *targets*.

Our example (**Ex: ex12/upload.rb**) starts with a customizable global variable set to the command to be used for file uploading.

The `DestWindow` class is a top-level window set up as a dnd destination capable of receiving text as well as uri targets (line 1-2). The `drag-drop` event triggers a call to the `Gtk::Drag` class method `get_data` to request for the target data. When the target data becomes available, a `drag-data-received` signal will be delivered.

```
$uploadCommand="/root/cmd/mput"

class DestWindow < Gtk::Window
  def initialize
    super("File Upload")
    @label = Gtk::Label.new("Ready To Upload")
    add(@label)
    set_default_size(500, 100)
    Gtk::Drag.dest_set(self,                              ## (1)
        Gtk::Drag::DEST_DEFAULT_ALL, [],
        Gdk::DragContext::ACTION_COPY)
    Gtk::Drag.dest_add_text_targets(self);
    Gtk::Drag.dest_add_uri_targets(self);                ## (2)

    signal_connect("drag-drop"){|w, dc, x, y, time|       ## (3)
      Gtk::Drag.get_data(w, dc, dc.targets[0], time)
    }
```

The `drag-data-received` event is connected to code to prepare for file uploading: obtaining file pathname from the uri data (line 4), splitting multiple files into an array (line 5), and scheduling each file for uploading (line 6) with an appropriate delay. The `GLib::Timeout.add_seconds` schedules any block of code to be called repeatedly every number of seconds. The repetition is stopped when the code block returns `false`.

```ruby
  signal_connect("drag-data-received"){
    |w, dc, x, y, selectiondata, info, time|
    f=selectiondata.data
    f = f.gsub(/file:\/\//,"")                    ## (4)
    files=f.split(/\r\n/)                         ## (5)
    @label.set_label(
        "Uploading #{files.length} files ...")
    delay=1
    files.each{|file|                             ## (6)
      ### scheduling repeated call to file_upload
      GLib::Timeout.add_seconds(delay){ file_upload(file) }
      delay=delay+delay
    }
  }
end  ## of initialize
```

Actual file uploading is performed by the `file_upload` method which invokes the given `$uploadCommand` and displays the resulting standard output through `@label`.

```ruby
def file_upload(f)
    result=`#{$uploadCommand} #{f} 2>/dev/null`
    @label.set_label(result)
    return false    ### stops repetition
  end
end  ## of DestWindow
```

The complete `upload.rb` can be found in the example code package.

12.10 For More Information

Ruby/GTK2 provides many widgets and other useful classes for GUI. The companion website provides a convenient *Ruby-GNOME2 Widget and Object API Search* function. For complete user guides, tutorials, installation, and API information go to the official site

`http://ruby-gnome2.sourceforge.jp/`

For more information on the Ruby language please visit

`http://ruby-doc.org/`

Glade is a rapid development tool for GTK, allowing you to interactively design a GUI by selecting widgets, edit their attributes, and arrange their layout. You can also connect event handling code. Glade allows you to save your work in XML files *filename*.`glade` that can be used to generate actual GUI code in C++ or Ruby/GTK2. Please go to `http://glade.gnome.org/` for more information.

12.11   Summary

The GTK+ is an object-oriented version of GTK (GIMP Toolkit), which is a graphics and GUI library built on top of X11. Through the Ruby scripting language and its GTK+ binding (Ruby/GTK2), GUI programming is made much easier.

A GUI program is event driven. It displays widgets in a well-designed layout for run-time user actions. The events produced by these actions lead to preprogrammed *event handler* code. A GUI program monitors events when not handling them.

Widgets in a GUI program form a containment hierarchy. The top-level window is the root container. Layout containers provide horizontal and vertical flow, row-column positioning in a grid, fixed positioning, and so on. Atomic widgets provide many familiar GUI features such as push buttons, selection buttons, radio buttons, text entry fields, and slider scales.

The `signal_connect` method of a widget is used to connect a signal for the widget to a code block that handles the particular event. In addition to basic events from the mouse and keyboard, semantic events represent widget-defined signals such as button clicking and text entry. Different event objects carry data related to particular events and are delivered to event handlers for processing.

A set of GUI examples shows how to write code, procedural and object-oriented, in Ruby/GTK2 that works under Linux, including a GUI front end for **tar**, a Tic-Tac-Toe game, and a drag-and-drop file uploader.

12.12   Exercises

1. Take the `click.rb` example in Section 12.1 and connect to the

   ```
   button-press-event or
   button-release-event
   ```

   instead of the `clicked` event.

2. Take the `volume.rb` example in Section 12.4 and modify it to use a 0.5 adjustment step and a vertical scale.

3. Take the `tz.rb` example in Section 12.4 and rewrite it using OOP.

4. Take the `api.rb` code in Section 12.4 and add the ability to display API for `Gdk` (for example, for `Gdk::EventButton`).

5. Take the `tz.rb` example in Section 12.4 and build a similar GUI front end for the **zip** command.

6. Write a picture viewer program that displays pictures given on the command line. A user may use mouse clicks to go to the next or previous picture.

7. Add a menu bar to the `upload.rb` example in Section 12.9 to provide an `Ask before upload` toggle option. If the option is set, the program will use a dialog window to ask the user to confirm or cancel each file upload.

8. Add GUI elements and move generation logic to the `tictactoe.rb` example in Section 12.7 to enable the program to make moves as a player to play the game with a user.

9. Write a Ruby/GTK2 program to display a pie chart based on percentages and colors given on the command line.

10. Create a simple text editor Ruby/GTK2 program.

Appendices Online

To reduce the volume of the book, the appendices are online at the book's website (`http://ml.sofpower.com`) where you can also find information updates and many other useful resources.

Appendix: Secure Communication with SSH and SFTP

SSH is a secure remote login program. It lets you log in and access a remote computer, often a Linux system, using your own desktop/laptop computer from home or anywhere you can access the Internet. SFTP is a secure file transfer program that allows you to upload and download files to and from another computer. See the appendix at `http://ml.sofpower.com/ssh.html`.

Appendix: Introduction to `vim`

Creating and editing text files is basic to many tasks on the computer. There are many text editors for Linux, but **vim** (**vi** iMproved) is a visual interactive editor preferred by many. See the appendix at `http://ml.sofpower.com/vimIntro.pdf`.

Appendix: Text Editing with `vi`

In-depth coverage of text editing concepts, techniques, and macros with the **vi** editor are provided. See the appendix at `http://ml.sofpower.com/vi.pdf`.

Appendix: Vi Quick Reference

Many editing commands are available under **vi**, and this quick reference card can be handy. See the appendix at `http://ml.sofpower.com/viQuickRef.pdf`.

Appendix: The `emacs` Editor

Rather than operating in distinct input and command modes like **vi**, **emacs** operates in only one mode: Printable characters typed are inserted at the cursor position. Commands are given as control characters or are prefixed by ESC or CTRL+X. See the appendix at `http://ml.sofpower.com/emacs.pdf`.

Website and Example Code Package

Website

The book has a website useful for instructors and students:

`http://ml.sofpower.com`

You can find the appendices for the textbook at the site. The site also offers a complete example code package for downloading, information updates, resources, ordering information, and errata.

Example Code Package

All examples in this book, and a few more, are contained in a code example package.[1] The entire package can be downloaded from the website in one compressed file, `MasteringLinux.tgz` or `MasteringLinux.zip`. The download access code is **LMCEP2010dL**.

The package contains the following files and directories

```
ex01/   ex03/   ex05/   ex07/   ex09/   ex11/   guide.pdf
ex02/   ex04/   ex06/   ex08/   ex10/   ex12/   license.txt
```

corresponding to the chapters in the book. You can find the descriptions for the examples in the textbook with cross-references to their file locations.

Unpacking

1. Place the downloaded file in an appropriate directory of your choice.

2. Go to that directory and, depending on the downloaded file, use one of these commands to unpack:

    tar zxpvf MasteringLinux.tgz

    unzip MasteringLinux.zip

    This will create a folder `MasteringLinuixCode/` containing the example code package.

---

[1]This example code package is distributed under license from SOFPOWER. The example code package is for the personal use of purchasers of the book. Any other use, copying, or resale, without written permission from SOFPOWER, is prohibited.

Bibliography

[1] Richard Blum. *Linux Command Line and Shell Scripting Bible.* John Wiley & Sons, Inc. New York, NY, USA, 2008

[2] Daniel P. Bovet and Marco Cesati. *Understanding the Linux Kernel*, 3rd Ed. O'Reilly, California, USA, 2005

[3] Arthur Griffith. *Gnome/Gtk+ Programming Bible.* John Wiley & Sons, Inc. New York, NY, USA, 2000

[4] Robert Love. *Linux Kernel Development*, 3rd Ed. Addison-Wesley Professional, Indianapolis, Indiana, USA, 2010

[5] Carla Schroder. *Linux Networking Cookbook.* O'Reilly, California, USA, 2007

[6] Ellen Siever, Stephen Figgins, Robert Love, and Arnold Robbins. *Linux in a Nutshell*, 6th Ed., O'Reilly, California, USA 2009

[7] Mark G. Sobell. *A Practical Guide to Linux Commands, Editors, and Shell Programming*, 2nd Ed., Prentice Hall, New Jersey, USA, 2009

[8] Mark G. Sobell. *A Practical Guide to Ubuntu Linux*, 3rd Ed., Prentice Hall, New Jersey, USA, 2010

[9] Tony Steidler-Dennison. *Run Your Own Web Server Using Linux & Apache.* SitePoint, Collingwood, Victoria, AU, 2005

[10] Paul S. Wang and Sanda Katila. *An Introduction to Web Design and Programming*, Course Technology/Cengage Learning, Ohio, USA, 2004

[11] Brian Ward. *How Linux Works: What Every Superuser Should Know.* No Starch Press, San Francisco, CA, USA, 2004

[12] Kevin Yank. *Build Your Own Database Driven Web Site Using PHP & MySQL.* SitePoint, Collingwood, Victoria, AU, 2009

Index